PESTS OF STORED FOODSTUFFS AND THEIR CONTROL

Pests of Stored Foodstuffs and Their Control

by

Dennis S. Hill

Institute of Biodiversity and Environmental Conservation,
Universiti Malaysia Sarawak, Malaysia

KLUWER ACADEMIC PUBLISHERS
DORDRECHT / BOSTON / LONDON

A C.I.P. Catalogue record for this book is available from the Library of Congress.

ISBN 1-4020-0735-3

Published by Kluwer Academic Publishers,
P.O. Box 17, 3300 AA Dordrecht, The Netherlands.

Sold and distributed in North, Central and South America
by Kluwer Academic Publishers,
101 Philip Drive, Norwell, MA 02061, U.S.A.

In all other countries, sold and distributed
by Kluwer Academic Publishers,
P.O. Box 322, 3300 AH Dordrecht, The Netherlands.

Printed on acid-free paper

To Hazel
And Liz
With gratitude

TABLE OF CONTENTS

PREFACE

This book is intended to serve as an introduction to the pests of stored foodstuffs of all types on a worldwide basis, and as a broad reference text. It is aimed at being complementary to the more detailed and more specific texts that are listed in the References. It does presuppose an adequate basic knowledge of entomology and zoology in the user. The stored products mentioned in the text are commercial products in the widest sense, including all types of plant and animal materials in addition to grain and prepared foodstuffs. Storage is viewed very broadly, from one day on a shelf to several years in a silo, or refrigerated store at -20°C. In many publications the produce surveyed has been restricted to stored grains, because of their obvious importance to human society, and because of the great quantities involved. For many different materials, of both plant and animal origin, there is a shortage of specific information, but it is to be hoped that this situation will gradually be rectified.

It should be clearly understood that any reference to animal pests is made in the strict zoological sense, and refers to any members of the Kingdom Animalia. There is a regrettable tendency in some circles to use the term 'animal' as being synonymous with 'mammal' – a habit to be deplored! There is definite emphasis on animal pests in this text, but micro-organisms are included where relevant. Most of the principal and general topics refer to all types of pests equally, and fungi and bacteria are broadly included as 'pests', but not specifically relating to individual crops and types of produce.

In recent years we have seen the establishment of supermarkets and a whole new approach to food retailing, and this led to a vast new industry – food processing and the production of convenience foods, including even ready-prepared meals packed and frozen. The food manufacturing industry has become a multi-billion dollar business in the USA and the UK (and Europe), and is now extending into Asia, Africa and parts of South America. Shopping malls and supermarkets have become worldwide phenomena, as have several of the multinational fast-food catering establishments (Macdonald's, Kentucky Fried Chicken, Pizza Hut, etc.). With these new styles of food retailing come new risks of food infection of which we need to be aware. For example, the present practice of meat bulking in hamburger production means that one diseased cow could end up as a thousand hamburgers sent to a hundred different retail outlets. One thing that is very clear is that there is a worldwide increase in the incidence of food poisoning. This is linked to a general increase in public awareness concerning food safety.

The need for a convenient, inexpensive text for both teaching and advisory (extension) purposes in tropical countries is self-evident in view of losses of foodstuffs during storage, and the urgent need for additional food sources. For teaching and training in the tropics, this book should ideally be used in conjunction with the training manual *Insects and arachnids of tropical stored products* (TDRI, 1984) whose illustrated keys are designed for use in laboratory practicals. In

Europe, *Common insect pests of stored food products* (Mound, 1989) should be used; a key is provided for each group of insects presented.

Nowadays in the UK, it appears that most of the universities and some colleges offer courses in 'Food Science', in a range of food-related degree courses. This development has taken place within the last 20 years. Stored products protection was active in the 1960s but was usually thought of as 'Post-harvest crop protection', and related mainly to grain storage. The Public (Environmental) Health Inspectorate (UK) and the Royal Society for Health (RSH) also run their own professional courses, and award their Diplomas and Certificates both overseas (e.g. Hong Kong) and in the UK.

It is hoped that this text will be of use for all these different training programmes, in both temperate (UK, USA, Europe, Australia) and tropical regions (S.E. Asia, India, Africa, etc.).

February, 2002

Dennis S. Hill
Institute of Biodiversity and
Environmental Conservation
(I.B.E.C.)
Universiti Malaysia Sarawak
94300 Kota Samarahan
SARAWAK

ACKNOWLEDGEMENTS

The book has its origins in a study of urban wildlife and domestic pests in Hong Kong, and I am greatly indebted to Mr Robin Wong, Office Manager and Entomologist of Bayer (China) Ltd. for his support and encouragement – without which the project would not have developed. Most of the drawings used were made by Karen Phillipps.

Assistance has been given by various colleagues over the years, and I am particularly grateful to the staff at the Pest Infestation Laboratory, Slough, England (later the Storage Department, Overseas Development Natural Resources Institute, Chatham, Kent, now Natural Resources Institute (NRI) of the University of Greenwich), especially B.J. Southgate. Professors J.G. Phillips and Brian Lofts successively were Head of the Department of Zoology, University of Hong Kong, where the initial work was done. The book was completed at the Institute of Biodiversity and Environmental Conservation, Universiti Malaysia Sarawak, and grateful thanks are owed to Director, Dr Fatimah Abang. Help was given by G.W. Chau (Hong Kong Urban Services Department), M. Kelly (Agricultural Development and Advisory Services, Ministry of Agriculture, Fisheries and Food, UK), Professor H. Schmutterer (German Agency for Technical Cooperation), I. Smalley and R.S. Parsons (Union Grain, Orby, Lincolnshire, UK). Thanks are due also to Bayer AG (Leverkusen), Degesch Co. (Frankfurt), Detia GMBH (Laudenbach), ICI Ltd. (Haslemere), ACIAR (Canberra), Food and Agriculture Organisation of the United Nations (Rome), Rentokil Ltd. (UK and HK Branch), and the Commonwealth Institute of Entomology International (London).

Other drawings were made by Hilary Broad and by Alan Forster. The photographs were taken by the author.

The first edition of this work was published by Iain Stevenson of Belhaven Press (London) (since acquired by John Wiley) in 1990 under the title 'Pests of Stored Products and their Control', and has been out of print for several years. However, it was thought that the work should be expanded to include a range of common internal parasites and micro-organisms, and that there should be more emphasis on prepared foodstuffs, resulting in a necessary change of title.

Two very useful sources of information were the book edited by Gorham (1991), and the eight-volume encyclopaedia edited by Macrae, *et al.* (1993): the publishers, the Association of Official Analytical Chemists, Arlington, Va, USA, and Academic Press (Harcourt Brace Jovanovich) are gratefully thanked.

INTRODUCTION

Many crop protectionists and agriculturalists have long been concerned by the relative lack of attention paid to the control of pests of stored products (foodstuffs), and the high losses during storage led to Ordish publishing his book, *Untaken harvest*, in 1952. Cereal grains are the most important crop produce in storage. Human civilizations have evolved around the different major cereal crops, and these are the staple crops on which we depend. In most regions there is only one annual cereal harvest, so some of the grain has to be stored for a minimum period of one year, and national famine reserves for 3-5 years. Seed grain must be stored until the start of the next growing season, 6-8 months on average. A similar situation prevails with dried pulses (beans, peas, etc.), which are the main source of protein in many human diets.

In a report to the Linnean Society in 1989, Professor Bunting said that the estimated harvest crop biomass for the world in 1984 was about 2,400 million tons, and of this some 1,800 million tons were cereals. Thus the average amount of cereal grains in storage worldwide must be about half the total, or 900 million tons.

Worldwide losses of foodstuffs due to insects and rodents have been estimated by the Food & Agriculture Organisation of the United Nations (FAO) as about 20% (figures ranged from 10% in Europe and North America, to 30% in Asia and Africa). Precise figures for storage losses of different types of crop produce are not readily available on an extensive scale, but most workers in tropical countries will have seen stores where infestations were so extensive that the entire contents of the store were ruined.

With the ever-increasing need for greater crop production in the third world tropics, it is important that accurate crop loss assessments be made at the different stages in production, and that appropriate control measures be taken. For example, in India the Ministry of Food and Agriculture estimated that in 1968 rats caused a grain loss of 10%, and yet in that year the government spent 800 times more on importing chemical fertilizers than was spent on rat control. Certainly up to about 40 years ago the post-harvest protection of crops was seriously neglected. But by the 1960s more attention was being given to stored produce pests, and scientists were recruited to this end by the various ministries of agriculture. The *Journal of Stored Products Research* was started in 1964, and the *Advances in Food and Nutrition Research* had been started in 1956. Since then, a number of new journals dealing with post-harvest studies and foodstuffs have appeared (see overleaf). In 1943, The British Pest Control Association was started, and it now represents more than 150 pest control servicing companies and pesticide manufacturers supplying the public health, commercial, industrial and domestic markets. Such associations are now present in most countries of the developed world. In 1987, in London, the Institute of Food Science and Technology (UK) was started, as the professional society for all those working in the British food industry. And recently, the European Federation of Food Science and Technology (EFFoST), and the International Union of Food Science and Technology (IUFoST) were established.

The major centre for food research in the UK is the Institute of Food Research at Norwich, on the campus of the University of East Anglia.

The seriousness of the international situation was finally recognized by the United Nations, and in 1975 FAO organized in Rome a special session on post-harvest food losses. They stressed the severity of the problem in tropical regions and they urged member countries to undertake research to reduce post-harvest food losses in developing countries by at least 50% by 1985. It was pointed out that there was a lack of agreed methodology for post-harvest loss assessment, and also that loss data were seldom correlated with financial costs, and in an attempt to rectify this situation, two further workshops were held in 1976, one in the United States of America (USA) and one in Slough, England. From these workshops came the publication *Postharvest Grain Loss Assessment Methods* (Harris & Lindblad, 1978). The situation now is that there is a great deal of information known about food pests, produce losses, and control measures, but in many tropical countries theory has yet to be put into practice to any real extent.

Interest in food technology has recently blossomed, and in the 1990s a great many major publications resulted, including such journals as *Trends in Food Science and Technology* (1990), and the *Encyclopaedia of Food Science, Food Technology and Nutrition* (8 volumes) (1993), and Academic Press launched the journal *Advances in Food and Nutrition Research* and the *Food Science Technology* series of reference books. The newly created Institute of Food Science and Technology (IFST) in London publishes the bi-monthly *International Journal of Food Science & Technology* and the quarterly *Food Science & Technology Today.* At the end of the References are listed the major international journals on food science, and it can be seen that most were started in the 1990s, and now a dozen or more are flourishing.

Most foodstuffs of an animal nature (fish, meat, eggs etc.) can only be kept for relatively short periods of time, and in general they require different storage methods, although if thoroughly dried they can be kept for months or even years.

The prepared foods industry is relatively young, but growing fast, and this demands a completely different approach with regard to storage and protection. The Supermarket system is clearly here to stay, and vast sums of money are involved. This would seem to be the main reason for the blossoming of all the new journals, encyclopaedias, and reference books dealing with all aspects of food science and technology. And now most of the UK universities have courses (and degrees) in Food Science, to prepare future staff for this burgeoning industry, as is also the case in Europe and North America.

In most countries, the study of pest infestations on farms or in on-farm stores is the responsibility of the agricultural entomologists working for the departments of agriculture, whereas in towns the food stores, mills, bakeries, markets, restaurants, and domestic premises are the purview of the public/environmental health inspectorate. Responsibility for regional stores often depends on whether the location is regarded as rural or urban.

CHAPTER 1

PRESENT TRENDS IN FOOD TECHNOLOGY

1.1 FOOD QUANTITIES

This has long been of major concern worldwide, especially throughout the third world tropical countries where a combination of land shortage and overpopulation has exacerbated the situation. Here traditional farming is based on smallholder plots where agricultural inputs (such as fertilizers and pesticides) are minimal or even non-existent, and yields are correspondingly low. The end result is that many millions of people in the tropics constantly suffer malnutrition, and some face starvation.

When weather changes cause the rains to fail, as in Ethiopia in 1984, then there is widespread harvest failure, and whole populations can die of starvation. At the present time, the world is experiencing major climatic changes, and there is a definite warming effect taking place. The precise reason for this is not known for certain, and several viewpoints exist (i.e. greenhouse (gases) effect; depletion of the ozone layer; natural interglacial climatic conditions). A measurable change is involved with the 'El niño' effect whereby the surface of the South Pacific is warmed and the prevailing surface air currents are reversed, leaving S.E. Asia short of rain and parts of South America flooded. Climatologists claim that these effects are very widespread, and extend to parts of Africa and the southern USA. The end result of the changes in the air-flow systems is that some regions are experiencing drought whereas others have torrential rain leading to widespread flooding. Thus in many countries agriculture has become even more unpredictable, and production losses have increased.

It has been demonstrated many times that tropical smallholder crop yields can be greatly increased – to a level comparable with temperate ones. The initial way was that used in the 1960s 'Green Revolution' in India and parts of S.E. Asia, which depended upon high levels of fertilizer on new high-yielding varieties of crop (rice and wheat mostly), and concomitant use of herbicides and other pesticides. The yields could be high, but the agricultural inputs were correspondingly high and the harvested crop was often more expensive and sometimes difficult to sell. A more recent approach is that of integrated crop management (ICM), which has the same overall rationale as IPM, or integrated pest management. In parts of South America and Africa, local ICM projects have been very successful, but they have needed a great deal of local research and effort. However, it would seem that the long term prospects for ICM holistic agricultural methods are excellent.

In temperate regions, particularly the USA, Canada, Europe, Australia etc., agriculture has generally become very sophisticated and efficient. Crop yields are sufficiently high that in some countries a land set-aside scheme operates to reduce over-production of wheat and other cereals. The regular huge surpluses of wheat and other grains in North America are still being utilized by Russia and the former members of the USSR for both human and livestock food, as their agricultural systems are still notoriously inefficient, and there is no way they can feed their populations on their own. European agriculture at its best is now extremely sophisticated, and advanced computer programs are making it possible to control soil fertility, harvest time and other aspects with great precision.

1.2 FOOD QUALITY

Now a major issue in many parts of Europe, North America and other countries. There are several aspects of concern, the first relating to basic quality. This can be expressed in terms of chemical constitution such as the nature of the types of carbohydrate and protein that make up the foodstuff. Vitamins and minerals are especially important, and a recent concern relates to fibre content. In general, people in the developed world are often eating unhealthily, using far too many 'convenience' foods and snacks including sweets, confectionary, crisps and cakes (cookies). Many of these food products have added salt (often used to generate thirst) and extra sugar. It should, perhaps, be mentioned that the extra sugar and/or salt added to canned and prepared foodstuffs is often included as preservative, and without this the storage life of the food would be shorter. In addition to the vast range of snack foods now available in stores worldwide, there is an equally large range of pre-cooked and frozen meals being consumed – they used to be referred to as 'TV dinners'. These ready-made meals are often treated with a wide range of additives, used to enhance flavour etc., but often the actual nutritive value of the meal may be quite low. But these frozen meals are very popular with certain members of society because of their convenience. In Europe, many people have insufficient exercise, and this, together with a diet lacking in fibre, can lead to chronic constipation and even bowel cancer – an ever-increasing urban complaint. Another alarming current tendency is for a large part of the population to be overweight, sometimes grossly so, and this is often attributable to an unbalanced diet combined with lack of exercise.

A major cause for concern that is on the increase in Europe at the present time relates to what are broadly classed as food contaminants, both deliberate and accidental. Pesticide residues in or on crops have led to spectacular cases of poisoning, both acute and fatal, and chronic and debilitating. All pesticides are carefully tested before receiving official government approval for the use against specific pests on certain specified crops, and usually at only a specified rate. Before registration, a chemical is rigorously tested by the manufacturer against a wide range of organisms, first under laboratory and then under field conditions. Finally, the World Health Organisation of the United Nations (WHO) approves an acceptable daily intake level (ADI) for humans, and this is typically low so as to be on the safe side. All approved pesticide chemicals have a recommended pre-harvest interval when used on edible crops, so that they will have degraded before there is any chance of their being ingested. But of course mistakes are made, and some farmers will deliberately apply more than recommended, so there are inevitably still occasional cases of pesticide poisoning from food crops. There are also cases of ignorance due to lack of research – for example, recent studies have found insecticide in the central core of carrots. It had previously been thought that the contact insecticides were limited to the outer skin of the root and so would be removed during food preparation (i.e. peeling). In the past worldwide there have been many thousands of cases of human death due to ingestion of pesticides on food crops, but nowadays this hazard is recognized and appropriate precautions are taken in most countries.

In many countries, medical statisticians have recorded collections of abnormal births in certain rural locations, and these foetal defects seem to be linked to local usage of certain pesticides. And another well-publicised case of insecticide poisoning has been the sheep farmers in the United Kingdom (UK) using the recommended organo-phosphorus sheep dip. This is a case of skin absorption and not connected with food, but sometimes the public fails to make the distinction. These and other similar cases have been the cause of some public concern in the UK, and undoubtedly are partly responsible for the increasing demand for 'organic' foodstuffs.

Many foodstuffs that are canned, frozen, or otherwise preserved have a wide range of additives applied to improve appearance, longevity, flavour, etc., and some of these can be regarded as undesirable if not potentially harmful to some people. And manufactured meals often contain quite a horrendous list of additives, the names of most of which are meaningless to the average consumer. Most of the additives are used for the benefit of the manufacturer and retailer, not for the benefit of the consumer. Many of them help to sell the product by making it appear more attractive, and others act as preservatives which give the product a longer shelf-life. In recent years in Europe there have been dramatic increases in certain bodily disorders. For example, in the last eight years the cases of asthma in children have doubled in the UK. Some of these disorders are attributed to increasing levels of pollution, but others are thought to be caused by certain food additives, and these include the following: migraines, aching joints, hyperactivity and attention deficit disorder in children, excessive lethargy, and irritable bowel syndrome. An indirect additive to meat, particularly beef produced in the USA, are various growth hormones, which can apparently persist into high quality steaks and similar cuts, and could theoretically present a threat to consumers. Perhaps careless use of systemic insecticides and acaricides could also result in contaminated meat. These topics are looked at in more detail in Chapter 7, under the heading of 'contamination'.

Pathogens and parasites are a constant source of concern. Beef and pork tapeworms are ever-present in countries where it is customary to eat raw or very lightly cooked meats, and trichinosis has long been an established problem in parts of Europe and the USA where pork sausages and smoked meats are regularly eaten. But these are parasites of long-standing that are a familiar problem in these countries. What is striking is the overall recent increase in cases of food poisoning resulting mostly from bacterial infections. This increase in numbers is partly due to the increase in the number of reported cases, but it is generally agreed that there are more cases these days. It is also widely agreed that probably the actual number of cases of food poisoning in the UK each year is about ten times the number reported. Food poisoning is looked at in more detail in the next section, and in Chapter 20. Occasionally, a new parasite appears, sometimes with a devastating effect, such as shown by the causal agent (a prion) of Bovine Spongiform Encephalopathy (BSE) that infects the spinal cord and other nervous tissues in cattle, and is thought to cause Creutzfeld-Jacob Disease (CJD) in man if ingested. First recorded from the cannibals in Papua New Guinea in the early part of the last century, it suddenly became newsworthy in the mid-1980s, when cases in the UK were reported. The basis of the problem seems to have been the practice of incorporating abattoir waste into protein diet supplements for intensive beef cattle rearing.

A very recent food scare is centred upon the use of genetically modified (GM) foodstuffs, which will be briefly reviewed later in the chapter. There are fears that the modified food might in some way be harmful to man.

Because of the ever-increasing public concern over food quality during the last few decades, there is now in most countries a series of centres and institutes concerned with food quality and aspects of food science. Important legislation in the UK includes such measures as the 'Food and Environment Protection Act, 1985', and the establishment of the Food Standards Agency (FSA) in 2001.

Concern over food quality led to the formation of the Codex Alimentarius Commission by the United Nations, more specifically as a joint venture between FAO in Rome and WHO, with its headquarters in Rome. All the countries that are members of the United Nations belong to the Commission. The Joint FAO/WHO Food Standards Programme publishes a series of guidelines relating to Standards for General Requirements (Food Hygiene) dealing with all aspects of food hygiene internationally, as well as reports and working papers. In most countries, the local contact is in the Ministry (Department) of Health.

1.3 FOOD POISONING

Food poisoning would appear to be on the increase, judging by reports in the national press, but in fact this is not completely certain because nowadays outbreaks are widely publicized and more cases are officially reported, but the main point is that every year there is an alarming number of cases of bacterial food poisoning, sometimes with lethal results. This will be discussed in more detail in Chapter 20. One surprising development is that several strains of what is usually a harmless human intestine dweller *Escherichia coli* have evolved with pathogenic properties, and they have been responsible for several major outbreaks which involved a number of deaths. A recent (June, 1999) statement from the UK Government stressed that poor hygiene by food handlers has led to an ever-increasing occurrence of food poisoning by species of *Campylobacter*.

Fungal infestation of food products is common but does not often result in food poisoning. Fungal spores are airborne, and so numerous and so tiny that they occur everywhere, and food contamination is usual. Control measures are centred upon the suppression or retardation of development of the fungi, and the utilization of the food product before the fungus has time to develop.

A major reason for the increase in cases of food poisoning is a change in human habits, linked to the proliferation of supermarkets. More people are eating out at restaurants and cafés as well as using fast-food outlets and takeaways than in previous decades. Much food in the United Kingdom is now prepared and handled centrally, usually in bulk, and then distributed to regional retail outlets. The more perishable goods are usually frozen and then carried in large refrigerated trucks maintained at about -20°C. Once in the supermarket, the shelf-life is usually regarded as about 3-5 days for raw meats, and 2-4 weeks or months for deep frozen, but only about 2 days for prepared dishes provided that they are kept in a refrigerator (at about 5°C). In the UK, the present domestic trend is for both partners to work, and this usually means that major household shopping is done only once a week with perishable foodstuffs being stored in either a refrigerator or domestic deep-freeze. Whilst frozen (at -20°C), foodstuffs are well protected, but while in a refrigerator, protection is more limited and for a much shorter period of time. Most household refrigerators do not maintain a temperature lower than about 5°C, and at this temperature biological processes are usually not halted but proceed slowly. For example, a Bluebottle trapped in a refrigerator is usually still mobile when the door is opened, albeit very sluggish, and bacteria and fungi continue to develop. If the door is opened frequently or is left open, the temperature inside will quickly rise as the cold air cascades out. Even in a deep freeze, food products are typically divided into three categories – those that can only be kept safely for one month, those for up to three months, and those for up to twelve months. Many household refrigerators are used carelessly, the door is often opened repeatedly throughout the day and often left open for long periods of time, and cooked foods may be stored while still warm. And uncovered foodstuffs can lead to cross-infection. People often forget that foodstuffs bought in a supermarket are already several days old when put on the shelf, and by the time they are purchased and maybe driven home in a hot car, pathogens may already be developing. Fungal spores and aerial bacteria are often ubiquitous in the atmosphere, and most foodstuffs are infected most of the time – and development will take place when condition permit. Most farinaceous products will yield a rich crop of fungi if kept long enough in storage, such as on a pantry or cupboard shelf. A loaf of bread, particularly one sliced, carefully sealed inside a polythene bag will develop mould patches in about five days at room temperature.

The other danger in this respect is the ever-increasing usage of prepared meals, pre-cooked at the site of manufacture, to be re-cooked or re-heated at home. It is so easy to keep these foodstuffs a little too long in storage, giving the pathogens or pests time to develop.

One basic problem in the food preparation industry is that some people are 'carriers' of various pathogenic organisms Typhoid is probably the best known example, and it can be transmitted by kitchen staff who display no symptoms whatsoever of the disease, but they infect the food they handle and the customers collapse with typhoid. The food retail industry is renowned for the rapid turnover of staff and for casual employment procedures. Also, in many retail outlets (hotels, restaurants etc.) the food handling areas and staff habits leave a great deal to be desired. But in most of the large supermarkets, the food preparation facilities are more closely supervised and less likely to have any contamination problems.

1.4 SUPERMARKETS

These have occupied a dominant role in retail marketing of foodstuffs for some 30 years in most western countries, and are rapidly becoming dominant throughout tropical Asia and Africa. Two important aspects of their activity are the packaging of most foodstuffs, and centralized production in bulk. Fresh foods are mostly packed in a polystyrene tray and covered with a plastic film, or items such as cheese are just wrapped in the film. These are then kept on chilled shelves at about 5-10°C.

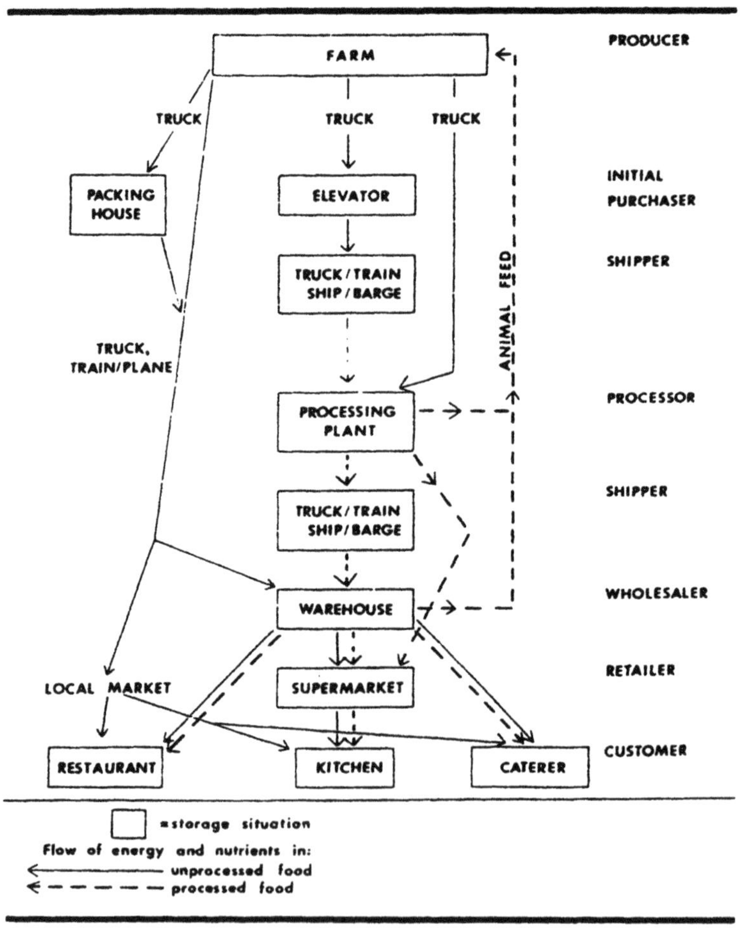

Figure 1.1. Main agricultural food-storage situations and flow of energy and nutrients from producer to consumer. (From:Gorham (1991) p. 39).

Dried materials such as nuts, raisins, potato chips (crisps), and other snack items are sold in airtight plastic packets that nowadays are often filled with nitrogen gas to prolong their storage life. Cornflakes, oats and other breakfast cereals are mostly still sold in cardboard boxes, as are some pasta products, but most are now enclosed inside a sealed inner bag – but they usually accommodate various insect, mite and fungus pests in small numbers. The insects and mites are invariably present as eggs which develop as conditions permit. If the products are kept in storage too long, then the pests develop. Most people have the experience of a cereal packet left at the back of the cupboard shelf for maybe a year, which on being opened yielded a small cloud of moths or a seething mass of beetles. This is common in the UK, but in the tropics the time required for such development is often only a couple of months. In the tropics generally, levels of food hygiene are far lower, and in the supermarkets the 5 kg and 10 kg bags of rice, in sealed plastic, are usually already lightly infested with adult rice weevils and other beetles.

Figure 1.2. A local 'Mini-Supermarket' on the outskirts of Kuching, Sarawak.

The other aspect of concern relates to time. The foodstuffs are prepared centrally and distributed to all the retail outlets by huge trucks. This all takes time, and then the produce is displayed on shelves and offered for sale. In order for all the produce to be sold, it will have to remain on display for some days, and so much corporate time and research goes into the study of food deterioration and preservation, so that shelf-life can be extended. The basic objective of the supermarkets is, of course, financial profit, but food safety is an important factor.

Present practice involves a number of separate stages which most foodstuffs have to go through in their journey from farm to consumer table, including storage in several places, and transportation from place to place. All this time gives opportunites for pest infestation, and for micro-organisms to develop. This is nicely summarised in a flow-chart in Gorham (1991; p. 39) (Figure 1.1).

In many countries, the development of shopping malls places a series of establishments in close proximity all under a single roof and sharing the same microclimate. In temperate regions, the malls are heated and air-conditioned, so conditions for pest and pathogen development are provided. In the tropics, the malls are cooled and air-conditioned, but this only retards insect pest development a little, while it encourages a concentration of human activity since the cooler conditions entice many people to come in out of the heat and humidity. Shopping malls are usually multi-storeyed affairs covering several hectares of land, and contain a supermarket, a couple of dozen shops, restaurants, cafés, and often a cinema. Most supermarkets are single-storeyed but cover 1-2 hectares in area (or are even larger), with a vast array of fresh, canned, dried, chilled and frozen foodstuffs. But in parts of Africa and Asia, the Mini-Supermarket is to be found – a much more modest enterprise offering a wide (but limited) range of produce often at very competitive prices (Figure 1.2), usually with a rapid produce turnover so pest problems do not often have time to develop.

1.5 IRRADIATED FOODS

Irradiated foods are causing some public concern as it might be thought they could present a health hazard. A considerable amount of research has been done recently without any evidence whatsoever of danger to man, so it is likely that irradiation of foodstuffs will become a standard method of pest control. A few years ago, the popular press was full of scaremongering articles about people's fears of irradiated foods, but these have now been replaced by articles on genetically modified foods. The topic is looked at in more detail in Chapter 4.

1.6 GENETICALLY MODIFIED (GM) FOODS

Food materials from genetically modified crops are the latest scare promulgated by the tabloid press, which has totally eclipsed scare reports about irradiated foods. The important point about this is that to date (October, 1999) there is no good evidence that there is any public danger from eating foods made from genetically modified (engineered) plants, and the press usage of the term 'Frankenstein foods' is totally unwarranted. On the other hand, our ignorance should not be taken as reassurance, and much careful research is needed on this topic. There is general agreement amongst most scientists that there possibly are environmental hazards that have not yet been investigated sufficiently.

Genetic engineering is being used in several different ways at present, including:

Herbicide resistance – soyabean is a particularly important crop as the bean flour is used in a vast range of human (and livestock) foodstuffs and other products. It has been possible to introduce a gene giving resistance to certain herbicides which greatly improves cultivation of a clean crop and increases yield. Weeds continue to be the major limiting factor in crop production worldwide on most annual crops, as is well recognized in most good IPM programmes, and herbicide resistance in the crop would be a most useful attribute.

Insecticidal chemicals – it is possible to incorporate genetic material of *Bacillus thuringiensis* (BTH) into some crops, which then makes them toxic to caterpillars which feed on the plant foliage. BTH is a well-known natural insecticide that kills caterpillars. Also, a range of chemicals that inhibit insect feeding can be taken from certain wild plants and incorporated into some crop plants.

Microbial fermentation – a quarter of the top 20 drugs used in human medicine are genetically engineered proteins produced in yeast cells and bacteria. Now there is the possibility of producing plant vaccines for treatment of enterotoxigenic *E. coli*, hepatitis B, cholera, and various viruses – these would be edible vaccines that would be cheap to produce and easy to administer, particularly in the third world.

Plant physiology – can be altered by genetic means to make a food crop more desirable, in several different ways. A spectacular case, some three years ago, was the modification of tomatoes to ripen more slowly, which gives them a longer shelf-life for both sale and processing. It is estimated that these new varieties yield 20% more tomato paste commercially.

There clearly is possible cause for concern in case careless use of GM plant materials, especially pollen, could have adverse effects on the flora of a region, or in other environmental ways. For example, in the UK, GM pollen from maize plants has been collected by foraging honey bees, and so it could enter human food chains. Pollen grains and aerial spores are major causes of human hay-fever, and the plant particles end up in the lung tissues. Both forage maize and oilseed rape are wind-pollinated (anemophilous) and produce large quantities of pollen for aerial distribution.

Genetic engineering is clearly a technique of great importance in aspects of pest management (e.g. production of hairy strains of cotton etc.). It must be stressed that this is basically little different from traditional plant breeding for crop improvement, in which we have been engaged for millennia, and which is the basis of all modern agriculture and livestock production. And for drug and medicine production, it is far too important a technique to be discontinued - although for production of food materials there are some techniques that may need very careful scrutiny. At the present time, the term is used too broadly and without discrimination, and this can be very misleading.

1.7 ORGANIC FOOD

This is foodstuff made from plants grown without the use of any pesticides or chemical fertilizers, so ideally it has no chemical contaminants. The high level of purity is achieved at the cost of some pest damage and a lower yield, so overall the food cost is higher than otherwise.

The advocates of organic food claim that it tastes better than that chemically produced; this can be demonstrated with fruit and vegetables, but produce freshness is an additional factor. Many consumers also believe that such farming is far less damaging to the environment and wildlife by avoiding the use of pesticides. And some of the worst environmental disasters in the UK have actually been through overuse of nitrate fertilizers and water contamination.

In the UK, there is a steadily increasing number of people wanting to buy organic produce in the supermarkets and farm shops, and gradually the Government is yielding to public pressure and is promising more financial support for aspiring organic farmers. And two UK universities are now offering BSc courses in organic agriculture.

CHAPTER 2

STAGES IN FOOD PRODUCTION

The period of time between the crop in the field and utilization by man can be divided into six separate stages for convenience. There is clearly often overlap between some of these categories depending upon the crop or produce concerned, but in general this categorization is useful and holds true in many cases. The first main departure from this simple sequence is transportation. With some crops the initial processing takes place in the field, but with others, transportation to a local processing station is required twice or three times, or even more, before they are sold. Packaging is the second event difficult to place in the sequence because it often occurs twice, at different times. Many products are shipped (exported) in bulk, often in boxes or sacks, and then re-packed into small bags for retail purposes. Rice is commonly shipped in large jute or plastic sacks containing a hundredweight (50 kg), and then re-packed into polythene bags for retail. The weave of the sacks is, of course, open to all types of infestation, and most rice shipments arrive infested with insects and mites as well as divers micro-organisms. The topic of packaging is of sufficient importance to warrant treatment in a separate chapter (Chapter 5). The sequence of stages is as follows:

- Harvest of produce
- Initial processing }
- Transportation }
- Storage } Post-harvest stages
- Final processing }
- Use or consumption }

2.1 HARVEST OF PRODUCE

This is the process of collection of the part of the plant, or animal, to be used as food by man. All the different parts of the plant body may be harvested, according to the type of crop, from roots, tubers, corms, bulbs, rhizomes, stems, buds, leaves, shoots, flowers, fruits and seeds. Some exceptional cases include plant parts such as anthers (saffron) or leaf petioles (celery). In most cases it is the fruits or seeds that are collected (cereals, pulses, fruit crops), but common vegetables include root crops (potato, carrot etc.) and leaf vegetables (*Brassica* spp. etc.). With animals to be used as human food, very often it is the entire body that is harvested (fish, fowls, cattle). With wild animals such as marine fish, the harvesting consists of commercial fishing using nets, lines and traps of different types. Basically, each type of plant produce or animal now has its own particular method of harvest, often involving specialised equipment.

Mechanisation has now become usual for most large-scale harvesting — it started with cereals and root crops and now includes top fruit such as apples and some bush fruits. This has involved extensive plant breeding to produce plants that are more suitable both physically and physiologically for such harvesting. An interesting case is the modification of the pea plant to produce a shorter, more upright habit with fewer flower trusses, so that maybe four trusses are ripe at the same time to harvest for freezing. Modern grain harvesting is now quite sophisticated, and the stems are cut in a swath of 10-20 feet, heads threshed and grains collected and blown into an accompanying truck for transportation to the local grain silo. Some harvesters also collect the straw into tied bales and drop them into the field for later collection. In large cereal production areas, several harvesters are used in concert, so that hectares of wheat are harvested in a matter of minutes rather than hours. But most smallholder crops are still hand-harvested; cereals, for example, may be hand-scythed and later hand-threshed.

One problem with mechanical harvesting is the likelihood of physical damage to the crop. On the whole, the earliest models of harvesters could be quite damaging, but the latest ones are surprisingly effective and cause little damage, and often leave little in the way of crop residues in the field. Grains and tubers/roots left in the field are often of particular importance to either wildlife or poor local villagers who glean the fields for spilled grain and overlooked tubers. Also, crop remains may constitute a reservoir of pests or disease organisms. Fruits and tubers damaged by cutting or bruising need to be discarded immediately, for they cannot be sold commercially (although they may be used as livestock food) and are soon a source of microbial infection. Most seeds have a testa that is strong enough to withstand physical pressure and protect them from damage. Evolution has led to a compromise between a strong testa for physical protection and one that can still be breached by the developing shoots. One result of this evolution is that only a relatively small number of insect pests (those classed as 'primary pests') are able to penetrate the intact testa of cereal grains, pulses and other seeds. But if the testa is cracked or scratched, the seed is then open to attack by the entire spectrum of arthropod pests that attack stored foodstuffs. It should be remembered that most of the insect pests are really quite small in relation to a cereal grain, with tiny mandibles, and the mites are microscopic.

In order to safeguard produce during storage, it is necessary to harvest the produce in such a manner as to minimize physical and mechanical damage, particularly bruising and cutting. With some insect pests it can be very important to ensure prompt harvesting as soon as the crop is ripe, to prevent unduly large field infestation – for example, both *Sitophilus* and *Sitotroga* lay eggs in or on ripe grains in the field, and the grain is infested before entering into storage. In parts of Ethiopia, both maize and sorghum crops are regularly infested by these two pests (see Figures 13.76 and 13.116), with 10-20% of panicles being infested. Some pests are only active in the field, but are a nuisance in storage, and the damaged produce is likely to be rejected. Thus, Codling Moth larvae damage the growing apples, and such fruit need to be rejected at harvest, which can be difficult with mechanisation.

2.2 INITIAL PROCESSING

Several aspects are involved immediately post-harvest, and these are usually cleaning, drying and, now, sometimes grading. With some foodstuffs, there would also be packaging as the final process (see Chapter 5).

2.2.1 Cleaning

This is the removal of pieces of plant debris, stones, soil particles, and the like. When cereal crops are hand-harvested, part of the stem is taken together with the panicle, and with legumes (pulses) the whole pod is collected. Thus the panicles and the pods have to be threshed in order to remove the grains and the seeds. In the rural tropics, this is still often done by hand, but mechanical threshing is becoming more widespread. The more advanced systems harvest and thresh and collect the grain/seeds as part of the whole process of harvesting.

With many vegetables, excess plant material that is not required for culinary use has to be removed. Thus nowadays the 'tops' (leaves) of carrots are usually removed, as are those of beets. Outer leaves of cabbage and cauliflower are now removed prior to sale, often in the field at harvest. Ware bulb onions are typically stored for months, and the traditional harvest method allowed the leaves to dry and wither naturally; the bulbs were usually lifted and then left lying on the soil surface to dry. The present practice is to mow the tops off carrots, beet and onions before immediate mechanical lifting.

Root crops and tubers destined to be sold in modern supermarkets, often in plastic packaging, are washed to remove soil particles. After washing, the root crops have to be dried before packaging. On washed tubers and roots, a little pest damage (such as by Carrot Fly larvae) becomes obvious and can quite unnecessarily result in produce rejection, even though the damage may really be quite slight. But very recently there has developed an interest in buying 'dirty' roots (potatoes, carrots, swedes etc.) as these are thought to be more 'natural'.

2.2.2 Drying

This is the removal of excess moisture in the produce, for if not removed it will cause fungal spores and bacteria to develop and moulds will result. Because of the ubiquitous distribution of bacteria and fungal spores in the atmosphere, they will inevitably develop whenever suitable conditions prevail; these are a moderate temperature, available moisture and a food substrate. In most food situations, the critical factor is moisture, including surface moisture (wet produce), moisture content of the produce, and atmospheric moisture (relative humidity). Produce that is actually wet at harvest has to be surface dried prior to handling, and this is usually done by field air-drying. Items to be kept for any length of time in storage have to be quite dry in themselves, and the storage area should be dry and with low atmospheric humidity. Drying is usually the single most important factor in the prevention of food deterioration, initially as field-drying and then later using special drying facilities such as large kilns or ovens where necessary. The problem of moisture is most encountered in connection with grain, and to some extent pulses in storage. At the time of harvest of wheat in the UK, the moisture content of the grain is usually in the range of 18-22%, although most farmers try to wait for harvest until the moisture content falls to 18-20%. In a dry summer, such as that in 1989, wheat grain in Eastern England was as low as 12-16%, which must be similar to conditions found in the dry tropics. The ideal level for grain storage is a moisture content of 10-12%. The common storage moulds need a moisture content of 15% or above for development. Traditionally, cereal crops were cut (harvested) first and then collected into 'stooks' (i.e. stacked in clumps) and left to air-dry. In Europe, the time of harvest is usually the latter part of the summer, which is also the dry season, so that air-drying the produce is quite feasible. When suitably dried, the sheaves were collected and the grain threshed so that it could be bagged and stored. Similar practices are followed throughout much of Africa and Asia for the local cereal crops. Nowadays in Europe and North America, with intensive agriculture and vast acreages of cereal monoculture, such harvesting methods are no longer feasible, and the combine harvester cuts the stems and threshes the grain in one operation, and the grain may be either bagged directly or, more usually, stored loose in bulk. If harvest takes place in dry weather, there are few moisture problems, but in rainy weather, with wet grain, there are major problems, and the grain has to be dried quickly to avoid mould development. Mechanised agriculture on a large scale is complicated and costly; harvesting is often done by commercial contract, and the process has to be conducted with little time flexibility, and often grain has to be harvested when really too moist for storage. Grain driers are expensive to build and to operate, so most if not all regional cooperative grain stores have a drier built in.

Most root crop and fruit harvests are usually timed to coincide with periods of dry weather so that the produce can be stored immediately after cleaning. In the tropics, vegetables and fruits may be harvested on a daily basis for immediate transportation to local food markets, and so there would be little time for moulds to develop. Similarly, vegetable crops for immediate sale need only be clean, and, in fact, moisture can be an asset for it enhances the 'fresh' appearance of the produce. Indeed, stallholders can sometimes be seen to spray water over their fruit and vegetables to encourage their purchase. But for storage, surface dryness is a requirement.

Other types of plant material are treated differently – dried mushrooms, dried herbs, spices and vegetables are completely dried before storage – usually sun-dried. Tobacco leaves are taken straight to the drying sheds after harvest. Coffee berries are sun-dried on the ground, and specialised crops such as tea are typically plantation-grown and then processed locally on the estate; the local smallholder harvest is also processed on the estate.

Dried fish and meats have to be dried immediately after collection, trapping or harvest, or else decay and putrefaction will commence. Deterioration can often be as much a result of natural enzymes causing tissue breakdown as of infection by destructive organisms such as bacteria and fly larvae. Drying as a means of food preservation is examined in more detail in Chapter 4.1.

2.2.3 *Grading*

Grading is a recent requirement in Europe for all food materials, both locally grown and imported that are for sale within the European Economic Community (EEC). Any damaged or infested foodstuff is immediately down-graded with a considerable drop in value. The Food and Drug Agency (FDA) in the USA is equally stringent with regard to food quality assessments.

2.2.4 *Packaging*

Packaging may be required as part of the initial processing. It is very important as a method of protection from pest organisms, and to keep the material clean, but is often of importance in making the material easier to handle. Most vegetable and fruit boxes are designed purely for handling purposes and to protect the produce from physical damage (i.e. bruising). This topic is of such importance that Chapter 5 is devoted to all aspects of packaging.

2.3 TRANSPORTATION

The traditional practice has been to collect the harvested produce in the field and to carry it to the farmstead. But some produce will go directly from the field to the local market for sale – this is particularly true for many vegetables. Other produce, such as wheat in bulk, will go by truck directly to a regional cooperative grain store for drying and storage. Some types will be taken from the harvest field to the farmstead for initial processing and sometimes on-farm storage for a while before being transported to either a regional store or a local market.

Transportation involves time and the danger of damage to the produce, particularly physical crushing, so the food materials have to be packed in such a way as to be conveniently handled as well as protected physically. The time factor is of importance in that development of pest organisms will continue during transport, especially in the cases of produce infected or infested in the field before harvest, and cross-infestation can take place during protracted transportation. The time factor is also of importance when considering endogenous physiological processes such as the ripening of fruits.

The transportation of agricultural and other produce internationally has long been the major factor in effecting distributing of pest species from one part of the world to another. This has been of particular importance in that many exotic pests are more damaging in their new locations owing to the lack of natural controlling constraints (e.g. predators). International transportation is now generally carefully controlled by phytosanitary legislation and international agreements. Unfortunately, in a few countries phytosanitary facilities are minimal or even non-existent.

A recent development of great relevance is the now universal practice of using 'containers' for international freight. The standard 'container' is about 12 feet long by 6 feet wide and 6 feet high (or 20′ x 9′ x 9′), and the end that opens can be sealed so that it is virtually airtight. The most recently developed containers have a gas tap through which fumigation can be effected, and they also have a coloured sticker (usually with a skull and crossbones insignia) on the outside to record very visibly details of the fumigation treatment. As the containers are airtight, they are also ideal for the use of phosphine-generating tablets, so food consignments can be fumigated during transportation without any loss of time.

Another trend is for global transportation by air freight of fresh fruit and vegetables. These are sold as either exotic items or else produce out-of-season in Europe, North America and other countries, and are very high-value commodities. Usually it takes about a week or less from field to market. Most produce is fumigated en route to prevent the introduction of pests. Further details are provided in Chapter 9.

Now that most food products are sold through supermarkets, in the UK most foodstuffs come from a central store or production area, and have to be transported rapidly throughout the country to the retail outlets; usually each place receives several such successive consignments each week. Even dry-goods have to be brought in on a weekly basis, for the supermarkets have very limited storage facilities for the extensive variety of products offered, and the more perishable materials, including frozen produce and prepared ready-cooked meals, have a short shelf life and have to be carried in huge refrigerated trucks. The UK fuel crisis in September 2000 made the public very aware of the new dependence upon road transport of foodstuffs, and many retail outlets quickly exhausted their supply of perishable (and other) foodstuffs.

2.4 STORAGE

It is usual to regard produce storage in two separate categories, first the on-farm storage of harvested produce, usually in small and relatively simple structures, and the larger regional stores of greater capacity and more permanent construction. The difference between these two types is greatest in the tropics where smallholder storage facilities may be rather crude and not too effective, and losses due to storage pests can often be high. In Europe and North America, farms are ever-increasing in size, by amalgamation, and stores of great size and complexity are being used. Some of these are owned by a single large farm (company) while others belong to regional cooperatives.

Traditionally, a main concern in agriculture has been for successful long-term storage of cereal crops, and to a lesser extent of pulses, as basic foodstuffs for both humans and livestock taken from a single annual harvest. In some scattered small societies, dried fish and meats were also required to be stored overwinter for months on end. Now, in Europe and many other places a vast range of prepared foodstuffs have to be stored, but mostly this is only for a short period of time. This topic is presented in more detail in Chapter 8. During long-term storage, pest populations can increase drastically, and damage to the produce can be extensive, and cross-infestation can result. It can be argued that store construction and management is the single most important aspect of post-harvest produce protection. Nowadays, storage includes a very wide range of facilities and structures, and can even include the packaging designed for long-term produce protection.

2.5 FINAL PROCESSING

Clearly, for some produce there is no real distinction between initial and final processing. In general, the term refers to the treatment of the produce (usually commercial procedures) used to make it ready for consumption. The most important type of processing has long been the milling of grain to make flour and meals for making bread, pasta and other farinaceous materials. The milling of grain involves the use of flour mills and stores for keeping the bagged flour and meals which may be kept thus for a long period of time before being sold. Thus the final processing of cereal grains is often a two-fold affair, involving first the flour mill and then the bakery where the bread is baked; or for pasta, the premises where the pasta is made. Pasta is made from Durum wheat and comes in a wide range of types and shapes, ranging from spaghetti, macaroni to sheet lasagna and small envelopes (ravioli) to be filled with minced meat or cheese; altogether, most UK supermarkets stock about 30 different types of pasta, and in Italy this total can be doubled. In many countries, pasta is now becoming a major staple, whereas traditionally it comes from Italy, although different types of noodles are made in China. Some countries manufacture their own pasta products, but large quantities still come from Italy, which involves both storage and transportation time. In previous years in Hong Kong, macaroni was imported from Italy loose, in bulk, in large cardboard cartons. Then it was locally packed into cellophane bags of one pound capacity, labelled and sealed. In many cases, there were *Sitophilus* eggs and larvae already inside the pasta, and, of course, there was air inside the bag, and soon adult weevils were to be seen inside the carefully sealed packages. Very often the careful, often airtight packaging takes place too late because the foodstuff is already infested with eggs and young larvae that are difficult to see.

Produce such as vegetables or fruit, pulses etc. are not usually subject to final processing, except possibly for grading and prepackaging of materials imported in bulk and to be sold in supermarkets. A small proportion of fruits and vegetables may be frozen, canned or made into soups, or otherwise cooked or preserved to different levels by different procedures.

2.6 USE OR CONSUMPTION

There is commercial pressure to produce foodstuffs that are visually or aesthetically attractive, or both. At the lowest level, this involves washing vegetables and prepackaging them into polythene bags. At the other end of the scale, the nutritional composition of the foodstuff is altered (improved!) by the addition of vitamins, mineral salts and the like. Chemicals can be added which serve as flavour enhancers (E461 or MSG), while others improve the colour, appearance, or taste, and some extend the shelf-life. There are some substances that are popularly regarded as undesirable and these may be removed from some foodstuffs, the most notable cases being the bran component of wheat, and the brown husk of rice. The Oriental preference for polished rice was initially emulated in the West, but recently there has developed an ever-increasing demand for brown (unpolished) rice. And now the cycle is completed by the sale of bran as a breakfast cereal or additive, totally without nutritional value, but it adds to the bulk of the food, which helps to maintain a healthy alimentary canal, especially the bowel region.

Two recent trends are, firstly, for various customer associations to discourage over-enthusiastic use of additives – some additives produce adverse effects on people who are sensitive to them, sometimes with serious consequences. Legislation now requires a precise listing of the entire contents, including all additives, in any given prepared foods. In the past, even such poisonous chemicals as strychinine, arsenic and formaldehyde were added in small quantities to some foodstuffs to improve their longevity. Secondly, most foodstuffs are cooked prior to being eaten – sometimes raw ingredients are used, but there is an ever-increasing tendency to use pre-cooked (or part pre-cooked) dishes bought from supermarkets, and these are often the source of food-poisoning bacteria because of inadequate cooking after purchase. Thorough cooking will invariably kill most of the parasites that attack man, but this does tend to diminish the flavour somewhat.

An alternative assessment of the flow process from producer to consumer (customer) is nicely illustrated in Gorham (1991; p. 39), and here in Figure 1.1.

CHAPTER 3

TYPES OF FOODSTUFFS

When foodstuffs and other materials kept in produce stores are considered in relation to damaging pest organisms, three broad categories emerge:

- Plant materials – fresh, dried and prepared.
- Animal materials – fresh, dried and prepared.
- Structural timbers, packaging materials and dunnage.

All types of produce of natural origin that is held for a while between production and consumption, in general produce stores, warehouses, markets or shops, or carried by ships and trains, may have to be considered. Initially, the materials and pests will be the concern of agricultural entomologists/crop protectionists, then the stored products entomologists, and finally they will come under the jurisdiction of the public health inspectorate.

In most produce stores, especially those in the tropics, there will be cross-infestation to some extent, and there will also be accidental infestation of foodstuffs by insects and other organisms from other sources. A notable case originating in Hong Kong was a shipment of brassières that arrived in Holland infested with Tobacco Beetle, and there was considerable chewing damage to the fabric.

In the past, many pest surveys have either been limited to cereals and grain stores, or else they only included the insect and mite pests that can be regarded as obligatory stored produce pests. As a result, published information on recorded pests of many of the less common commodities kept in storage tends to be sparse.

Most of the plant species referred to here are grown as a number of different sub-species and varieties (i.e. cultivars); for example, rice has a known 1,000 varieties within a series of subspecies; most of the cereals exist as hundreds of different varieties. And the plant breeders continually hybridize in their quest for particular characteristics, so the total numbers of different varieties are legion. Each variety of crop plant will have its own susceptibility or resistance to particular pest organisms.

A recent development of great importance with regard to pests has been the introduction of new soft-grain varieties of maize into Africa to replace the original flint varieties that are so hard and difficult to prepare for food. But the old flint varieties are resistant to insect attack and store well, whereas the new softer varieties are devastated in storage by insect pests.

Other conspicuous examples arise from the fact that some varieties of carrot are seldom attacked by Carrot Fly, and some apples are rarely found with Codling Moth larvae.

3.1 PLANT MATERIALS – FRESH

These are materials that are generally not stored for very long, and so most of their pest problems are brought into the store with them from the harvest field. There are exceptions – potatoes and ware onions are stored in the UK for up to eight months, and apple crops can be kept for several months. Many types are now regularly frozen or canned, and some fruits are sold as juice in bottles. If carefully harvested, produce that is frozen or canned usually avoids pest problems, but can have problems of their own, such as Syrphid pupae and small snails in frozen peas, and ants in canned pineapples.

The main categories of fresh plants used as food are as follows:

Fruits - apple, avocado, banana, cherry, citrus, currants, coconut, grape, guava, lychee, mango, melons, nectarine, peach, pears, papaya, pineapple, raspberry, strawberry.

Vegetables – asparagus, aubergine, beans, brassicas, capsicums, carrot, cucumber, marrows, onions, parsnip, peas, tomato, sweetcorn, turnip, swede.

Roots and tubers – cassava, ginger, potato, sweet potato, yam.

Miscellaneous – some herbs and spices are sold fresh.

3.2 PLANT MATERIALS – DRIED

It is not feasible to list all the different types of plant material that are kept in produce stores and shops throughout the world. The list of plant materials to be found in a large herbalist's or Chinese medicine shop would be very extensive and probably not very useful here. In recent years, there has been a great upsurge of interest in herbal medicine, and many species of wild plants are now harvested and dried for this purpose.

An interesting new development is seen in the London-based company 'Fruits of the Nile', which operates in Uganda, and local smallholder farmers are drying finely sliced fruits in the sun which are then flown to Europe to an ever-increasing market. The most popular fruits are banana, pineapple, mango and papaya.

The main types of plant materials can be categorized as follows:

Cereals (large grain) – maize (corn).

Cereals (small grain) – barley, buckwheat, millets, oats, rice, rye, sorghum, wheat.

Cereal products – animal feedstuffs, biscuits, bran, bread, flour, meal, pasta.

Pulses – beans, chickpea, grams, groundnut, pea, pigeon pea, soya bean.

Roots and tubers – cassava, ginger, potato, yams.

Nuts – almond, Brazil, cashew, groundnut, coconut, hazel, macadamia, pecan, pinenut, pistachio, walnut.

Dried fruits (whole) – apricot, currants, date, fig, prune, raisin.

Dried fruits (sliced) – banana, mango, papaya, pineapple.

Oil seeds – castor, cotton, copra, groundnut, mustard, rape, sesame, safflower, sunflower.

Beverages and spices – chillies, cinnamon, clove, cocoa, coffee, soya sauce, spice leaves (bay, rosemary, sage etc.), spice seeds (coriander, nutmeg, pepper etc.), tea, tobacco.

Miscellaneous – bean curd (tofu), book bindings, cotton, cork, dried seaweeds, dried vegetables, flower bulbs and corms, mushrooms and various fungi, seeds for propagation, yeast.

3.3 ANIMAL MATERIALS – FRESH

Most meat is sold dead but fresh, and periods of storage are brief, except for frozen meat which may be kept at -20°C (deep freeze) for months or even years. Some meats are likely to act as host for a range of parasites, such as Protozoa, helminthes and Nematoda as these occur naturally in the food chains, and some are particularly susceptible to bacterial infections. Beef carcases are traditionally allowed to hang for some days under cool conditions to allow natural enzymes to start tissue breakdown. This tenderizes the meat and makes it easier to chew as the connective tissue breaks down, and also many people claim the flavour is enhanced. The UK practice of hanging game animals in an outhouse for a week or so before consumption originated in part for the need to start tissue breakdown to tenderize the otherwise very tough muscle tissues. The other factor was that in the UK, the traditional roast grouse enjoyed in London restaurants had to be transported by road from Scotland, where they were shot, and this journey usually took a week by coach, so the birds were 'ripe' by the time they reached their destination, and so for the afficionados the tradition of eating 'ripe' game meat was established.

Fish and shell-fish (Crustacea, Mollusca) are either kept alive or else stored on ice for a rapid sale in the fresh fish-markets. Once the animal dies, decomposition starts rapidly, and retail value plummets. One consequence is the ever-increasing market for frozen prawns as well as the more traditional tinned shell-fish (crabmeat, shrimps, clams etc.), and tinned marine fish (herring, mackerel, sardines, tuna and salmon).

The main categories of fresh animal products are as follows:

Meats – beef, mutton, lamb, pork, veal, goat, rabbit; bones and milk can be included; 'bush-meat' in some countries.

Poultry – chicken, duck, goose, turkey, quail, pigeon, ostrich, eggs.

Game animals – game birds (grouse, pheasant etc.), hare, pig, venison, and a wide range of tropical game animals (antelope, deer etc.).

Fish – Marine fish – a very wide range of species is caught and eaten.
– Freshwater fish – trout and carp are especially important, but other species are also eaten. Many species are now reared commercially. Caviar, some other fish eggs.

Shell-fish – Crustacea – prawns, shrimps, crabs, lobsters, crayfish etc.
– Mollusca – clams, mussels, cockles, oysters, snails (including the Edible Snail), squid, octopus.

Miscellaneous – Oriental markets include frogs, turtles, lizards, snakes, wild birds, civet cats etc.

3.4 ANIMAL MATERIALS – DRIED

Some of the meats are traditionally sun-dried, but mostly they are smoked and dried. Such meat, if kept in dry storage, can last for many months. Smoking can adequately preserve the outer meat surface, but parasites and pathogens can survive in the middle of the pieces of meat if there is no proper cooking prior to consumption. Meat pastes and similar concoctions are often left with an air-tight seal of solidified fat on the surface in the jar which makes an efficient barrier to micro-organisms. Bacon, ham, salami and other sausages have the outer surface smoked and covered with a layer of salt as preservative. Many also have an outer layer of cloth, by which they are hung from hooks in the kitchen ceiling rafters. Cheeses are usually wrapped in a cloth and allowed to develop a hard protective rind. Pressed (corned) beef is cooked, pressed and canned (tinned), and can then be kept in storage for years if need be.

Dried fish have played a very important role in human history as a source of food both for long journeys (Vikings etc.) and long winter periods (Asia, North America), but nowadays they are mostly used in Africa and parts of tropical Asia. Oily fish, if not too large, can be preserved by drying (curing), the oil helping to preserve the flesh. Trout and salmon are either flattened (if small) or filleted (if large) and smoked or sun-dried.. The North American 'Indians' and some Inuit tribes in Alaska and thereabouts (including parts of Siberia) were totally dependent upon dried salmon for their winter survival. Herring and mackerel dry well, especially if smoked. Dried cod (split, salted and sun-dried) was the mainstay that permitted the exploratory pursuits of the Vikings 2,000 years ago, and later the seafaring Portuguese explorers. In Africa, sun-dried cichlids (especially *Tilapia*) from Lake Victoria, Malawi and the other Rift Valley lakes are the protein source for several local tribes. Ideally, the fish and some shell-fish are subjected to some form of curing before being smoked and dried. If salt is available, the fish can be soaked in brine for a while, or else fine salt crystals can be rubbed into the cleaned flesh. An alternative to salt in some regions has been a mixture of rum and molasses. One traditional English delicacy is potted shrimps, where cooked, peeled shrimps are packed in small pots under a layer of butter. The Royal Navy and many British merchant ships relied on salt pork (raw pork pieces in barrels of brine) to feed the sailors on long voyages.

The main types of meats dried or otherwise preserved include:

Meats – beef, mutton, pork, venison, bacon, ham.

Meat products – biltong, pressed (corned) beef, salami, sausage, pork pie, haslet, meat paste, dried milk.

Poultry – chicken, duck, dried eggs, dried bird nests (S.E. Asia).

Fish - Marine fish – salmon, cod, herring, mackerel etc.
- Freshwater fish – trout, *Tilapia* and other Cichliidae, eels etc.
- fish products include fishmeal, fish paste, dried shark fins.

Molluscs – abalone, clams, mussels, oysters, octopus, scallops, squid.

Miscellaneous – animal feedstuffs, bones and bonemeal, cheeses, Chinese medicines, leather, oyster sauce, fish sauces, shrimp paste, skins and hides, wool.

For many years, canning (USA) or tinning (UK) has been used to preserve a wide range of fruits and meats including fish. Sardines (pilchards etc.) occur in vast shoals at several points in both the Atlantic and Pacific Oceans, as well as salmon, and local industries evolved to exploit them. The fish are caught, cleaned, cooked and canned locally, and then exported to all parts of the world. A similar industry was established in Europe using herring, and later tuna were added to the list of species extensively canned. Similarly, the vast cattle ranches in the Argentine solved their distribution problem by the development of canning, and pressed (corned) beef became a worldwide commodity.

A very recent development in food preservation is the now widespread use of refrigeration, both commercially and domestically, so that perishable foodstuffs can be safely kept for weeks or months. This has revolutionized how we live and feed ourselves with a greatly increased diversity of diet. All types of meat, fish and shellfish are regularly available frozen in most supermarkets, and in the UK, to eat Pacific Mussels (*Perna viridis*) and Tiger Prawns is now commonplace.

3.5 STRUCTURAL TIMBERS AND PACKAGING MATERIALS

Some of the animals found here (in stores), especially rats and mice, can be of prime importance when they are introduced accidentally to virgin islands and other lands. Some insects can emerge from timbers or packaging, and damage produce in storage. Other insects may be quite harmless to the produce, but their appearance may be alarming. Rice Water Weevil was accidentally introduced into Japan from the USA (Florida) in rice straw dunnage.

The main categories of materials are as follows:

Wooden timbers – mostly pine and other softwoods, but some hardwoods are used for building construction.

Bamboos – in parts of tropical Asia these feature prominently for construction.

Wooden boxes and crates – may be made of any cheap local wood.

Cardboard boxes and cartons – these offer little resistance to many animal pests which can gnaw holes in them. The large drums made of compressed card and with metal lids tend to be very successful for transportation of valuable produce.

Sacking – sacks and bags made of Hessian, jute and other 'natural' materials are still widely used for handling grain. Plastic materials may be used, but if entire, like polythene bags, there may be moisture condensation, and holes can be gnawed in the plastic. Woven plastic allows air circulation and insect pest entry. Some rice sacks are now made of woven plastic strips and lined with polythene (Figure 5.6).

Straw dunnage – in the past, this has often been the source of rats, mice and various insects and micro-organisms being introduced to all the isolated Pacific islands as well as other continents.

CHAPTER 4

FOOD AND PRODUCE PRESERVATION

4.1 DRYING

Drying is the first technique used by man to preserve foodstuffs, and undoubtedly goes back to Stone Age times. Later, the objectives of dehydration of produce evolved, and drying became an important stage in the chain of food production. It provides extended storage life, ease of handling, and enhancement of quality. Clearly, initially, the scale of operation was small, and the quantity of moisture removed was not great, but nowadays some establishments are on a huge scale where many tons of produce per hour are dried, and a great deal of water can be removed.

In the majority of foods, the water content is of several types. Firstly, if the produce is wet, there is free *surface* water, which almost always has to be removed. *Free* water is that which fills the cavities of the structure, trapped by capillary forces. The total quantity of free water is directly correlated to the pore structure and porosity of the product. *Bound* water is inside the cell walls, and attached by the hydrophilic properties of the cell, and clearly requires a great deal of heat energy to break the bonds. Finally, there is *water vapour* which, mixed with air, occupies the empty holes that are unsaturated by free water. The specific water content of cellular structures when saturated but without having any free water within any cavities is known as the hygroscopic moisture content. Above this level, the product is non-hygroscopic, but below it is hygroscopic.

The hygroscopic behaviour of a material is a key factor in the exact assessment of the drying process. One complication with the use of salt (NaC1) in the curing/drying process is its strong hygroscopic properties – in moist air, it will quickly deliquesce.

4.1.1 Air-drying

Air-drying is generally only successful in a steady flow of air (wind) with a fairly low relative humidity. The wind will both lower the temperature (usually rapidly) and the moisture content (quite slowly), and there will be structural changes and some deterioration of certain components of the foodstuffs. The air supplies the external surface of the product with energy, and is also responsible for the removal of the water vapour. Clearly, the wind (air) has to have a low relative humidity for the drying process to be effective. Because the process is slow, the time required for drying can be quite long. Rain has to be avoided, so the usual practice is to build open-sided sheds with an overhanging roof to keep off the rain. Air-drying of tobacco leaves is practiced in southern and central Africa very successfully. Once the sheds are built, this is an inexpensive method of drying the tobacco leaves.

In olden days in the UK, the usual practice at cereal harvest was to cut the stems and then to bind them in bundles which were placed upright in groups (stooks). The bundles of stems were then left in the field of stubble for some days or weeks in order to air-dry sufficiently before being threshed for the grain. In parts of the tropics, cereal crops are left in the field to dry before harvest.

4.1.2 Sun-drying

This usually involves the spreading out of the material in the sunshine so as to permit free air circulation, and the result is often a mixture of air- and sun-drying. But the energy input is provided by the solar radiation, so that the temperature of the product can be quite high (rather than being cooled by the wind). The overall process can be quite rapid in sunny weather in the tropics. The main drawback is that solar radiation is an intermittent source of energy, and the sunlight will modify some properties of the dried product (colour, flavour, vitamin content etc.). But in rural areas of the world, sun-drying of produce is very important, particularly since it is free, renewable and non-polluting.

The most commonly used form of direct drying is termed 'open-air sun-drying', whereby the produce is exposed to sunlight. In Africa, smallholder coffee is still traditionally sun-dried by being laid out on mats on the ground, as also are shelled maize and sorghum. Rice in South East Asia is usually dried in the sun after threshing. The energy for the drying process is provided by the radiation transfer from sun to produce. Limitations to this method are wetting by rain, and contamination by dust, dirt and insect pests. The solar radiation concerned with sun-drying is in the wavelength range of 0.3-3μm, and corresponds to the range between ultra-violet (UV) and near infra-red.

Recent improvements include the manufacture of drying cabinets with several trays for the produce, and a solar panel for collecting the solar radiation. A transparent cover will protect the produce from rain, and the cabinet usually has a ventral slit for ambient air intake and a dorsal aperture for the humid air to exit. This is the type of simple solar drier developed by Tropical Wholefoods (London) for the production of dried sliced fruits (pineapple, banana, mango, papaya) for the Ugandan 'Fruits of the Nile' company, which exports the dried fruits to Europe.

The colour of some foodstuffs is affected by sunlight, particularly the UV radiation. This can have an unfavourable effect on green vegetables and fruit when the ascorbic acid (Vitamin C) content is considerably reduced. Alternatively, some products such as dates, grapes, cocoa beans and prunes require sunlight for development of full flavour and the correct dark colour.

Meat and fish may be spread out on a flat surface, usually a horizontal wooden lattice, which permits free air circulation and receives maximum solar radiation, or may be hung vertically from a wooden framework where air- and sun-drying may be equally important. Meat has to be cut into thin strips for rapid drying. Fish are cleaned (guts removed) and may be left intact if small, or split longitudinally and flattened. With meat and fish it is vital that the drying process is rapid, otherwise the flesh is likely to be fly-blown, and there will be the start of putrefaction by the natural enzymes present in the tissues. Clearly, such drying is only feasible in the hot, sunny tropics or in other places during a hot, sunny summer period. Air-dried freshwater fish are the most vulnerable, while marine fish are rather salty, and this deters some insect pests. Fish such as eels, salmon and trout have a particularly oily flesh which generally preserves well, although it is usually smoked.

The flattened dried ducks so familiar in the Orient appear to be partially preserved by the natural thick layer of body fat under the skin. Air-dried octopus and squid resist pest infestation quite well without salting, for the flesh is tough and rubbery, and dries hard. These animals are displayed in the windows and stalls of food shops, often just hanging from hooks. The more delicate molluscs, such as abalone and scallops, are traditionally air-/sun-dried, and then stored in large glass jars with screw-top lids where they are well-protected, for they are of very high value. Oysters and mussels are usually smoked and dried.

Indirect solar drying is generally more energy efficient than direct sun-drying, but often relies on the use of a very large surface area of energy collector. The heated air passes up through the perforated trays in the drying cabinet, and exits at the top. When developing a drying cabinet using solar radiation, it is necessary to optimize the surface area of the solar collector or the drying time. The choice is closely linked to the type of produce to be dried, and its moisture content. If the drying process is too slow, deterioration and spoilage may occur. Conversely, under certain temperature conditions, overheating or overdrying may occur. A higher temperature level generally results in a shorter drying time.

4.1.3 Fluidized bed-drying

This is an industrial process whereby air is made to flow through a bed of solid particles supported on a grid or perforated plate. Frictional drag will cause a pressure drop across the bed. The drag, and hence the pressure drop, will increase as the air velocity increases, and when the frictional force overcomes the weight of the particles, they move apart, and the bed expands and begins to behave like a fluid (giving rise to the term 'fluidized bed'). Food particles suitable for this method of drying should be in the size range of 20 μm to 10 mm, and should be uniform in size and shape, and should not be too sticky. Cereal grains fluidize readily over a wide range of moisture levels.

4.1.4 Spray-drying

Nowadays, this is the most commonly used method for drying food liquids and slurries. The foodstuff liquid is converted into a spray or fine mist and then mixed with heated air. The result is very rapid drying, converting the liquid droplets into powder particles. The particles are small in size, 10-200 μm, and so have a very large surface area for drying. Drying times are usually short, 1-20 seconds, and there is also a significant evaporative cooling effect, so the surface temperature of the droplet remains low, and heat damage to the product is minimal.

A large range of commercial driers are now in use worldwide, and the main food products that are spray-dried include milk (especially skimmed milk), whey and buttermilk. Icecream mixes, cheese and butter are other dairy products that can, with care, be dried. Whole eggs, egg yolk and egg white can be dried – usually for commercial use. Instant coffee is spray-dried on a vast scale, and sold in supermarkets worldwide, and to a much lesser extent tea can be produced for sale in an instant form. Whole fruits and fruit juices as well as various vegetables can also be spray-dried into a powder form for industrial food preparations (soups etc.). Apart from tomatoes, the other fruits and vegetables need to have additives to facilitate drying – often liquid glucose is used. Meat and fish extracts and purées can also be dried.

4.1.5 Kiln-drying

This relies on either the use of large ovens or kilns (some are termed 'furnaces', but these are usually for the very high temperatures needed for metal melting or the combustion of materials). Commercial ovens are basically very similar to domestic ovens but larger. They can be heated by gas, electricity, or combustible fuel such as oil, coal, wood or charcoal. Many ovens rely on convection for hot air circulation, but they can also be fan-assisted.

There are many different types of produce-drying that can come into this category, ranging from small simple arrangements to large complicated and sophisticated equipment with total automatic computer control. For small quantities (of seed, for example), a wooden box with an internal perforated tray and a couple of light bulbs or a small paraffin lamp underneath works very effectively, and is widely used for drying insect specimens and plant herbarium sheets. In the UK, ware onions and some other vegetables are dried of excess moisture (prior to storage) in a long enclosed barn with a heater at one end and a huge fan at the other, so that warm air is pulled through the 10-20 foot layer of onion bulbs (Figure 8.10).

Grain usually goes into tall silos for temporary holding (Figure 8.13) before being passed through a tall drier (Figure 8.14). The grain falls through a stream of hot air in the drier, and this is repeated until the required level of dryness is achieved. The dried grain is then carried away on a conveyor belt (Figure 8.14) to be stored in large storage bays (Figure 8.11) or large, individual holding silos. In the example illustrated here, at Orby in Lincolnshire, UK, there are six silos for initial holding, and one double drying tower, with six very large storage bays behind them. The structures are all joined by conveyor belts on which the grain is carried. Each storage bay has a capacity of 3,000 tons, and it has the floor grooved by ventilation channels which run the full width of the bay at 10 foot intervals. These are connected to ventilation outputs on the exterior (Figure 8.12), and air can be forced through the grain pile from the exterior on cold dry days; the air can either be sucked in through the outlets into the floor grooves and then up through the grain, or, alternatively, the outlets can be used to suck the air down through the grain pile and out to the exterior via the floor grooves. In the UK, this forced air circulation is usually done on cold dry days, and the effects may last for several months. The whole system is under computer control, and can be completely automatic – there are temperature and humidity probes placed strategically throughout the system. For more details, see Chapter 8. At harvest time, the grain usually has a moisture content of 18-20% in the UK, so it has to be dried to 14-15% before storage. Here the regional store has its own permanent drying unit, for grain usually needs drying before storage, and during a prolonged period of wet weather the grain could get damp again and require another session of drying. But in drier regions it may be more useful to have smaller mobile drying units that can service the local farming community and their produce. In Lincolnshire in 1990, after warm dry summer weather, the wheat harvest had a moisture content of 12%, and the driers were not needed at all; but a couple of years later, the grain had a reading of 23% at harvest, and the drier ran night and day for several weeks..

4.1.6 Freeze-drying

Essentially a recent development used mostly for the preservation of soft biological specimens in museums, such as green plants, caterpillars, and other soft-bodied or small animals – animals up to the size of a rat can be freeze-dried. Modern museum displays are now usually incredibly life-like because of this process.

Freeze-drying in the food industry (also called lyophilisation) is the process of removing water from a product by freezing it quickly and then subliming the ice to vapour using a vacuum chamber. Freeze-dried foodstuffs have a long shelf-life and low weight, and are generally of high quality and sought after for culinary use.

But despite the success of this technique in biological laboratories, in food technology the cost is very high, and there have been scale-up problems. The result is that it is used mainly for expensive instant coffee, soup mixes, some Mollusca, herbs and military rations (Dalgleish, 1990).

4.2 SALTING

Once really dry, meat, fish and plant materials are usually very resistant to bio-deterioration and store well, but the process of drying is fraught with risk. Because of the difficulty of air/sun-drying meat and fish without it becoming fly-blown or infected with bacteria, there has long been a tendency to apply salt (sodium chloride) when drying flesh. The addition of salt does also make the flesh more palatable once the excess has been removed.

Salting is done either by dipping or a short soak in a brine solution (saturated salt), and then allowing to dry, or else salt crystals are rubbed into the flesh, which is then allowed to dry. In a few cases, salt crystals may be poured over a layer of fish, and then further layers of fish and salt added to fill a box-like container.

The surface of the meat or fish dries more rapidly, and the salt layer deters most insects as well as fungi and bacteria, which will not develop on a substrate of high salinity. As long as the ambient air remains relatively dry, salted fish and meat keeps very well; but salt is hygroscopic, and at high air humidity levels it will absorb water vapour and will deliquesce. A few insect pests will attack salted fish (e.g. *Dermestes frischii*), but this is quite rare because salting is a strong deterrent.

Highly salted beef and fish may be eaten without cooking, but the heavily salted flesh usually has to be soaked in water (to remove excess salt) and cooked.

Salt in the form of brine is used as a medium for pickling.

4.3 BOTTLING

This is a surprisingly old system of food preservation, and it consists basically of heating or cooking the fruit and vegetables and storing the hot produce in sterilized wide-necked glass jars. The glass lid has a rubber ring seal and a spring/clip system for closure, and the lid is closed while the contents are still very hot. In this way, the produce is sealed inside an air-tight container without any living pest organisms included, and will keep for months or years. The bottle or jar is usually topped up with either sugar syrup, vinegar or brine, according to the product being preserved.

The origin of bottling was probably on isolated rural farms, where there was a need to preserve harvested fruit for use throughout the year. It is now developed commercially for the storage of jams and sauces mostly, with screw-top lids.

4.4 CANNING

It is likely that this idea developed from bottling, and was more suitable for small-scale commercial enterprises, and, of course, this led to large-scale canning factories for fish (sardines, salmon etc.) and meat (processed beef etc.). For the sardine, pilchard and salmon fisheries, canning factories were built in the isolated coastal regions where the fish were found, and the finished product was then exported throughout the world.

The can or tin is made of thin sheet mild steel plated with tin in order to prevent rusting, usually cylindrical in shape, with a base pressed on to the bottom of the cylinder, and a lid pressed on to the top after filling. As with bottling, the empty can is sterilized, and the cooked meat or fish placed inside before sealing. Often, the liquid preservative/filler used is vinegar, or salt solution for fish and shellfish, and sugar syrup for fruit. But nowadays, especially in Scandinavia, small tins of preserved herring, mackerel fillets, sardines, smoked oysters and mussels, and the like, contain vegetable oils, mustard, curry, tomato and other sauces.

If the cans are properly made and sterilized, this is a very effective form of food preservation, but sometimes cans are substandard or are physically damaged during handling, then leak and become contaminated. Most cans will contain some live bacteria, which will eventually develop and spoil the food, so most cans have a shelf-life of 1-3 years. Some of the worst cases of microbial food poisoning have been attributable to cans of crabmeat, shrimps, and meats that had deteriorated and become infected.

Some canned produce is quite expensive, and so the presence of dead insects, or bacteria, inside the cans can result in whole consignments of produce being rejected. Tinned pineapple with enclosed fruit flies, and corned beef with ant bodies inside, are quite a regular occurrence.

4.5 PICKLING

Another ancient method of food preservation is pickling, where the produce (usually flesh, but sometimes fruits or vegetables) is steeped in a preservative liquid in a suitable container. The original idea was purely for food preservation (raw beef, pork, fish, fruit, vegetables), but the fish and some of the fruits and vegetables developed an appealing flavour which was sufficiently attractive for this treatment to become a standard method of food preparation. The most successful examples include pickled herrings, eggs, onions, walnuts, and some fruits.

Four main types of solution are traditionally used in pickling, with water as the basic liquid. Should the chemical concentration be (or fall) too low, the preservative value of the solution is reduced, and biological contamination or decay will result. Dilution will automatically result when the food material is added to the solution, particularly if the food is fresh and succulent, so care has to be taken to keep the solution sufficiently concentrated to act as a preservative.

Brine, a saturated solution of salt (sodium chloride), is often used for preserving pork and beef, and sometimes fish and eggs. The salt successfully repels insects and mites, and inhibits development of bacteria and fungi as long as the concentration remains high. The solution should ideally be kept saturated. This was the standard method of meat preservation for ships' crews (pork in barrels of brine) until as recently as 1925. Some meats and fish are steeped in brine for a while before being hung out to dry.

Pickling in vinegar probably started in antiquity with the use of sour wine – where the alcohol has been converted by bacterial action to acetic acid. The foodstuffs traditionally pickled in vinegar are fish, onions, gherkins (small cucurbits), eggs, some other vegetables (cauliflower, cabbage etc.). But acetic acid can be invaded by bacteria if the concentration falls, and moulds will grow on the surface quite rapidly. Some fly larvae can tolerate weak solutions of vinegar. Because of the added flavour imparted to the food by the many different types of vinegar used, pickling has remained a very popular form of preparation, especially for some types of fish and vegetables. Pickled raw herring (and mackerel) is a national dish in Scandinavia (as is pickled red cabbage in Germany – sauerkraut – and Russia; and pickled onions, gherkins, eggs, pigs' feet and walnuts are well-known appetizers in Europe and North America. In the UK, the popular dish of fried fish and chips would never be so popular without vinegar on both the chips and the fish. A method of food production involving the use of vinegar is the production of chutneys and vegetable pickles.

Alcohol is effective for storage, but is generally expensive, and only really reliable at concentrations of 40% or more (e.g. bottled peaches in brandy etc, especially popular at Christmas). Its use is mostly limited to museum collections and Chinese medicines, where the items preserved are of high value. For edible items, ethyl alcohol is the usual chemical preferred because the other commercial alcohols are toxic. The major problem with alcohol is that on exposure of the liquid, it is the alcohol component that evaporates most rapidly. Dilute solutions are no protection against most micro-organisms, endogenous enzymes, and many insects.

Sugar syrup, as long as it is saturated, will inhibit bacterial development and fungal spore germination. But syrup is attractive to many different insects, although only a few insect larvae are able to survive actual immersion in a saturated sugar solution. Fruits such as peach, pear, cherry, strawberry, raspberry, pineapple and mango are the main types of food which are preserved in syrup. Most of these are now alternatively tinned in natural juice, often with apple juice as the sweetener. Similar preservative methods involving sugar include the making of jams, and also candied or crystallized fruits (and ginger). Many types of confectionery are relatively safe from microbial activity, but are attacked by a number of different insect pests in storage and in domestic situations.

4.6 SMOKING

This is actually a method of part-cooking as well as preservation, and the meat is said to be 'cured' when the operation is completed. Often salt is used in the process, and sometimes materials such as rum and molasses are used – this will have a preservative effect, but will also be a flavour enhancer. The use of salt is commonly seen in the curing of bacon and hams. The process takes some time, and essentially it dries the flesh somewhat and at the same time covers the outer layer with a concentrated mixture of tars, phenols and other chemicals that have a strong inhibitory effect on pests and micro-organisms. Sometimes fish and poultry are smoked at a higher temperature, in which case the flesh is partly cooked.

The usual meats for smoking include the oily fish – salmon, trout, eel, herring and mackerel, as well as pork products, venison (deer flesh) and some poultry. In rural communities in both Eurasia and North America, it was the practice to have a smokehouse downwind of the house for domestic production. The type of wood burned in the smokehouse can influence the final flavour of the fish or flesh, and it is common to see on commercial products labels which stipulate that the fish etc. was smoked over hickory or oak chips.

Some very perishable, easily contaminated shellfish such as oysters and mussels are very successfully smoked and dried; some are stored dry in large glass jars, but others are immersed in edible oil and canned.

4.7 FERMENTATION

A few food products are made by a fermentation process, usually under anaerobic conditions which inhibits pest organism development (except for a few bacteria), and some of the by-products are inimical to most living organisms. An ancient system first recorded by the Romans consists of digging a hole on a sandy beach, lining it with wet seaweed, and filling it with small dead ungutted fish; finally, a layer of seaweed is placed on top, and covered with a thick layer of sand. After 6-12 months of underground fermentation, the foetid paste is dug up and sealed in air-tight jars, and is used sparingly as a flavouring in Oriental cooking. A similar process involves the use of heavily salted shrimps or small prawns. This had been a widespread practice throughout the coastal regions of the Orient for aeons, and was used by the Romans two millennia ago. A very successful use of fermentation has long been practiced in the Far East, where soya beans (soy beans) are fermented into bean curd (tofu – Japan). This legume is the richest natural vegetable food known, and has manifold culinary and agricultural uses. The other product of this fermentation is soy sauce, used as a flavouring in Oriental cooking, and which in turn can be used to make oyster sauce.

4.8 ADDITION OF PRESERVATIVES

In the past, it was a common practice to add various chemicals to some foodstuffs in order to preserve them. As might be expected, these chemicals have to be general biocides in order to be effective in preventing pest development in the produce. Traditional additives included small quantities of arsenic, strychnine, formaldehyde, and sodium nitrate. Some of these are accumulative poisons in man, and repeated small doses can eventually be lethal, so their use was gradually discontinued. However, the practice was essentially so useful that prepared meats (sliced ham etc.) are still treated with both preservatives (such as sodium nitrite and potassium nitrate) and anti-oxidants (ascorbic acid, sodium ascorbate), but in carefully measured proportions thought to be quite safe. Various types of animal feedstuffs and other foods have selected antibiotics added to minimize microbial activity.

Addition of pesticides could be included here, but this topic is dealt with in more detail in Chapter 20. It should perhaps be stressed that there is more and more legislation in the EEC countries and elsewhere to prevent, or at least minimize, food materials being adulterated with pesticides. The two main concerns are the danger of chronic chemical poison accumulation in the human body, and contamination of natural food webs. But for many years it was the practice in the USA to add malathion to grain for export to reduce as far as possible insect pest populations and damage to the grain. Similarly, food for export was regularly fumigated with ethylene dibromide and others.

Section 4.13 includes some of the types of additives that can be used to ‘improve’ foodstuffs.

4.9 HEAT TREATMENT

The object of heating food materials is to raise the temperature of the food to a level lethal to the pest organisms. The ultimate heat treatment is obviously cooking; the use of high temperatures below this level is seldom feasible, except for smoking, curing and milk pasteurization, because at such temperatures there would be explosive development of pests, particularly bacteria.

Hot water dipping treatment of bulbs and some fruits is done very successfully to kill some nematodes, bulb mites, and fly larvae. Kiln treatment of timber successfully kills most timber insect pests.

If heating is to be used successfully to kill certain pest organisms, it is most important to ensure that the central region of the produce has reached the required temperature for the necessary period of time. Tapeworms and various Nematoda are transmitted by meat which has not been sufficiently cooked so as to kill the larvae. The centre of beefburgers or thick steaks is often left quite raw when the outside might be charred. In point of fact, we often prefer to cook our meat this way for the sake of flavour and taste, but we should be aware of the risk we run. But meats such as chicken, where there may be a risk of *Salmonella* or other bacteria, should be cooked thoroughly to ensure their destruction.

4.10 COOLING AND REFRIGERATION

The rate of development of biological organisms depends primarily on temperature and food availability, within acceptable limits of atmospheric or food moisture. In a food store, food for pests is unlimited, but pest development can be curtailed by lowering the ambient temperature. There are basic differences between produce /food stores in the tropics and those in temperate regions. In Europe and North

Figure 4.1. Chilled food display cabinet in supermarket, with assortment of meats and vegetables; Kuching, Sarawak.

America, in unheated stores the winter is sufficiently cold that all pest development will come to a standstill. Most temperate insect pests survive there only because they are able to enter a state of diapause, and to wait for the warmer weather in the spring. The accidentally introduced tropical pests will usually die during the temperate winter, but they do survive in heated stores and domestic premises. In the tropics, continuous pest development in stores usually occurs, and at high temperatures pest development is very rapid. Many insect pests might only produce one generation per year in Europe, but could have 5-10 generations in the tropics.

In North America, the wheat production zone extends from Texas up into Canada, and the grain is stored in large terminal elevators – these are massive, regional bulk storage centres. After initial treatment, the grain is stored here. During the winter, the grain can be cooled below 10°C by forced aeration, and this prevents any insect propagation.

The lowering of the ambient temperature by even a few degrees can have a significant effect on the activity level of individual insects and the rate of an insect pest population increase, and thus on the damage level to the produce. This approach is being used in some integrated pest management programmes, with considerable success (see Chapter 21). Chilled food materials of all types have an extended storage and shelf-life, and part of the new practice of supermarket retailing involves extensive shelf storage of chilled foodstuffs.

Refrigeration is the ultimate form of cooling; firstly, the temperature is lowered to the point of freezing (0°C), and then lower in some cases. As the temperature falls, the activity of the animal pests decreases and then finally ceases. But death may not occur until a much lower temperature is reached. For complete cessation of biological activity, a temperature of about -20°C is recommended – this is the temperature used in domestic deep freezes and most commercial freezers, but some commercial food stores are kept at -30°C. Most microbial activity ceases at about -10°C, and this temperature will kill most, though not all, insects and mites; they will be inert, but will recover when the temperature rises. Most biological enzymes need a temperature lower than -18°C for inhibition. This is why food freezers are kept at -20°C.

The literature on the effects of low temperatures on stored food pests is somewhat confusing. As the temperature falls, the activity level slows and finally ceases, but typically death does not occur until a much lower temperature is achieved. Similarly, the duration of a cold period is important because many organisms can survive a short period of low temperature, but not prolonged exposure.

Some foods can be easily frozen and thawed without any damage, but most foods contain some free water, and have to be frozen with care. Slow freezing will result in ice crystal formation, which destroys cell structure. But rapid freezing can supercool the water and turn it into non-crystalline 'glass'. Commercial quick-freezing is usually done by cold air-blasting. Cryogenic freezing using liquid nitrogen is very effective but expensive. Similarly, freeze-drying is very successful with smaller biological items, but is expensive and is not widely used for food materials. Most fruits are not suitable for freezing, and neither are some vegetables, and these are stored at temperatures of around 3-5°C to avoid ice crystal formation.

Figure 4.2. Frozen food cabinet (chest style) in supermarket; Kuching, Sarawak.

An ordinary domestic refrigerator is typically kept at about 5°C, but the basal region (where vegetables are kept) may be higher. And if the door is opened too often, the interior of the fridge will be warmer as the cold air is lost. At these temperatures, some insects can still develop, and many micro-organisms can grow. Thus, while an average domestic refrigerator is a vital component of most households, and also a range of food retail outlets (cafés, restaurants etc.), it must be emphasized that the protection it affords perishable food materials is limited – it does not kill many pests, it merely slows them down. But the domestic deep-freeze at -20°C does give a high level of protection to most stored foodstuffs. The combination of domestic refrigerator and freezer makes the average household food preparation régime relatively simple, but, of course, it is subject to the vagaries of human behaviour, and we all make mistakes which can lead to food deterioration and poisoning. And the whole food industry is now concentrating on prepared foodstuffs and frozen raw materials.

There is an unfortunate tendency to assume that foodstuffs in a refrigerator are being protected indefinitely, but this is not so – the micro-organisms on the food are all developing, albeit rather slowly, and the food is spoiling. Adult flies often gain access to a fridge interior even if the door is only opened briefly, and after an overnight stay can usually manage to fly slowly away when the door is opened.

As mentioned in Chapter 20, the tropical insect and mite species can be killed quite easily by refrigeration – most tropical species cease development at about 18-19°C. It must be remembered, though, that death results from a combination of low temperature and time. And some species will have a diapause, usually in either the egg or pupal stage, so that although larvae and adults may die because of the cold, the population will survive through the diapausing individuals.

4.11 CONTROLLED (MODIFIED) ATMOSPHERE

This approach started initially by sealing foodstuffs (grain mostly) into air-tight stores – if the container was filled to the brim, the amount of air included was quite small, and the oxygen was soon used up by the biological organisms there, and there was a concomitant increase in carbon dioxide. So, after a while, when most of the oxygen had been used, the atmosphere in the store was a mixture of nitrogen and carbon dioxide, and quite inimical to life. But obviously there will be some initial damage by pest organisms while the oxygen lasts.

Thus, the next development was to replace the air in the produce container with a repressive gas such as carbon dioxide. This is not toxic, but has a suffocating action which kills the pests quite effectively, and, being heavier than air, it can be used to fill a container easily as it displaces the air from base to top.

But the latest development is the use of a totally inert gas such as nitrogen. Nitrogen is relatively easy to obtain and use, and can be purchased worldwide in steel pressurized cylinders. Small packets of nuts, biscuits and the like are now mostly packed with nitrogen for a far greater shelf-life, and the food is fresher when the packet is opened. Figure 4.3. shows a packet of 'snacks' manufactured in Malaysia, packed in a nitrogen atmosphere.

It could be said that the ultimate in controlled atmosphere is vacuum-packing, where there is no atmosphere at all – this is mentioned in section 5.9.

4.12 IRRADIATION

A relatively recent development in food preservation is irradiation using sources of X-rays (usually cobalt-60) or γ-rays (caesium-137). The main objective is to preserve food from bacterial activity, which causes more spoilage than any other pest organisms; but a treatment that will kill bacteria will kill all the other pests and parasites. Most countries have had a national food irradiation research project which started in the late 1960s, and they have experimented with the protection of a wide range of foodstuffs in addition to grain (wheat and rice especially). Most countries have also been particularly concerned with using radiation to inhibit onions and potatoes in store from sprouting. Irradiation slows down the natural process of ripening, usually by the control of growth hormones.

Figure 4.3. 'Party Snacks' manufactured in Malaysia, "Nitrogen packed for freshness."

Extensive testing in the UK, USA and other countries has shown that irradiated food appears to be quite unaltered and quite safe for human consumption, except that there is some destruction of Vitamins A, B, C and E. However, in most countries, the general public has not been enthusiastic about the idea of irradiated food – many people think that the food may become radioactive! Stored grains can have insect pest populations destroyed successfully, and various meats, fish and other foodstuffs have a greatly increased shelf-life. In Hawaii, fruit growers are interested in a scheme to irradiate papaya fruits for export to kill fruit fly larvae instead of using a hot water dip, and mangoes are also being irradiated to kill insect pests.

But the consensus of professional opinion inclines to the view that, in the immediate future, widespread food irradiation is unlikely because of public concerns, but the food industry and governments are definitely interested in large scale foodstuff treatments, and they recommend further studies.

4.13 FOOD IMPROVEMENT

Since time immemorial, it has been a common practice to add things to foodstuffs, usually to improve them in some way – mostly preservatives and flavour enhancers. Nowadays, we have reached the slightly absurd point where more than 3,000 additives are available to food manufacturers. Within the European Economic Union, all these additives are allocated an E-number for quick reference purposes.

The two main aims have remained the same, namely, to preserve the foodstuffs (increased shelf-life) and to enhance the flavour.

Preservatives include sodium nitrite, potassium nitrate etc.

Flavour enhancers – the most frequently used is the Oriental monosodium glutamate (MSG) (E 645), and the fermented fish/shrimp paste (belacan) can come into this category, as well as condiments.

Colouring – the chemicals being used are mostly synthetic aniline dyes. Yellow is commonly the result of tartrazine, which is now shown to cause hyperactivity in some children. Because of these chemical side-effects, there is a renewed interest in the use of natural cochineal produced from the body of the Cactus (Cochineal) Mealybug, which gives a range of colours from orange to dark red.

Anti-oxidants include ascorbic acid and sodium ascorbate.

Sweeteners – sugars; a wide range of sugars are used. Saccharine was the first of the artificial sweeteners, now replaced to some extent by 'Sweetex', and more recently by atrazine ('Nutrasweet'), but the latter has been shown to have some adverse physiological side-effects in some people.

Bleaching agents are widely used with some flours and meals to produce a white product.

Emulsifiers and stabilizers are added to certain manufactured products such as ice cream and synthetic creams, to bulk up the product and stop it separating.

Condiments such as salt, pepper, mustard and cayenne (chilli) are typically added to prepared foods to improve the flavour. In the tropics, the use of chillies as a flavouring/spice for both meat and vegetable dishes is very important.

Bread has long been subjected to 'improvement' during its manufacture. The earliest methods of grain-grinding using stones resulted in a brown-coloured flour with bran and germinal areas included – a healthy, but deemed by some unattractive, colour. So, sometimes the flour was bleached to make it whiter, and the modern milling process removes both bran and germ, which also promotes whiteness. During World War II, by government decree, bread flour had chalk powder added to increase the whiteness – this had the real benefit of adding calcium carbonate to the diet, which resulted in children having stronger teeth and bones. Nowadays, the cycle is completed, and there is a great demand for stone-ground, wholemeal (brown) bread. Some manufacturers even add extra bran and wheatgerm to their bread. Bran and other plant fibres are added to both bread and flour, and to cereal products, to provide roughage for the human diet. Roughage is now known to be necessary for intestinal health, and to avoid constipation, which is one of the scourges of modern living.

Dried fruits traditionally are treated with sulphur dioxide (SO_2 E.220) as a preservative, and then given a fine coat of mineral oil to make them glossy and minimise their sticking together. There have been some medical problems in the use of mineral oils, so generally their use is now replaced by various vegetable oils.

CHAPTER 5

PACKAGING

5.1 INTRODUCTION

Produce is packaged for several different reasons: partly to make the material easier to handle, to protect it from physical damage and to keep it clean, and as a method of protection from pests. Most packaging is used to contain the produce for shipment and distribution, but is also used for containment on retail premises, such as the storage of dried abalone, scallops etc. in large glass jars in Chinese fish shops. Different types of packages will have different properties according to their prime purpose. Vegetable crates and some fruit boxes are designed purely for handling purposes and to safeguard the produce from physical damage (bruising etc.). At the other extreme, as already mentioned, the storage of dried abalone and scallops in large glass jars is purely for protection against insect pests and micro-organisms.

The use of packaging in modern food distribution systems provides a potentially effective tool in the management of food pests. There is a need for systems that will protect food from infestation from the time it is packed until the package is opened by the consumer.

There are two main problems with packaging. One is that often produce may be free of insects and mites in the larval and adult stages, but when packaged there are often eggs present which will eventually develop under favourable conditions given sufficient time. Secondly, there is the ever-present likelihood of cross-infestation taking place. Warm, humid environments favour both insect development and adult activity for cross-infestation, so these problems are likely to be more serious in tropical regions and during the warm summer months in temperate zones.

Some products are more likely to become infested during storage than others, and these serve as foci for infestation of other products stored nearby. Dry pet food is seldom packed in insect-resistant containers, and is a suitable medium for several species of insects to develop. Clearly, this food can then serve as a reservoir for insect pests when stored in grocery warehouses containing other edible foodstuffs. Packaged foods can also be infested during shipment on ships, rail-cars and trucks, and during storage at retail outlets. Length of storage, infestation severity, quality of packaging, and ambient conditions will all affect the ease and severity of cross-infestation.

Stored products insects can be regarded according to their infestation activity, as either penetrators which can bore through packaging materials to gain access to the foodstuffs, or the more common invaders which can only gain access to packages through an existing hole or opening. Of the latter, many species are sufficiently tiny that the small openings left in many packages as currently designed are sufficiently large for them to pass through. Some of the 'invader' species are able to 'penetrate' if sufficiently hungry (e.g. *T. castaneum*).

The more important species of penetrators are *Rhizopertha dominica, Lasioderma serricorne, Stegobium paniceum, Plodia interpunctella, Cadra cautella, Trogoderma, Dermestes, Corcyra cephalonica. Tenebriodes mauritanicus* larvae are strong penetrators but seldom found in processed foods (Highland, in Gorham, 1991; pp. 345-350).

5.2 PAPER BAGS AND CARTONS

Paper bags were one of the earliest forms of packaging, typically for the retail trade as much of the produce (rice, flour, oats, meals, sugar etc.) came in sacks (jute or Hessian) and was ladled out into paper bags at purchase. The paper was quite thick, but offered little resistance to the insect penetrators, and being only loosely closed, allowed all small insects access. Later, when produce packaging became the norm, it was customary to buy cereals, flour, oatmeal and the like in small (mostly 1lb in weight) paper or card cartons (called 'consumer-sized packages'), with a folded and partially glued base and top. In thickness, the carton material was twice or three times the thickness of the paper (bags). Against the penetrators, this increased thickness offered little real resistance, but the invaders were limited to the folded top and bottom of the carton to seek openings. The latest development in paper cartons is to have an internal cellophane or plastic bag for extra protection, the bottom usually sealed and the top either just folded or folded and sealed, and this gives good protection against insects.

Multi-wall paper shipping bags are large (10-20 kg size), with the closures either sewn or taped (or both). Generally, those only sewn are liable to insect penetration, but can be made insect-resistant by using a folded-over, stepped-end closure sealed to the opposite face of the bag with a heat-activated adhesive. These bags are often given an insecticide coating (permethrin or synergized pyrethrin) on the outer ply. Another type of bag, used widely in the US international aid programme for processed cereals, is a multi-wall paper shipping bag, treated with synergized pyrethrins and has paper tape(similarly treated) heat-sealed over the stitching on both ends of the bag. Sometimes an additional foil over-wrap is also provided.

In general, paper and cellophane are the least resistant of all flexible packaging materials to penetrating insects.

5.3 PLASTIC BAGS

The earliest 'see-through' bags were probably cellophane, but on their own were not too successful, and were easily penetrated by insects. A wide range of plastic films have been tested against insect pests, both separately and as mixed films. Many of the combination films were quite successful, and especially some that incorporated aluminium foil. Almost all of the flexible polymer films, or combinations of films, can be penetrated by one or more insect species. Polyethelene as an over-wrap on paperboard cartons was usually successful provided the film was quite thick (5-mil or more). At the present time, cereal grains (rice etc.) and many other types of produce are sold in stout polythene bags with heat-sealed tops and bottoms.

Figure 5.1. Rice packed in 10 kg plastic bags for retail; bag heat-sealed top and bottom; Kuching, Sarawak.

They are very convenient to handle, and the enclosed produce is clearly visible for examination. In the tropics, the rice is usually lightly infested with rice weevil, but nowadays adults are seldom present in the bags, although eggs are usually. If the rice is left in the bag too long, then developing weevils can be seen through the polythene, but if the rice is sold promptly, then weevils do not develop enough to cause any problem.

As summarized by Highland (1991), in general, plastic films vary in susceptibility to penetration depending on thickness, on the basic resin from which the film is made, on the combination of materials in the film, on the nature of the package, and on the species of insects concerned.

Figure 5.2. Rolled oats packed in a sealed plastic bag; imported in bulk from Australia and packed locally in Sarawak.

Figure 5.3. Plastic bag containing frozen prepared mixed vegetables; heat-sealed and airtight; Kuching, Sarawak.

5.4 CARDBOARD BOXES

The British cardboard box is known in the USA as a 'corrugated paper case' because, basically, cardboard is just several layers of paper with a corrugated insert. It is the standard container for shipping foods (as well as a wide range of manufactured goods) produced in the USA and Europe. These cartons are seldom resistant to insects, although they can be so made using well-sealed tape over the flap junctures, or it can be over-wrapped with a heat-shrunk polythene film.

In the 1960s, pasta (macaroni especially) was imported into Hong Kong from Italy in bulk, loose in large cardboard boxes (0.5 cu m), and was packed into small plastic bags nicely sealed for local sale. But the pasta was already infested by *Sitophilus* eggs, and soon larvae and pupae were to be seen in the bags. If the pasta was sold quickly and eaten soon after purchase, all was well; but if a packet was kept on the shelf for several weeks, it soon became a crawling mass of weevils.

The present trend is to pack bulk food produce inside a thick polythene bag (or liner) inside a large cardboard carton some 3-5 mm thick, and this gives good protection for several months, even in the tropics.

Figure 5.4. A carton containing 'Coffee-mate' sealed inside a foil and plastic bag, heat-sealed and airtight; Kuching, Sarawak.

The ultimate form of cardboard container is now the cylindrical drum about one metre tall and half a metre in diameter. It is constructed of pressed cardboard up to 1 cm thick, with a sealed metal base and a round metal lid that clips on to the top of the drum with a pressure spring lever. These drums are virtually airtight, and when new are certainly insect-proof, but they are rather expensive and not very large, so are not economic to use for cheaper products. In S.E. Asia, they were most frequently used for animal feed concentrates and supplements. For retail purposes, a common practice is to make a small cardboard carton into which is placed the produce inside a sealed foil and plastic bag which both keeps it fresh and keeps pests at bay.

5.5 SACKS

Formerly throughout S.E. Asia in the 1960s, it was customary to export and import rice in large sacks (1cwt) made of Hessian or jute (burlap in the USA), and to this day it is still the practice for carrying quantities of maize, sorghum and wheat in Ethiopia and parts of eastern Africa. These sacks are the right size and weight to be carried by an experienced dockyard worker in S.E. Asia, or a larger version to be carried on the back of a donkey in Ethiopia. These sacks are quite suitable for the purpose of carriage and handling, but are an open invitation to insect pests, for the weave is sufficiently loose to allow insect entrance. When such sacks are used for overseas shipment, they are often the source and or recipient of insect pest cross-infestations. Of course, in Europe the standard hundredweight or half-hundredweight sack (nowadays 50 or 25 kg) is still used for potatoes and some other root crops, but it is now being replaced by multi-layered paper sacks that do not allow free air circulation, and so potatoes in domestic storage suffer from storage rots only too frequently.

Sacks made of woven plastic strips are available and are successful in replacing Hessian in that they allow free air circulation, but are equally open to insect infestation.

Figure 5.5. A selection of 50 kg rice sacks, with rice from five different countries, for sale in local mini-supermarket; Kuching, Sarawak.

5.6 SEALED JARS

These are glass jars with a wide neck and usually a screw-top metal or plastic lid. There is also an insert of waxed card or plastic inside the lid to make an air-tight seal.

Figure 5.6. A 50 kg sack of Thai rice; sack is made of woven plastic strips, but lined with a polythene bag and sewn top and bottom.

There are typically two main types of jars. The small (0.5-1 lt.) ones are for retail of fancy food items such as roasted nuts, pickles, chutneys, jams and other preserves; because the jar is made of glass it is relatively expensive and also heavy, and present practice is to retail dry items such as roasted nuts in foil bags instead. Large glass jars are widely used for the storage of dried molluscs such as abalone and scallops and many types of Chinese medicines, and they do provide air-tight storage. These jars are typically 1-2 gallon capacity and also have screw-top lids.

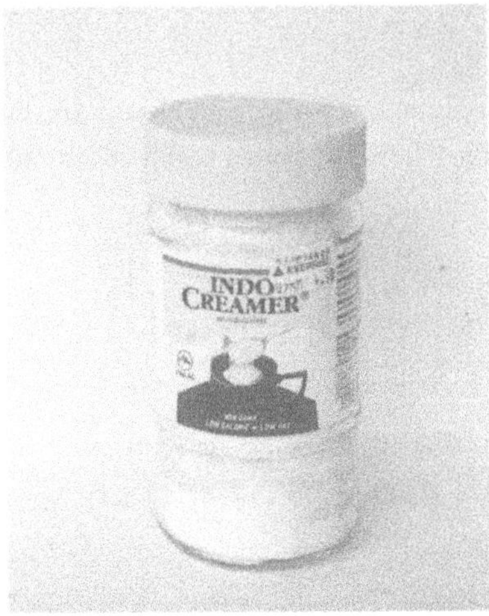

Figure 5.7 (a) Glass jar of coffee creamer with plastic screw-top lid; Kuching, Sarawak.

Figure 5.7.(b) Glass jar of peanut butter with plastic screw-top lid; Kuching, Sarawak.

One of the earliest methods of food preservation by 'bottling' was with the use of wide-necked 'Kilner' glass jars with a solid glass lid held by a metal spring, and made air-tight with a renewable thin rubber ring. The fruit (or vegetables) was cooked separately in a pan and then poured in to the heated jars, and the lids sealed while it was all still quite hot (and sterile), and once sealed, the fruit would stay edible for years. This treatment was the on-farm/domestic precursor to canning.

Earthenware jars, and sometimes glass jars, are used to keep seeds safe until needed for sowing. In many rural areas, seed jars are earthenware pots of 1-5 gallon capacity that have the top closed with clay or wax to make an airtight seal. It is more usual for the seed jars to be 1-2 gallon capacity, and the larger jars used for storing pulses, dried roots and tubers etc. Any society that can make large earthenware jars has a definite advantage in terms of seed and foodstuff preservation. In parts of the tropics, large dried gourds are used for seed storage because, having a narrow neck, they are easily sealed. Large sealed earthenware jars can be stored either inside the dwelling place, or if the hut is on stilts (Figure 7.1) as in parts of S.E. Asia, the jars can be stored underneath. Sealed gourds are often stored hanging from the eaves of the building. With air-tight storage, experimentation has shown that when the oxygen content falls to 2%, *Sitophilus* weevils will be asphyxiated.

In tropical Africa and Asia, large earthenware jars of 5-10 gallon capacity, with a removeable lid, are used for short-term storage of grains, pulses, dried vegetables and the like for daily food production. These are kept in the hut, or the kitchen area, for ease of access.

5.7 BOTTLES

These are typically also glass, but with a narrow neck and usually used for storage of liquids. The narrow neck makes it easier to keep air-tight with either a small screw-top or else a bung of some sort inserted into the neck. Although usually for liquids, bottles are similar to narrow-necked gourds and can be used for seed storage. Bottles are commonly used for storage of sauces (soya, oyster etc.) which if left exposed would undoubtedly attract flies, especially *Sarcophaga.* One advantage with the use of glass bottles and jars is that with care they can be used repeatedly, many times.

5.8 CANNING (SEALED TINS)

This has long been an acceptable method of preservation of foodstuffs. It was probably started on a local scale by isolated farmers in the USA where they needed a long-term method of food preservation. Starting with fruits and vegetables, the logical sequence was then to can meats. Once the technique was established, canning factories were built on the coast of Alaska to use the migrating salmon, and in California and Chile for the pilchard/sardine shoals. By World War II, canning factories were established in South America for the production of pressed (corned) beef from the cattle herds on the pampas.

Figure 5.8. Can of baked beans; Kuching, Sarawak. Many modern cans now have a ring-pull top for ease of opening.

Now, a tremendously wide range of produce is canned — fruits, vegetables, fish, and meats make up the bulk of the produce. Traditionally, the food material is either cooked prior to canning or else is cooked in the open can which is sealed by the lid clamped on or soldered while the produce is still hot. When the food cools, the air pressure in the can falls until a near-vacuum is achieved. When done carefully, this is a very effective method of preservation, and most foods will last for years.

But there is some air trapped in the can (tin), and there may be some spores or propagules which over the course of time will develop and propagate, so the canned meats will not last forever. If the seals on the can are not air-tight, there is a serious danger of microbial contamination, and this can develop quite quickly. In the UK there have been more than a few cases of *Salmonella* poisoning, and most of the more scarce cases of Botulism have arisen from either badly canned meats, or canned meats that had been kept for too long.

Most micro-organisms are killed by the high temperatures involved in cooking, but the spores of some moulds (e.g. *Byssochlamys* spp.) are heat-resistant, and may cause spoilage of canned fruit.

In the USA, *Geotrichum candidum* is the main mould contaminant of processing equipment, and its presence in canned fruit is usually an indicator of unsanitary conditions within a processing plant.

Certain products are sometimes at risk from particular local pests — in Hawaii, canned pineapple may occasionally have enclosed dead fruit flies (*Drosophila* spp.), and corned (pressed) beef from the Argentine may contain ant bodies.

Some products are now dry canned and at the same time vacuum packed. Sealed tins of about five gallon capacity, resembling the well-known jerry-cans used for carrying petrol, are being used for consignments of high value commodities such as spices, shelled cashew nuts, macadamia nuts, pistachio nuts, from countries such as India and Indonesia where pest infestation levels are typically high. If packed clean, the use of these tins affords maximum safety for the produce. The lid is only about 4-5 cm in diameter, and can either be a screw-top or else be soldered shut. If the tin is filled there is little air space, but nowadays either the tins can be vacuum packed or else an inert gas such as CO_2 or N_2 can be introduced.

5.9 VACUUM PACKING AND MAP

Some expensive commodities of small size, such as cashew, macadamia and pistachio nuts etc. are retailed vacuum packed inside small tins. This is a very successful method of food protection, but it requires quite sophisticated equipment, and is an expensive process. It is really only applicable to items that are both small in size and of high value.

In recent years a great deal of progress has been made in this field. The main objective is to remove oxygen from the foodstuff because it causes oxidation reactions, which damage vitamins, pigments, fatty substances and flavouring substances which are often catalyzed by enzymes, and it also permits growth and activity of pest organisms (especially aerobic bacteria, moulds and yeasts).

This type of packaging started in about 1960 in conjunction with the development of plastics technology, and is now one of the most widely used systems of small-scale foodstuff protection for retailing purposes.

For wet packaging in chilled conditions, the air is flushed out of the package using steam; the container is usually rigid, and the atmosphere is replaced by steam which, after sealing, condenses; the food is usually perishable, and has to be stored in a chilled state.

Figure 5.9. Pack of ten chicken sausages, vacuum packed and frozen; Kuching, Sarawak.

For dry packaging, a vacuum pump is used to suck out the air from a plastic bag — these packages are usually flexible, and after evacuation are sealed hermetically and collapse on to the packaged product once the package is exposed to atmospheric pressure. Several different plastics are regularly used to make shrink bags and pouches for produce containment. The latest development uses thermoformed packages, made by two rolls of plastic — the bottom one is unwound and heated on a warm plate and formed into cavities where the food is placed; inside a vacuum chamber, the second roll of film is sealed onto the base. Thermoforming is used for the packaging of many types of perishable foodstuffs, both consumer and industrial units, up to the size of a whole ham.

The other way to remove oxygen from a package, or at least to reduce it to a low level, is through the modified atmosphere packaging (MAP). This is another relatively new technique used to extend the shelf-life of certain types of food, mainly those that are fresh or chilled. These include raw and cooked meats, fish, poultry, fresh pasta, fruit, vegetables and various bakery products. MAP techniques are based on the principle that modifying the atmosphere in the food package changes the growth characteristics of the microbial flora present. Oxygen, nitrogen and carbon dioxide are the gases generally used in MAP, and the plastic package may be permeable, semi-permeable, or may provide a barrier to gases either leaving or entering the package.

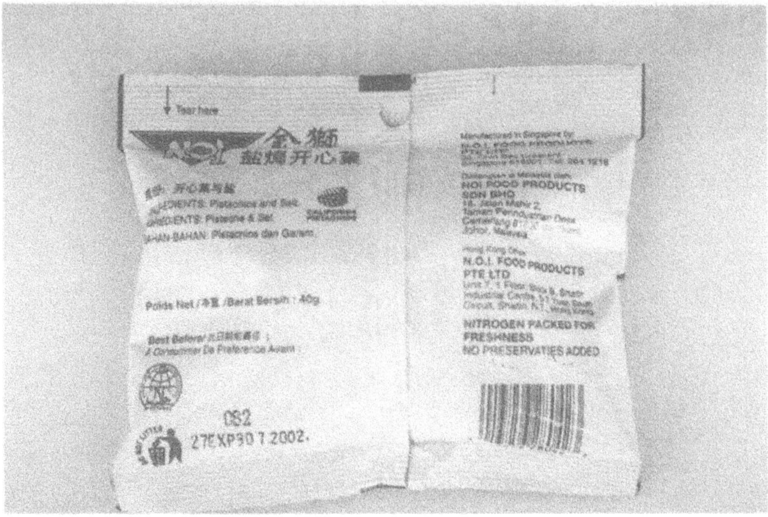

Figure 5.10. Pistachio nuts sealed inside a foil-lined plastic bag in an atmosphere of Nitrogen.

Figure 5.11. Prawn crackers sealed inside a partly clear plastic bag and packed in Nitrogen, giving an open shelf-life of almost one year; Kuching, Sarawak.

With some packaging, a high level of oxygen is used (80%), others are totally without oxygen; a high level of carbon dioxide is effective against many micro-organisms. It seems that each type of produce to be packaged has its own requirement in terms of gas, type of plastic membrane, and temperature for storage, both to maintain its appearance, flavour and keeping quality and to suppress the micro-organisms that normally infest it. But some organisms thrive under conditions often associated with MAP, such as low oxygen, so care has to be taken with its use. The use of nitrogen as the filler gas appears to be very successful in the suppression or micro-organisms, and now is used very widely.

5.10 POLYSTYRENE TRAYS WITH A PLASTIC FILM (SKIN PACKAGES)

As part of the supermarket approach to food retailing, there is much use of small polystyrene trays with a cover of clear plastic film. The result is an air-tight display pack with moisture retention that will initially keep the produce moist but will rapidly lead to decay following bacterial and fungal development in the saturated atmosphere. It is reported that in the USA, the use of porous plastic film lowers the enclosed humidity level. Such packaging is used for expensive perishable commodities (fresh meat, fruits, vegetables) as part of the final processing procedure, in consumer or family units, and the produce is only kept thus for a few days and usually under chilled conditions.

Figure 5.12. (a) Skin package on polystyrene tray kept in the chilled cabinet of a supermarket; Kuching, Sarawak.

Figure 5.12. (b) Skin package on polystyrene tray kept in the freezer chest of a supermarket; Kuching, Sarawak.

Clearly, it is important to use the correct type of plastic film so that the moisture is retained but condensation avoided; this will maximize both produce protection and storage shelf-life.

This is a type of thermoforming, where the top of the film conforms tightly to either the product shape or to the edges of the tray. Rigid trays are generally made of either polystyrene or polyvinylchloride.

CHAPTER 6

TYPES OF PESTS

A stored products/food pest can be defined as 'any organism injurious to stored foodstuffs of all types (especially grains and pulses), including vegetables, fruits, seeds and diverse types of plant and animal materials stored for human purpose, including a wide range of prepared foods'. Storage time can be as long as three to five years (for some National Crisis/Famine Stores), or as short as a day or two — infestations can take place very rapidly, especially those of micro-organisms.

Foodstuffs on a worldwide basis start with cereal grains and pulses in storage as the most important, partly because of the quantity involved and for the length of storage time, which may be years in some cases. But they also include processed grain (flour, meals and a wide range of farinaceous materials), vegetables, fruits, nuts, oilseeds, spices, stimulants (coffee, tea, cocoa, tobacco), dried fish and meats. Most of these commodities are dried or processed, and some are stored for long periods of time (months), but some vegetables, fruits, meats and fish are fresh, and are only stored for a short time (days or a week or two at most).

In many parts of the world, there is an ever-increasing food industry manufacturing prepared foodstuffs including complete meals, such as 'TV dinners', which only require heating in a microwave oven. Such gastronomic delights used to be the purview of the western temperate countries, but more and more prepared foodstuffs are being made in parts of Asia, and Africa as well as South America. Most of this food is mass-produced in a few locations, and then has to be distributed throughout the country, through the major supermarkets. Some are sold fresh, most are chilled (to about 5°C), and others are frozen (-20°C usually). All these prepared foods have possible pest problems connected both with their production (mostly hygiene) and length of storage prior to sale and use.

Many of these pests are also referred to as 'urban pests' or 'domestic pests' because of their association with Man and human dwellings — they are also termed 'synanthropic species', and in a few cases the association has been so long and so intimate that these species cannot be found truly in the wild any more.

There are, however, several different ways in which the pest organisms can be viewed according to different circumstances, as shown below.

6.1 SYSTEMATIC APPROACH

This is the grouping of the organisms according to their official scientific status, by Class, Order, Family, so that the related species are presented together. This is the most basic approach, and is of importance since similar species with similar habits are grouped together, and generally close relatives have a similar biology, although there are a few striking exceptions. The following table lists the organisms concerned in a systematic sequence.

SYSTEMATIC LIST OF PESTS OF STORED FOODSTUFFS

Insecta (Orders 1-10)

(1)	Thysanura	- Lepismatidae	- *Lepisma, Thermobia*
(2)	Orthoptera	- Gryllidae	- *Acheta domesticus*
(3)	Dictyoptera	- Blattidae	- *Blatta, Periplaneta*
		- Blaberidae	- *Pycnoscelis surinamensis*
		- Epilampridae	- *Blatella germanica* & spp.
(4)	Isoptera	- Rhinotermitidae	- *Coptotermes, Reticulitermes*
		- Termitidae	- *Macrotermes, Odontotermes*
(5)	Psocoptera	- Liposcelidae	- *Liposcelis*
		- Trogiidae	- *Trogium*
(6)	Hemiptera	- Aphididae	- several
		- Pseudococcidae	- several
		- Coccidae	- several
		- Diaspididae	- several
		- Reduviidae	- several (predators)
		- Anthocoridae	- several
		- Lygaeidae	- several
(7)	Coleoptera	- Histeridae	- predators
		- Scarabaeidae	- several
		- Buprestidae	- several
		- Elateridae	- *Agriotes*
		- Dermestidae	- many species
		- Anobiidae	- several
		- Ptinidae	- several
		- Bostrychidae	- several
		- Lyctidae	- *Lyctus*
		- Lophocateridae	- *Lophocateres*
		- Trogossitidae	- *Tenebroides*
		- Cleridae	- *Necrobia* etc.
		- Nitidulidae	- several
		- Cucujidae	- *Cryptolestes*
		- Silvanidae	- *Oryzaephilus, Ahasverus*
		- Cryptophagidae	- *Cryptophagus, Henoticus*

		- Lathridiidae	- several
		- Mycetophagidae	- several
		- Tenebrionidae	- many species
		- Anthicidae	- several
		- Cerambycidae	- several
		- Bruchidae	- *Acanthoscelides, Bruchus, Callosobruchus* etc.
		- Anthribidae	- *Araecerus*
		- Apionidae	- *Apion, Cylas*
		- Curculionidae	- many species
		- Scolytidae	- several
(8)	Diptera	- Mycetophilidae	- *Mycetophila*
		- Sciaridae	- *Sciara, Pnyxia*
		- Cecidomyiidae	- *Contarinia*
		- Scenopinidae	- *Scenopinus*
		- Phoridae	- *Megascelia*
		- Syrphidae	- *Eumerus, Merodon*
		- Lonchaeidae	- *Lonchaea*
		- Drosophilidae	- *Drosophila*
		- Psilidae	- *Psila*
		- Anthomyiidae	- *Delia*
		- Muscidae	- *Musca*
		- Fanniidae	- *Fannia*
		- Calliphoridae	- *Calliphora, Lucilia, Sarcophaga*
		- Agromyzidae	- several
		- Piophilidae	- *Piophila*
		- Tephritidae	- many species
(9)	Lepidoptera	- Tineidae	- several species
		- Heliodinidae	- *Stathmopoda*
		- Yponomeutidae	- several
		- Oecophoridae	- several
		- Cosmopterygidae	- *Pyroderces*
		- Gelechiidae	- several
		- Tortricidae	- several
		- Pyralidae	- many
		- Pieridae	- *Pieris*
		- Noctuidae	- several
(10)	Hymenoptera	- Braconidae	- *Bracon*
		- Evaniidae	- *Evania*
		- Chalcidoidea	
		- (Pteromalidae)	- several parasites
		- (Trichogrammatidae)	- *Trichogramma*

		- Bethylidae	- *Cephalomia*
		- Formicidae	- many
		- Eumenidae	- *Eumenes, Odynerus*
		- Vespidae	- *Vespa, Polistes*
		- Xylocopidae	- *Xylocopa*
		- Apidae	- *Apis, Trigona*
(11)	Aranaeida	- several	- many
(12)	Acarina	- Phytoseiidae	- *Blattisocius*
		- Eriophyidae	- *Philocoptruta*
		- Cheyletidae	- *Cheyletus*
		- Tarsonemidae	- several
		- Pyemotidae	- *Pyemotes*
		- Acaridae	- several
(13)	Gastropoda	- Limacidae	- several
	(Mollusca)	- Helicidae	- several
(14)	Trematoda	- Fasciolidae	- *Fasciola*
			- *Clonorchis*
			- *Paragonimus*
(15)	Cestoda	- Pseudophyllidea	- *Diphyllobothrium latum*
		- Cyclophyllidea	- *Taenia solium*
			- *Taeniarhynchus saginatus*
(16)	Nematoda	- Trichinellidae	- *Trichinella spiralis*
		- Ascaridae	- *Ascaris lumbricoides*
		- Anisakidae	- *Anisakis*
		- Heterodidae	- *Meloidogyne*
(17)	Protozoa	- Zoömastigophora	- *Giardia lamblia*
	(= Protoctista)		- *Pentatrichomonas hominis*
		- Amöebidae	- *Entamoeba histolytica*
		- Emeriina	- *Sarcocystis hominis*
(18)	Aves	- Passeriformes	- *Passer* etc.
		Columbiformes	*Columba, Streptopelia*
(19)	Mammalia	- Muridae	- *Mus, Rattus* etc.
	(Rodentia)	- Cricetidae	- *Microtus* etc.
		- Hominidae	- Man
(20)	Fungi	- Moulds	- *Aspergillus*
			- *Claviceps purpurea*

			- *Penicillium*
			- *Fusarium*
			- *Alternaria*
			- *Trichothecium*
			- *Mucor*
			- *Rhizopus*
			- *Cladosporium*
		- Yeasts	- *Saccharomyces*
			- *Candida*
			- *Torula*
			- etc.
(21)	Bacteria (= Prokaryotae)	-	- *Bacillus*
			- *Campylobacter*
			- *Clostridium*
			- *Escherichia coli* (strains)
			- *Listeria*
			- *Salmonella*
			- *Streptococcus*
			- *Vibrio parahaemolyticus*
			- *Yersinia enterocoliticus*
(22)	Other Micro-organisms	- Protein prions	- BSE (vCJD)

This list does include a few species that are predacious on the pests, and a few that attack or inhabit the buildings or packaging and so are not directly pests themselves, but they are to be found in stores and warehouses/godowns, and their presence requires some explanation; moreover, they can be responsible for contamination of produce.

6.2 TRUE STORED PRODUCTS PESTS

The other classification of pests depends upon the degree of dependence on this way of life. A few species fall nicely into the following categories, but there is a large number that are difficult to categorize because of the subjective nature of the categories, so this approach is rather limited in its usefulness.

6.2.1 Obligatory Pests

Sometimes called *permanent resident pests*, these are the species that are generally only found in stored produce or in domestic situations. They have evolved for a sufficiently long period of time in this lifestyle that no populations are to be found in the wild. These pests include *Tribolium castaneum, T. confusum*, and to some extent *Lasioderma* and *Acarus siro.* Also in this category are usually placed the species that are not yet quite so closely evolved with the stored products. These are the species that are found in the wild in small numbers on crop plants (and sometimes wild hosts) but in the stores they breed up into large populations, and they will continue to breed indefinitely in the stores.

Some of these insects, particularly the grain pests, although they may sometimes occur in quite heavy field infestations, would in fact find it quite difficult to survive as 'wild' populations. Presumably, in their early evolution as cereal grain pests, they must have survived in the wild, but in present times it is clear that the field populations are completely dependent upon the store populations. The species in this category would include *Sitotroga, Ephestia, Cadra, Sitophilus, Acanthoscelides, Rhizopertha*, food mites, and various other Coleoptera (Dermestidae, Tenebrionidae etc.) Clearly, these categories as here used are not too sharply defined, and there will be many species that either are intermediate to, or overlap with, the following category.

6.2.2 Facultative Stored Produce Pests (Temporary, Casual, or Visiting Pests)

These are species which live elsewhere but invade stores for food, and may stay to live there for a while; some species appear to be evolving into permanent stored produce pests, and are really intermediate between these two categories. Potato tuber moth comes into this category, for it is predominantly a field pest but can have several generations on stored tubers, although then it needs to go back out on to a growing plant to revitalize the population. *Callosobruchus* species are in a similar category, as are some Tenebrionidae. The usual 'invading' pests would include rats, mice, birds, cockroaches, ants and the like.

6.2.3 Temporary Pests (Accidental, Field Pests)

These are basically field pests, to be found on growing crops, and carried into food and produce stores, to some extent by accident. The life cycle is normally completed in the field, and often before harvest takes place. But for those larvae (sometimes pupae) taken into the store, development will proceed, but the population stops at the adult stage, and breeding will not normally take place in the store. For example, the grain and pulse pests that will only oviposit or develop in young moist seeds; with the dried seeds in storage, the adults will either not oviposit, or should they do so, no development is possible. Some of the common and obvious examples of this category are:

Bruchus pisorum (Pea Bruchid) in peas.
Pectinophora gossypiella (Pink Bollworm) in cotton seed.
Cryptophlebia spp. in madacamia nuts, oranges and other fruits.
Cydia pomonella (Codling Moth) in harvested apples.
Ceratitis, Dacus spp. (Fruit flies) in a wide range of tropical and subtropical fruits and vegetables as larvae.

6.2.4 *Accidental Stored Produce Pests (Field Pests)*

These are simply field pests that have been taken with harvested produce into storage, and usually no further development will take place in the store, but the larvae (or whatever) will die and their bodies and frass will contaminate the produce, and there may also be damage evident which will reduce the value and quality of the produce. Examples are aphids and caterpillars on vegetables, slugs in holes in potatoes, and slugs and snails on cereals or peas for freezing. In recent years in the UK, there has been a new problem with the corpses of *Sitona* weevils in harvested peas, both dried and for freezing.

The commercial importance of these pests lies mainly in their contamination of produce so that there is a reduction in quality, or grading, or saleability; and also, to some extent, their damage (if evident) has a similar deleterious effect. The produce involved in this category is usually fresh fruit and vegetables for which the storage period is typically a matter of days rather than weeks.

6.3 FEEDING HABITS AND DAMAGE CAUSED

On the basis of feeding behaviour, and to a lesser extent on type of damage caused, some six categories can be designated. It will be obvious that some pests do not readily fall in to any of these categories, but it is sometimes a useful approach for the cereal grain pests.

6.3.1 *Primary Pests*

The usual criterion is that these pests (usually referring to insects) are able to penetrate the intact protective testa of grains and seeds, and so they are of particular importance and especially damaging. Some of these species show a preference for feeding on the germinal region of the seed, and so they are responsible for a loss of quality and nutritive value in cereal grains for food, and a loss of viability (germination ability) in seeds. The larvae of *Sitotroga* come into this category, and so do the adults and larvae of *Sitophilus, Rhizopertha, Trogoderma*, and the larvae of *Callosobruchus* etc., and, of course, rats and mice. It should be remembered that the great majority of stored produce pests are quite small, and so have tiny mandibles in relation to a grain of maize or wheat, or a bean seed.

6.3.2 *Secondary Pests*

These are only able to feed on damaged grains and seeds, where the testa is cracked, holed, abraded or otherwise broken, either by physical damage during harvesting, or by too rapid drying, or by the prior feeding of a primary pest. Some of these species will be found in grain stores together with primary pests such as *Sitophilus* or *Sitotroga*, but often the secondary pests are to be found in greatest numbers on flours, meals, and other processed cereal products. The more notable of the secondary pests that are quite common on grains and seeds include the larvae of *Ephestia, Cadra, Plodia*, and the beetles *Oryzaephilus, Cryptolestes, Tribolium* and the other Tenebrionidae.

6.3.3 Fungus Feeders

A large number of insects found in food stores are actually fungivorous and are feeding on the fungal mycelium growing on damp produce. Moisture is usually a major problem in most produce stores and on domestic premises, so moulds are a constant problem. Some species are partly fungivorous and partly a secondary pest, and some start by feeding on the fungal mycelium and then continue on the foodstuffs or seeds – this commonly happens with some Psocoptera and some beetles. Several families of Coleoptera (e.g. Mycetophagidae) are commonly referred to as fungus beetles, for the fungivorous habit is frequent.

6.3.4 Scavengers

This represents a large group whose members are basically polyphagous and often omnivorous, and they feed on the general organic debris in the store. Some of these species are not permanent residents in the store, but are either occasional visitors or casual pests. They can be very important and damaging to certain types of produce. Included here would be silverfish, crickets, cockroaches, some ants and beetles, and on a larger scale, rats and mice could be regarded thus.

Most of these species are essentially phytophagous and feed upon the detritus in the stored grains and foodstuffs, but some species feed on animal materials, and others are quite omnivorous and feed on debris of both plant and animal origin.

6.3.5 Specialized Plant Feeders

Some of the phytophagous species will only feed on certain types of plant materials, and in some instances they are quite food (host) specific. Examples include:

Oilseeds and copra: *Necrobia rufipes* (Copra Beetle); but sometimes the insects will eat the fat on bacon and hams.

Pulses: beetles of the family Bruchidae (Seed Beetles); many species show host specificity to a particular genus of legume.

Tobacco: not usually a preferred host because of the nicotine and alkaloids in the dried tissues, but several insects, including *Lasioderma serricorne* and *Ephestia elutella*, will feed and breed on both stored (cured) leaf and cigarettes.

Dried fruits: these are characterized by having a high sugar content, and attract *Carpophilus* (Sap Beetles), *Ephestia* spp., Date Moth, and others.

6.3.6 *Animal Materials*

These food materials come in two forms, the one being protein (flesh), usually dried, and the other keratin (and also insect chitin) in the form of hair, fur, wool, skins, hides and horns. Fresh flesh is attacked mostly by Muscoid flies, but dried meats and dried fish are equally likely to be attacked by Dermestid beetles. A few plant protein products, such as bean curd and soya sauce, may be infested by the fly species that feed on animal proteins.

The species that feed on keratin include the larvae of Tineidae (Clothes Moths) and Oecophoridae (House Moths), and some Pyralidae, and many beetles in the Dermestidae (Hide and Carpet Beetles etc.). Some Psocoptera are very damaging to museum collections where they eat out the interior of dried insects and other specimens. Some mites are protein eaters and are to be found infesting cheeses and bacon.

6.3.7 *Predators and Parasites of Stored Produce Pests*

When surveys of produce stores are made, it is usual to collect all the insects and other animals found there, and the final collection can be quite a mixture of species. Along with all the different pests, there will be species that are predators or parasites of the pests, and clearly these are to be regarded as highly beneficial and not to be harmed in any control programme. The different animals likely to be encountered in this category are included in the previous list, and mentioned in Chapters 13 and 14.

6.4 ORIGIN OF STORED PRODUCE PESTS

The prehistoric origins of storage pest species are speculative, but there are records from Roman and Egyptian times, although the actual origins must be far older than that, stemming from the earliest days of human agriculture. There are references to grain storage and stores in the Bible (e.g. Genesis 41:56), but not to specific pests. The earliest, identifiable written references in the United Kingdom date from about 1825, but in the USA there is information from the previous century.

True storage pests fall into two simple categories – those that breed in seeds, fruits and the fleshy parts of plants (weevils, bruchids, pyralid moths etc.), and the omnivorous scavengers. (Howe, R.W. in Gorham, 1991; pp. 13-16).

6.4.1 *Phytophagous Species - Pests of Ripening Seeds*

In the wild, some fruits and seeds remain on the plant for a long time after ripening; with cereals, the grains stay in the panicle to a variable extent – with some cereals, the seeds (grains) are shed within weeks of ripening, but with maize and some varieties of sorghum etc., the grains may be retained indefinitely. But, of course, these cultivars are not the ancestral forms; with wild sorghums, the grains are not shed too readily but do eventually fall, being dislodged by wind action.

With legumes, there is much variation in pod dehiscence – some species have dispersal mechanisms that relate to atmospheric moisture, and the pods explode and the twisting valves throw the seeds a considerable distance. With other legumes, pod dehiscence is a more gradual and gentle matter, and insects can have easy access to the dried seeds in the pod.

Grains and seeds are not such desirable items of diet when dry, and most insects feed on them when they are young, soft and succulent. But presumably during the course of evolution the stored produce insects have adapted and evolved to feed on the older, harder and drier seeds, for which there is less competition. These insects would initially have occurred in the wild in small numbers on the wild hosts that were the progenitors of crop plants. Then, as early man started to collect wild grasses to store over winter, and later started to sow the seeds to grow a crop which led to overwinter storage of the grain in greater quantity, he constructed the forerunners of produce stores and established an evolutionary niche for the insect pests. Since then, these insects have increased in numbers and have become more and more adapted for life in the produce store.

A nice evolutionary series of species that shows this adaptation is present in the Bruchidae. *Bruchidius* is found mostly on forage legumes (clovers and vetches) but only as field pests; *Specularius* is another genus of field pests, but it is reported that some have apparently been bred on dried pulses in laboratory experiments. *Bruchus* are also field pests, but when carried into stores in harvested pulses, they continue their development there, but do not breed in the store. *Callosobruchus* are equally abundant in field crops and in produce stores, and they usually attack the pods at an older stage than do the *Bruchus* species; but they do not attack dried closed pods. *Caryedon* only attacks groundnut pods after harvest while they are drying in the sun. But *Acanthoscelides obtectus* is a highly adapted stored produce pest, and typically feeds on dried beans in storage, although it does still often start as a field infestation. It is clearly capable of continuous breeding in stores without any apparent need for field generations.

A similar pest showing partial adaptation to store life is *Phthorimaea operculella*, in that the larva starts as a leaf miner and eventually bores down the stem in to the potato tuber. Several generations can develop in the stored potato tubers, but the population soon declines in size and vigour, and they need to revert to a growing plant again.

The other major pests that belong to this category include *Sitotroga cerealella, Ephestia* spp., *Cadra* spp., *Araecerus, Sitophilus* spp., *Stegobium paniceum* etc.

6.4.2 Detritus Feeders and Omnivorous Scavengers

Detritus feeders occur in forest leaf litter where they feed on decaying plant material, fallen fruits etc., and animal faeces and remains. This habitat niche is occupied by many cockroaches, a number of Tenebrionidae, Thysanura, crickets, and some weevils.

Birds' nests, animal lairs and caves contain the plant material used in the nest construction, together with animal fur and feathers, epidermal flakes, faecal matter and food remains – a rich source of food materials for scavengers. Small caves are often used for animal lairs. Large caves are typically inhabited by bats in the darker recesses, and birds such as hirundines (swallows and martins) and swifts in the illuminated parts, and the droppings and corpses of young etc. are a rich substrate for cockroaches, Tenebrionidae and mites. Nests are typically inhabited by a broad range of insects and mites, including fly larvae (Diptera), some moths, many beetles, and a number of mites (Acarina; Acaridae).

6.4.3 Fungus Feeders and Some Secondary Pests

In the wild, most fungivorous species are to be found either on tree bark or in leaf litter. On tree bark, there are some fungi growing alone and many are present as lichens in a state of mutualism with symbiotic algae. Most of the process of litter decay is actually conducted by fungi, but typically the mycelium is not obvious to the unaided eye, and most detritivores are actually feeding on fungal mycelium even though they are ingesting pieces of cellulose (decaying leaf material or wood).

In storage situations mould is a major problem, and these insects and mites will feed on the fungus, and then in some cases they progress to eating the damaged produce. During the course of evolution, a few species will adapt to feeding on the produce itself without any intermediary feeding on the fungi. Many Coleoptera and the store Psocoptera will come into this category.

At the present time, the distribution of produce stores, warehouses, mills, markets and pantries (kitchen stores) is so extensive that man has created a completely new habitat for pest organisms – quite worldwide, abundant and permanent, and the process of trade and commerce will ensure a continuous cross-infestation and pest population replenishment.

6.4.4 Sap and Fruit Feeders

Ripe or drying fruits have a very high sugar content, as shown by dried grapes (currants, raisins, sultanas), dates and the like, and these do occur naturally in the wild, but prunes and dried apricots tend to be the product of human agency.

Similarly, concentrated plant sap can have a high sugar content, as shown in the extreme cases of sugarcane, sugar-sorghum, and sugar-maple trees. The Nitidulidae (Sap Beetles) as a family have adapted to such a diet, and several are serious pests of dried fruits in storage, together with some Pyralidae (Lepidoptera). It appears that after long association with cereals in produce stores, some of these moth species have also evolved to include cereal grains in their diet.

6.4.5 Pests of Unknown Origin

A few common stored produce pests are normally never to be found in the wild (unless adjacent to a store), and they are presumed to be of ancient and unknown origin as stored produce pests – some *Tribolium* spp. and *Lasioderma serricorne* are such species.

6.4.6 Animal Feeders - Animal Corpses and Cadavers

In the extreme situation, a dried cadaver, after the flesh has been eaten, is attacked by a series of beetles and moth larvae that feed on the dried fragments of flesh and the keratin of the skin (epidermis) together with hair, fur, feathers, hooves and horns – these are species specially adapted to feed on dried keratin, such as beetles in the family Dermestidae, and moth larvae in the Tineidae, Oecophoridae and some Pyralidae.

A fresh corpse in the wild (particularly in the tropics) will normally be eaten in the main by scavenging mammals (Carnivora etc.), and scavenging birds (vultures, raptors, crows etc.) which will remove all the flesh and soft parts, although fragments of flesh will adhere to the skin and bones. The most adapted of the scavengers such as the Spotted hyena will eat the skin, hooves and most of the smaller bones if hungry, but such efficient scavengers are not widespread on a worldwide basis. In some situations, large scavengers are scarce or even absent, and it is then up to the insects to dispose of bodies.

The usual process of decomposition of a dead animal body can be viewed as follows:

- Fresh corpse: larvae of *Calliphora, Lucilia, Sarcophaga, Musca* (Diptera).
- Putrefying corpse: *Fannia*, Sepsidae (Diptera).
- Cadaver (dried or mummified): *Dermestes, Anthrenus, Attagenus, Necrobia* (Coleoptera); *Piophila* (Diptera), *Tinea*, *Ceratophaga* etc., some Pyralidae (Lepidoptera).

The corpse will suffer initial putrefaction as the internal organs break down and liquefy to a large extent, partly the result of the endogenous enzymes; there will be a combination of muscle (flesh) breakdown and liquefaction, and desiccation as the flesh dries. The precise sequence of events will be dictated partly by the ambient conditions of temperature and humidity and by other factors, such as the nature of the substrate.

When the corpse is fresh, the bulk of the material is flesh (muscle), although the internal organs are quite extensive, and the invading insects are usually Muscoid flies whose maggots eat the flesh and soft parts. After initial putrefaction when some tissues are liquefied, the larvae of *Fannia* often predominate as they prefer a semi-liquid medium.

Once the soft parts and the flesh have been eaten, the cadaver tends to be dried and desiccated, and most of the remaining tissues are skin, tendons, cartilage, hair or fur, claws, horns, hooves and bones (basically keratin and bones). Then the feeding insects are more specialised and include certain moth larvae (Tineidae and some Pyralidae) and beetles of the family Dermestidae, and also a few other beetles from different families. It is not uncommon to find termites breaking down dried bones and skulls in parts of Africa.

The initial breakdown of soft parts, putrefaction and liquefaction, tends to be quite rapid under hot or warm conditions (a week or so, or even a matter of days), but the final breakdown of the dried tissues may be quite prolonged, and can take weeks or even months depending on the extent of the insect population. Sometimes in the tropics, an animal corpse will decompose very rapidly.

An interesting case was seen here in Kuching (Sarawak) in January, 2002.
A dead dog by a roadside storm drain was not attacked by vertebrate scavengers, but was fly-blown immediately after it died. After four days (Figure 6.1), it was reduced to a bag of seething third instar maggots (Calliphoridae). After only nine days, all that remained were clean bones and the ring of hair stuck to the concrete pathway (Figure 6.2) – even the skin, ligaments and cartilages had been eaten (the skull had fallen into the drain). It appeared that all this destruction had been done by the fly maggots, and that the usual cadaver fauna of beetles and moths was not involved. The midday air temperatures reached 32°C, but on the exposed concrete, the surface temperature would have been much higher (up to 40°+) when the sun shone. The RH was high (c.90%), and there were some periods of rain, and clearly putrefaction was very rapid.

Figure 6.1. Dead dog after four days – a sack of skin filled with large maggots (Calliphoridae); Kuching, Sarawak.

Figure 6.2. Same dog corpse after nine days, all skin and soft parts eaten, just clean bones and hair remaining.

6.4.7 Birds' Nests and Animal Lairs

These inhabitants are often the more omnivorous and scavenging species that feed on the mixture of plant materials used for the nest, fragments of skin, fur and feathers, together with pieces of food and faecal matter. Typically, a few blood-sucking species are to be found in nests where they take blood from the young animals in the nest. Sometimes there will be a corpse or corpses in the nest when fly larvae may predominate for a while. Several species of Dermestidae are recorded in the wild in nests and lairs, as also are cheese and bacon mites, flour mites, and some other pest species. It is now thought that this route must have been the one taken by many stored products pests, both insects and mites.

6.4.8 Under Loose Tree Bark on Insect Remains in Spiders' Webs

At first sight, such a niche or habitat would appear to be very limited, but in practice dead trees with loose bark are common in forests, and our normal climax vegetation has been forest for millennia. Several eminent entomologists have been of the opinion that many of the important stored produce pests have evolved from species feeding on the dried remains of insects in spiders' webs (under loose tree bark). Spiders are fluid feeders and they suck the haemolymph (blood) from their insect prey, and to some extent, through the agency of venom and intestinal enzymes, they can cause partial digestion (liquefaction) of internal tissues. But the husk of the insect body left by the spider will contain much of the original muscle tissue which will dry and shrivel. Some of the Dermestidae and some Tenebrionidae, as well as several other beetles, are thought to have evolved in this manner.

It should be remembered that just as there are no major biochemical or dietary differences between animal protein and plant protein, so there is not much basic difference between keratin (vertebrate skin, hair, feathers, nails, horns etc.) and the insect chitin that forms the arthropod exoskeleton.

6.4.9 Internal Parasites

With pests such as the internal helminth and protozoan parasites, the human involvement as a host appears to be that of the top carnivore in a natural food chain. By eating the intermediate host (usually with inadequate cooking), we take in the larval stage, which then grows to maturity in our body as it would in the body of the alternative final host.

CHAPTER 7

TYPES OF DAMAGE

7.1 FOOD SPOILAGE

Human food is obtained from plants and animals, and immediately after harvest or slaughter the spoilage process begins, and if steps are not taken to prevent spoilage, the food will become useless to humans. Food spoilage is caused by many factors - physical damage, eating by animal pests, contamination by pest bodies and faeces, chemicals, enzymes, temperature injury and the actions of micro-organisms (causing decay, altering taste and smell, producing toxins etc.).

Microbial contamination is the main source of spoilage because fungal spores and bacteria are ever-present in the atmosphere. They are minute in size, and are carried on air currents; most are found worldwide, and in certain regions (cities, farms, factories etc.) the aerial populations are very large, and any food will almost immediately be infected upon exposure to the air. Processed foods are clearly most at risk, as opposed to dried grain and pulses in bulk stores, and foods given a severe preservation treatment such as canning or drying are not easily damaged by micro-organisms. Thus the processing industry assigns foods to different spoilage potential categories based upon the type of preservation processes applied, and the consequent shelf-life of the product.

- Highly perishable; no preservation process (fresh poultry, some breads, most fresh meats); shelf-life 3-5 days.
- Semi-perishable; lightly processed (pasteurised milk, some cheeses, bacon); shelf-life 1-2 weeks.
- Semi-shelf-stable; two or more preservation processes (vacuum-packed, cooked meat product); shelf-life 1-3 months.
- Shelf-stable; highly processed (canned vegetables, tinned fish and meats, dried milk etc.); shelf-life 1year or more.

Thus the type and extent of microbial spoilage of foodstuffs will be influenced by the processing (or lack of it) which the food receives

7.2 PRE-HARVEST FACTORS

A point of particular importance is that the cereal plants of many local land-races and indigenous varieties have evolved, over long periods of time, various defences against phytophagous animals, and a major feature in cereals has been the evolution of hard grains. Some of the old varieties of maize grown in Africa are known as flinty-grain types. Although such grain is not easy to grind or break up, it is naturally resistant to many insect and other animal pests. Many of the new improved high-yielding, hybrid varieties of maize now being introduced are soft-grained; this makes them ideal for cooking purposes, being easy to grind, but such grain is devastated in storage by weevils and other pests. This point applies to some extent to all cereal grains, and to some other seeds, but it is often most pronounced in maize.

7.3 DIRECT DAMAGE

This is the most obvious and typical form of damage, when the pest eats part of the seed or fruit etc. The damage is usually measured as a weight loss or a reduction in volume; this is mainly because of historical precedent in that the early studies involved grain in sacks, and weight loss is the most convenient measurement. But, of course, this is not totally accurate as there is an accumulation of insect bodies (both alive and dead), exuviae, faecal matter and fragments (frass).

With cereal grains and pulses, damage assessment studies have included the percentage of grains damaged and the percentage weight loss, so that empirical data may establish a direct correlation, and future sampling be more useful as a result. Data accumulated indicated that with cereals and pulses, the percentage weight loss is directly related to the percentage of damaged grains (but not constantly related); it is always less than the percentage of damaged grains since the grains are seldom entirely consumed. In wheat infested with *Sitophilus* weevils, the observable weight loss is about 30% of the percentage grains damaged; but in sorghum it is about 20%.

A personal experience involved flour mite infestation of a proprietary tropical fish food, where the sealed carton contained a final bulk of living mites, mite corpses, exuviae, frass and faeces to about one-third of the original food volume (the food material had been entirely eaten).

Eating of dried meats, cheese, fish etc. by insects is often disproportionately damaging, for fly larvae and adults practice enzyme regurgitation as well as having enzymes in their saliva, and their feeding usually encourages putrefaction of the meat. Therefore, quite a low infestation level can induce putrefaction.

Further information is available in the publication *Postharvest grain loss assessment methods* (Harris & Lindblad, 1978).

7.4 SELECTIVE EATING

A number of different insects (both beetles and moth larvae) and also some mites, show a feeding preference for the germ (embryo) region of seeds and grains. This is the region where proteins, minerals and vitamins are to be found, whereas the bulk of the cotyledon(s) is just starch. A cotyledon can be bored or eaten to some extent without affecting the viability (germination) of a seed, but if the embryo is attacked, the seed is destroyed. Bean Bruchids can make several holes in a bean seed, in the cotyledons, without destroying its ability to germinate (Figure 14.64). So not only do these pests seriously affect the viability of stored seeds, but when attacking good grains, they seriously reduce the nutritional quality of the grain. Such selective preference is shown by larvae of *Ephestia* and *Cryptolestes*, and also adult *Rhizopertha* and others. Sometimes, quite a low level of infestation can cause a disproportionately large extent of damage.

7.5 CONTAMINATION

In stored cereal grains for local consumption, the effects of contamination are not too serious. In the Orient, most people expect a certain level of *Sitophilus* and *Tribolium* infestation of their rice, and their food preparation processes automatically include cleaning the grain.

But cereals for export (especially to the USA and EEC countries) are rigorously inspected for pests. Contamination is usually visual in that insects, or rat droppings, can be seen and recognised. Live insects, dead insects, larval and pupal exuviae, together with faecal matter and food fragments (often termed 'frass') are the main visual contaminants in stored produce. With rodents, the droppings and sometimes the footprints are most obvious. With both rodents and insects, signs of damage could be included under the broad heading of visual contamination, especially when holes and gnawing marks are evident.

Contaminant odours are usually associated with cockroach infestations, and the smell of rat urine is quite unpleasant and very distinctive. Rodent droppings and urine contamination of foodstuffs can be important from the point of view of human hygiene, as disease pathogens can be transmitted. Certain ants, cockroaches and some other insects have also been recorded distributing pathogenic micro-organisms in domestic premises, and the list of diseases transmitted by synanthropic muscoid flies is quite horrendous.

Some fungi can apparently enhance the nutritive value of some foodstuffs, but their visual effect is one of produce deterioration (spoiling), and a few species of micro-organisms produce mycotoxins and other toxins that are in fact lethal to man and other vertebrate animals (such as the aflatoxin in mouldy groundnuts).

Produce for export is often rigorously inspected before being accepted by the country of purchase. In the USA, the Food and Drug Act is very strict about insect and rodent contamination of foodstuffs being imported, and apart from the more obvious insects, a search is made for rat and mouse hairs, which are very durable (although very small) and can be specifically identified by microscopic examination.

In the countries of the developed world, consumer tastes have become more demanding, especially as the supermarket trend has increased together with the practice of pre-packing produce in transparent wrappings. The end result is that consumers will usually reject produce that shows the slightest sign of pest contamination. This means that, to the producer, any obvious sign of produce being contaminated will represent a serious loss financially, for the produce will either be rejected or will be downgraded. Contamination by only a very small number of pests, and sometimes even by only a single insect, can be sufficient to have a consignment of produce rejected. Thus, produce contamination losses can be totally disproportionate to either the infestation level or to the damage level, and can be very serious economically.

Food contamination is a regular occurrence, usually the result of carelessness or accident. Two years ago, I purchased a pound of minced beef from a very reputable local butcher and found the body of a *Sarcophaga* fly in the middle. But in 2001, in the UK, the 'Daily Mail' has run a series of spectacular articles showing first a European Tree Frog in a sealed bag of fresh salad, then a baby Mole in a cabbage, and finally a House Mouse (sliced) in a sliced loaf of bread from Sainsbury!

The larvae of moths in the family Pyralidae all produce silk for the construction of the pupal cocoon, and some also make extensive silken galleries in which the larvae live. The silk webbing can clog machinery, and generally is a great nuisance. It is seen most commonly on the outside of sacks in large grain stores in the tropics, but it can be a serious problem in temperate regions. The species that produce the most silk are in the genera *Plodia* and *Cadra.*

7.6 LOSS OF QUALITY

As just mentioned, selective eating of the germ region of seed results in an overall loss of quality in the grain, both in respect of germination and dietary value.

Dried skins with holes made by feeding beetles are severely affected in terms of quality and value. One fur, or fur coat, in storage damaged by a clothes moth caterpillar suffers a vastly disproportionate loss in value.

Dried fish and meats are sometimes unsaleable with quite low levels of damage inflicted by insects and other pests.

Both direct damage to the produce and contamination will cause an overall loss of quality of most types of produce, especially that for export or sale, for it may be down-graded, and could even be rejected, and this would represent a considerable financial loss.

Loss of international reputation can be very important – if produce from a particular country is regularly found to be infested, then this information spreads internationally, and other countries will tend to assume contamination, and produce will either not be accepted at all or prior fumigation will be required.

Microbial infection can lead to major changes in smell and taste, and the end result can be quite unpleasant.

The European Community has published quality standards for fresh fruit and vegetables – a retailers' guide, obtainable in the UK from MAFF (1989) (now DEFRA) offices. Basically, 20 types of vegetables and 13 fruits are classed, according to their quality, as

Extra Class - excellent quality, and usually only very specially selected produce.
Class I - good quality produce with no important defects.
Class II - reasonably good quality, but a little deficient in shape and colour, maybe with small blemishes.
Class III - marketable quality, but not up to standards of the higher classes

At that time, Class III category produce was not marketable.

Clearly, pest damage of any obvious sort would result in produce downgrading, probably to Class III, which could only be sold in times of shortage of supply throughout the community.

A recent development in the UK has been the establishment of the Food Standards Agency (FSA), an independent official agency concerned with all aspects of food safety and quality in Britain. This was formed in early 2001, and as yet is still getting established.

In S.E. Asia, there has been recent concern over the quality of foodstuffs being exported to both the USA and EU, and in a local newspaper article (Sarawak Tribune, 21 November, 2001), the World Trade Organisation (WTO) has emphasised the need for companies producing and exporting food products to follow HACCP principles to ensure that foods produced are safe, nutritious and correctly labelled. And the newspaper reported that, in Malaysia, 36 companies in the food business are now HACCP certified, and more are working towards this goal (Rahman & Hill, *in press*).

7.7 CROSS INFESTATION

A major problem in produce storage and shipment is that of cross infestation. It happens in two main ways – either clean produce is brought into a dirty store (i.e. one already infested with insects, mites etc.), or else a clean store containing uninfested produce receives a consignment of infested produce, and then the insects spread into the previously uninfested material. Many of the serious pests are sufficiently polyphagous to spread easily from one foodstuff to another. In some of the oriental godowns (warehouses), it is common to find up to a dozen different edible consignments stored together in one large building, so the likelihood of cross infestation will be high.

Some of the large regional stores, or the godowns and warehouses in ports, are in actual fact never properly cleaned, for they are never empty – there is continual removal of produce and introduction of new material into the store, and there can be continual cross infestation. With most cargo ships, especially bulk grain carriers, the vessel is empty at the port of destination, and cleaning and fumigation can then be carried out. Similarly, most on-farm stores are empty just before harvest time, and there is an opportunity to remove and destroy produce residues and lingering pest populations.

Sometimes physical damage is done to packaged produce without actual eating. In 1965, a shipment of brassières from Hong Kong to Holland was heavily infested by adult *Lasioderma* beetles which had eaten their way through the cardboard cartons into the lingerie; a few of the brassières were nibbled, but the wrappings were holed, and the presence of dead beetles was a cause for concern.

One particular problem with rodent infestations is that the animals often gnaw holes in sacks and cardboard boxes and other soft containers, and this leads to produce spillage. In the western world, much grain is stored loose in bulk in large silos, but in tropical countries, where such silos are not readily available, most of the rice and other grains are still stored in sacks - for an obvious reason, since each sack contains about either 50 kg or 100 kg in weight, according to whether man or donkey is the local beast of burden.

With pests such as the internal helminth and protozoan parasites, the human involvement as a host appears to be that of the top carnivore in a natural food chain. By eating the intermediate host (usually with inadequate cooking) we take in the larval stage which then grows to maturity in our body, as it would in the body of the alternative final host.

7.8 HANDLING AND PHYSICAL DAMAGE

Many fruit and vegetable crops are particularly susceptible to physical damage, especially from mechanical harvesting equipment. Damaged individuals are almost always more liable to be infected by pathogens and certain insect pests. The keeping qualities of such fruits or roots are greatly impaired, and they will act as foci of infection for the rest of the crop.

Post-harvest handling can be crucial, for some of the more delicate fruit crops such as peaches, pears, grapes etc., but nowadays they are packed into preformed plastic trays which cushion them individually.

7.9 TEMPERATURE INJURY

Heating of bulk grain is one curious aspect of temperate zones that can be important. In bulk grain the air is stagnant, and in areas of high insect density the air will become heated as a result of the insects' metabolism, and 'hot spots' will develop. The moisture from the insects' bodies will condense on the cooler grains at the edge of the hot spot, and this water will cause caking. There will also be fungal development, and the moisture can cause some grains to actually germinate. All this biological activity generates more heat and produces more water, so the hot spots enlarge, and temperatures will increase. Germination of grains in storage is not uncommon in the tropics, and cases of spontaneous combustion in grain stores are recorded from all parts of the world. Clearly, grain stored too moist is particularly at risk in such situations. Drying, followed by cooling and ventilation of bulk-stored grain, as well as temperature monitoring, is an important aspect of long-term storage.

Some types of produce (especially fruits) can be damaged by both high and low temperatures during storage, and especially by freezing. But the use of cold to kill tropical pests and heat to kill temperate pests is now a widespread practice, provided that the produce is not affected by the temperature. A great deal of research has been carried out on the cold storage of foodstuffs, especially with fruit and vegetables (see Snowdon, 1990); such treatment, clearly, will increase the shelf-life of fresh produce, and although it will not actually kill most temperate pests (+ 5°C) it will slow down their metabolism considerably.

7.10 PHYSIOLOGICAL DETERIORATION (CHEMICAL)

Harvested and stored plant and animal tissues suffer spoilage due to catalysis by enzymes. The end result can often be a softening of tissues, increased respiratory rate, protein destruction, degradation of collagen in fish fillets etc. - usually a loss of quality of one type or another. Sometimes this change can be desirable, as in the hanging of fresh beef for a few days to encourage the conversion of muscle into meat, and the hanging of game animals to tenderize them. In the case of heat-treated foodstuffs, where enzymes have been destroyed, nonenzymatic chemical reactions can play a part in food spoilage. For example, loss of Mg from chlorophyll (under conditions of low pH and high temperature) results in brown discoloration of green vegetables. Maillard browning is a very complex reaction between reducing sugars and free amino groups of amino acids or proteins –it is a non-enzymatic process generally affecting colour, flavour, nutrition, and possibly also safety of foodstuffs, usually caused by temperature changes. When reducing sugars react with amines, an extremely complex mixture of compounds is obtained, controlled to some extent by the reaction conditions, namely, the reaction time, temperature, pH and the concentrations of the reactants. The food industry is very interested in this process because of the desire to produce and control the characteristic aromas obtained in cooking (frying, baking, roasting or grilling). It is now possible to synthesize some of these aromas. Sometimes the colour resulting from non-enzymatic browning is desirable (bread crust, beer, coffee, meat crust), but at other times it is spoiling. However, the process is still not understood very well – in outline only.

Enzymatic browning is the production of pigments in plant materials injured mechanically or subjected to temperature changes. Cell disruption leads to the release of phenolytic substrates and enzymes , and in the presence of oxygen leads to oxidative production of coloured quinones. This brown coloration is obviously detrimental to the quality of the foodstuff, and is exacerbated by the wounding of plant tissue, especially during industrial processing. Heat treatment (blanching) is the simplest and most direct method of enzyme deactivation.

Oxidation of food components is the formation of acidic residues, often termed oxidative rancidity. In addition to flavour loss, colour and nutritional quality can also be affected. Lipids and proteins are the most susceptible food elements, and light is important as a causal agent as riboflavin and chlorophyll are particularly photosensitive. Vegetable and fish oils are highly susceptible, as is milk.

7.11 MATURATION, RIPENING AND SENESCENCE

Most fleshy fruits reach physiological maturity at an unripe state unsuitable for immediate consumption. The ripening process that follows, either on the tree or after picking, is the result of several physiological and biochemical processes – these are revealed in a series of changes in colour, texture, aroma and taste, and this leads to the physiological state at which the fruit is considered commercially edible. Many fruits are able to ripen after being picked from the tree (plant) provided that physiological maturity was reached before picking. The transition from maturity to senescence is often associated with a sudden rise in respiration rate and a synthesis of ethylene. Infestation of fruits by some pests, such as insects or micro-organisms, can result in a dramatic increase in the rate of ripening; often, infested apples on a tree can be recognized by their premature ripening. Some fruit disorders during storage were thought to be microbial in origin, but are now known to be physiological.

7.12 PATHOLOGICAL DECAY

Decay is the term used to describe the process whereby micro-organisms feed on a food item, utilizing part of the material as a source of nourishment for themselves and physically replacing it by their proliferating body (a mycelium for a fungus). Some of their by-products are just waste products, equivalent to an animal's excreta, but others are secondary in that their role is not understood and they may be very toxic (e.g. mycotoxins). The micro-organisms cause the food material to undergo localized discoloration, often a softening of the material, especially in fruit and root crops, and often culminating in liquidization. The metabolic by-products, both primary and secondary, may be toxic to animals and humans that ingest them, but typically they have both an unpleasant odour and taste. Fungal infections usually end up with a dense cover of fur-like mycelium of distinctive coloration - green, blue, yellow, brown or black - often in circular patterns, as the fungus grows radially (e.g. mushroom and toadstool 'fairy-rings').

CHAPTER 8

STORAGE STRUCTURES

Where food storage structures are concerned, there are now at least three different systems. The simplest is found in Africa and Asia and parts of Europe, for the smallholder farmer growing basically for his own family. Secondly, there is the large scale temperate farmer growing only one or two crops on a large scale (100-500 ha). Also, there will be the horticultural growers with vegetables and fruit on a somewhat smaller scale (10-100 ha). In some locations (e.g. Lincolnshire, UK), vegetable production is now large-scale and farms up to 500 ha can be found. Of course, in USA and Canada the 'large' farms are many times this size, both for 'truck'crops and cereals. Nowadays, with the development and proliferation of supermarkets, there is the relatively new industry of prepared foodstuffs or 'convenience foods' – these can include complete meals manufactured by or for the supermarkets by specialist firms which are sold either chilled for immediate consumption, or frozen, when they can be kept frozen for a month or more. Each supermarket has its own sources and suppliers, usually just the one for each product, and the produce has to be transported all around the country to the various retail outlets. Many of the supermarket products are dry, packaged or tinned, and so do not need special treatment for storage and transport. The supermarket foodstuffs flow-chart will lead both directly to domestic consumer outlets and to commercial ones such as cafés and restaurants, where it is finally prepared and consumed by the public.

8.1. TROPICAL ON-FARM AND RURAL DOMESTIC STORES

Most of the world still practises agriculture on a smallholder basis, where a single family cultivates from 0.5-2.0 hectares, mostly for food crops, although there may be some communal cultivation. Traditionally, each farmer will also have a cash crop as the source of money that he needs; this can range from coffee or tea to extra cereals, potatoes or vegetables if there is a suitable market nearby. In countries such as Malawi, with an extensive lake, fishing may be the source of revenue.

The on-farm stores are generally large enough to store a ton or two of produce – basically, grain and pulses to feed the family for up to one year, until the next harvest. Siting and construction are very much related to local conditions (weather etc.) and building materials available, and in recent years, sad to say, the ever-increasing likelihood of theft. Cash crop produce is generally only stored on-farm for a while (often for initial processing), and traditionally has its own store.

Vegetable cash crops are common throughout S.E. Asia, China, Japan and parts of Africa and India, near cities, where there is a daily demand for fresh vegetables.

With a vegetable or fruit cash crop, there is generally no storage as the produce is taken to market straight after harvest to be sold fresh. Coffee as a cash crop has to be dried on the farm, but tea is usually taken straight to the local plantation for treatment. Tobacco may be treated in either way.

8.1.1 Pots

In Africa, many dwelling places are built on the ground with projecting eaves for protection from the torrential rain. But in parts of S.E. Asia, it is the practice to build huts on wooden stilts with a concrete base to each leg. The elevation is done to raise the floor well above the ground level to keep it dry during the monsoonal rains, and the concrete leg base is to prevent termite destruction of the basal structural timbers. Such huts thus create a dry, sheltered area underneath, which can be used for produce storage (Figure 8.1). The legs (stilts) also make it easier to control ant access to the dwelling. More modern wooden houses in Malaysia are still built on stilts (Figure 8.2), but these are made of Bornean Ironwood (*Eusideroxylon zwageri* – Lauraceae) which is termite-resistant, and they stand embedded in a concrete platform as foundation, but nowadays the underfloor area is not often used for produce storage, unless in a very rural location.

Figure 8.1. Rural dwelling place in Malaysia elevated on wooden stilts with concrete bases, to protect the structural timbers from termite attack; the area underneath is protected and dry and used for storage.

Large earthenware pots, up to 1 m tall and 0.5 m wide, have been used since ancient times for storage, and at the present time are widely used in the tropics for shelled legumes and grains, and if provided with an adequate lid, they can be very effective stores. Pots can be stored underneath huts built on stilts, or outside the hut but sheltered by the eaves, or else inside the hut where they are more protected.

Figure 8.2. Wooden guest house on UNIMAS campus with stilts embedded in concrete to protect from termite attack; Sarawak.

Seed grains and pulses are usually kept in small earthenware pots sealed with clay, and are mostly kept inside the dwelling place. Gourds are also used for small quantities of seeds.

Small quantities of dried roots, tubers, and sometimes also pulses are often kept in baskets of woven grass, banana or palm leaf fibre, suspended from the eaves of the house.

The larger local stores are generally not very effective at keeping pests at bay, and they often suffer considerable pest damage. Generally, farmers are not happy at the idea of changing their traditional grain and food stores. But if the basic design can be retained, and if the local materials can be used, and if definite improvements result, then many farmers can be convinced. In different parts of Africa and Asia, there are many projects trying to tackle this problem under the heading of 'Appropriate Village Technology'; results to date have been rather mixed, for a number of different reasons. On elevated stores, the fitting of rat guards on the stilt legs is usually accepted. Another innovation is to construct a sealable mud brick container that can be rendered over with concrete and made air-tight. The use of metal bins is being promoted in some areas. Air-tight construction permits the use of phosphine-generating tablets for fumigation, as well as giving complete physical protection.

8.1.2 Storage Huts

Traditionally in most parts of the tropics, the main on-farm store is a small hut with wickerwork walls and a thatched roof. The floor may be of beaten earth, or it may be of wooden slats raised off the ground. The idea of the wickerwork walls and slatted floor is to allow free circulation of air, and the thatched roof keeps off the rain. In regions of heavy rainfall, the roof eaves are very extensive so that the overhang prevents undue wetting of the walls. But, of course, if there are spaces for air circulation, there is clear access for insect and rodent pests. Stores such as these are quite suitable for keeping maize on the cob and sorghum panicles prior to threshing, for this is a period of drying. At Alemaya University of Agriculture in Ethiopia, short-term stores were constructed for maize cobs out of wire netting, wooden posts, a concrete base and a metal roof.

Figure 8.3. Traditional small on-farm grain store as found in the Highlands of Ethiopia, made of wickerwork covered with a mud/dung mixture and a grass-thatched roof.

The more protected stores have the walls plastered with a mixture of earth (clay) and cow dung, and the roof covered with corrugated sheet-iron, and have a well-fitting small wooden door. Sometimes the walls are made of adobe (sun-dried clay bricks).

The size and construction of these stores will depend in part upon the nature of the produce to be stored. Maize on the cob, unthreshed sorghum panicles, seed cotton, copra and tobacco leaves are bulky crops, and a large store is usually needed, but only for a short period of time (on-farm storage of such produce is usually for a short time). Such stores are often large, but of simple construction and open for free air circulation (Figure 8.5).

Figure 8.4. Two small temporary on-farm stores for freshly harvested maize cobs, made of wire netting with a concrete base and sheet-iron roof; Alemaya University of Agriculture, Ethiopia.

Figure 8.5. Large open shed for temporary storage of maize cobs; Alemaya University of Agriculture, Ethiopia.

Shelled maize, other grains, shelled pulses, dried cassava tubers etc. are foodstuffs in a more concentrated form, and smaller stores suffice on the farms; the food is also to be kept for a longer period of time. Stores for such produce are often about 1.0-1.5 m high, and some 0.75 m wide, and typically are raised off the ground on short wooden stilts, and the roof is steeply sloped to aid rain runoff. Access to the grain is often by a small hatch at the base of the store (Figure 8.7). In the many agricultural reform programmes in Africa, such stores are being made out of sheet iron, or are having the walls rendered with concrete, so that they can be made almost air-tight for effective pest control (Figure 8.6).

Figure 8.6. Two better quality maize stores made of corrugated sheet iron, with a concrete base; weatherproof and ratproof; Alemaya University of Agriculture, Ethiopia.

Figure 8.7. A row of small maize stores solidly constructed of corrugated iron, with a sheet iron roof, concrete base, and small access door at the bottom; Malawi.

8.1.3 Underground Stores

In the dry regions of Africa, India and the Middle East, underground pits have been used for storage of grains for literally thousands of years. Basically, the pit is a hole in the ground with the entrance sealed in some way. In its crudest form, it is just a hole in the ground with a wooden lid. The more developed pits are lined and sealed, and the lid can be more or less air-tight. Soil fungi and bacteria will have access to the grain, and there will be water seepage unless the soil is very dry. Modern refinements to this method include sealing the earth wall with a layer of concrete, and waterproofing the surface with a sheet of polythene film, and the lid can be a carefully fitted hinged metal plate.

Figure 8.8. Experimental site comparing long-term sorghum storage in two underground pits and two traditional above-ground stores; Alemaya University of Agriculture, Ethiopia

The shape of the pit is either cylindrical, about 1 m wide and 2 m deep, which is typical of Eritraea and the Sudan, or like a conical flask, so that the entrance is narrower and more easily sealed, being only about 0.5 m in diameter; this is the usual shape in most of Ethiopia. Previously it had been usual to locate the pit just outside the family hut, situated for preference on an elevated site just in case of a rainstorm or a flash-flood. But theft is becoming a major hazard in most countries, and in Ethiopia at the present time about half the pits are now located inside the hut for safety. Concrete and polythene sheeting is often not available, so a large range of materials are used either to try to waterproof the earth wall or to keep the grains from touching the damp earth. The lid usually consists of pieces of wood plastered with cow dung and covered with earth. Underground pits are usually only opened every couple of months, so that pest control measures might be more effective.

When the pit is emptied annually, it is usually cleaned of debris and then fumigated by having a fire at the bottom for a while. Figure 8.8 shows a trial site at Alemaya, Ethiopia, where grain stored in two underground pits is compared with that in two traditional above-ground stores.

8.2 EUROPEAN ON-FARM STORES

Most European farms are now larger than in previous times, for larger units are more economically viable, and there has been much amalgamation of farms in recent years. The earlier stores on the farms were brick-built with a tiled roof, and were not pest-proof at all; nowadays some of these stores are being adapted and modified to meet present needs. New on-farm stores are being built larger and more secure, and for more specialised functions. At the present time in the UK, there tends to be agricultural specialisation, and some of the large farms grow mostly grain and rape, whereas other large farms grow mostly vegetables on a field scale.

Figure 8.9. A somewhat old fashioned English equipment barn, modernized for produce storage by concreting the open bays and adding a new large entrance.

Thus, some on-farm stores are designed for grain, rape and pulses, although most of these crops are now stored regionally in co-operative silos. On-farm vegetable stores are now quite sophisticated. Many are built of concrete with metal walls and roof (Figure 8.10), and are of large capacity. Some have a section which is designed for drying or cooling produce, and have an enormous fan positioned by a heating unit (or a cooling unit, or both) so that hot air or cold air can be driven through the bulk-stored vegetables. If the air is cool, then cabbages can be stored for a while if market conditions require it. The modern method for onion (ware crop) harvesting is to mow off the tops (leaves) in the field while still green and fleshy, then to lift the rows of bulbs on to the soil surface where they are left a while for the soil to dry. Then the bulbs are scooped up and riddled to remove soil and stones; the cleaned bulbs are then taken into the store, and hot air is blown through the pile of bulbs which is about five metres deep; there are also air ducts along the floor in the concrete. After a few days or a week, the bulbs are dried enough for sale and overwinter storage.

Figure 8.10. A series of new vegetable storage sheds, of metal construction and large capacity; one designed for onion drying etc., with large fans and heating and cooling facilities; Boston,Lincolnshire, UK. An on-farm facility for intensive vegetable production – mainly brassicas and onions

8.3 EUROPEAN DOMESTIC STORES

There is so much variation in the structure of domestic stores for agricultural produce in the developed world that generalisation is difficult. But they are invariably built of brick with a tiled roof, if they are among the older ones, whereas many of the modern buildings tend to be of concrete with a metal roof, and sometimes the upper parts of the walls are also metal. Many of the structures are rat-proof, but few are insect-proof. Most of the older structures were not built as food stores, and are modified dwelling places. The food materials are not often stored locally for long, and the cool weather helps to keep insect pest populations to a minimum, but even then there are occasional outbreaks.

8.4 REGIONAL STORES

These are large stores owned by either the government or a local co-operative, because of the size and costs involved, but some are private commercial ventures. Some are located in towns and some in ports, but in Europe some are rural and sited either in the centre of an area of intensive agriculture, or sometimes sited conveniently near to a railway terminal. In Figures 8.11-8.17 is shown a modern co-operative storage system in Lincolnshire, England. The two main structures each contain three large bays (each bay holds 3,000 tons of grain).

Figure 8.11. Union Grain Co-operative, Orby, Lincolnshire, UK; a six-bay storage facility for grain, pulses and rape seed, with a total capacity of about 20,000 tons.

At the side are six holding silos, each of which can take 250 tons. These are connected with a small cleaning unit, which is in turn joined to a double drying tower which can dry 60 tons per hour in a continuous flow system. The wheat comes from the fields by lorry, and on arrival the load is weighed on the weighbridge, and the grain is sampled. It is tested for moisture content, grain weight and temperature in a small automatic computerised machine (the Dicky-john) (Figure 8.16) – another machine can test the grain for protein levels, oil content and moisture level (Figure 8.17), and also gives the result on a paper print-out.

The grain is destined for milling if the quality is good, but if the protein content is low, or grain weight low, then it will be used as feed.

Figure 8.12. Side view showing the ends of three bays and ventilation outlets; basic structure is of steel walls and asbestos roof.

Figure 8.13. Six holding silos, each with a capacity of 250 tons; receiving pit is in the foreground.

Figure 8.14. The double drier unit, connected by conveyor belts to the holding silos on the right and the cleaning unit on the left, with door behind tractor; the receiving pit is in front of the cleaning unit, the metal doors at present closed.

The lorry-load of grain is then dumped into the receiving pit in the ground, and carried on conveyor belt to the appropriate holding silo according to its nature and final destination. Later, grain is taken from the holding silo by conveyor to the cleaning unit, where debris is removed. After cleaning, the grain is carried to the top of the drying tower; the grain falls down the tower in a continuous flow at a suitable temperature for drying. After about six passes through the tower, the grain is dry enough for storage, and is taken by conveyor again into one of the large bays that hold either 3,000 tons or 3,500 tons. The wheat usually comes in with a moisture content of 18-20% and is dried to about 14-15%. The duration of the drying is correlated with the initial moisture content, of course. The temperature of drying can be important, for high temperatures can have a deleterious effect on the grains. Wheat for milling is usually dried at about 80°C, but wheat for feed can be dried at 120°C; beans and peas should not be dried at higher than 60°C or the testa will crack. Along the floor of the bays are ventilation gullies covered with a steel mesh, and ambient air can be passed through the grain during the period of bulk storage, and this keeps the grain temperature reasonably low. Some of the grain is usually in storage for up to ten months; during the storage time, probes are left in position in the grain to monitor temperature and humidity, and are connected to the central computer. When the time comes for grain to be taken out of storage, it travels by conveyor to one of three large hoppers for loading on to lorries (Figure 8.15).

Figure 8.15. Three large hoppers for direct loading on to lorries when the grain is removed from storage.

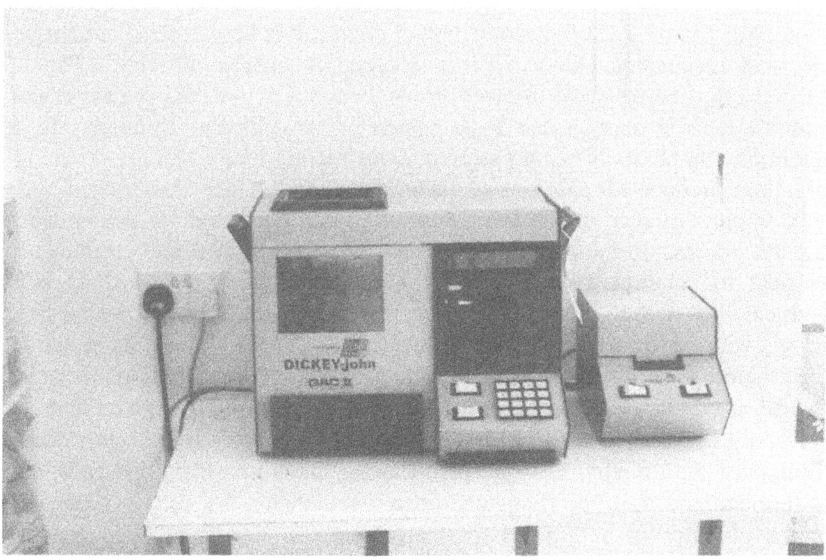

Figure 8.16. The Dicky-john computerized laboratory equipment for directly measuring grain moisture content, grain weight and temperature, with printer to the right.

Figure 8.17. Technicon grain sampler for direct reading of grain protein content, oil and moisture content.

The produce usually handled by this co-operative is mostly wheat, but some barley, rape, peas, and field beans (*Vicia faba*) are included; the equipment could be calibrated for a wide range of other crops.

Such a system could be regarded as ideal, from the points of view of management and produce protection, but, of course, it is very expensive to build and equip, and maintenance in a tropical situation is always difficult. There is a concerted effort being made in many tropical countries to develop a system of co-operatives to best support the local system of smallholder farming. In these programmes, emphasis is being placed on constructing regional stores where all the smallholder produce for sale can be accumulated and, if need be, treated. A very nice example was seen in Malawi a few years ago, designed for use with maize, wheat and pulses. In the main building was a unit for mechanical cleaning, then a drier next to equipment for bagging. The grain was stored in three 200 ton cylindrical silos next to the main building. The silos were of metal with a concrete base, and were air-tight, with gas pipes attached so that the grain could be fumigated while in storage if need be. There was also a disinfestation chamber adjacent, where a loaded lorry could be driven in and the whole thing fumigated in the event of a heavily infested consignment. Thus, any produce known to be infested could be fumigated with methyl bromide before being taken into the store, and grain for long term storage could be fumigated once it was inside the silo. For purposes of sale, the grain could be bagged for convenience of handling. The system was extensively mechanized, and, overall, was an excellent example of how a regional storage system should be organized.

In some parts of Africa, the regional co-operative-type stores are so successful that local farmers sell all their grain to the regional depots, and later buy back what they need when they need it, thus obviating any necessity for them to struggle with pest control in their own small stores – the extra cost is more than covered by the savings. In Europe, there are now very few farmers supplying their own kitchens as used to be the case at the turn of the century; most farmers, as already mentioned, tend to be specialized, and the current practice is to sell all the produce and buy food at the local supermarket.

Centralised stores for long term grain storage are sometimes underground. In the tropics, the two main physical problems for foodstuffs storage are the high ambient temperature and the high humidity which is prevalent in many regions; then there is the problem of easy access for pests. By building large underground stores, the temperature extremes are avoided, and access for pests is reduced to the entrance of the store. In parts of eastern Africa, large underground stores, called Cyprus bins, are used for long term grain storage. These bins can be sealed and made air-tight, either for fumigation purposes or for faunal suffocation. Such stores are usually government-owned, and used only for long term storage, for pest control is only effective so long as the bins remain sealed. An interesting account of a project in Israel in the Negev Desert was given in *New Scientist* (1988). Large stores were constructed underground and filled with grain. At night, when the air temperature in the desert is very low, the stores were opened and cold air was pumped through the grain. The stores were then kept sealed during the heat of the day. This daily cooling of the grain was very successful in depressing pest activity, and three-year grain storage was reported to be very successful.

8.5 REFRIGERATED STORES

Two forms of refrigerated store are usual: for frozen meat (carcasses etc.), large stores, often sited partially underground, are kept at a temperature between -20°C and -40°C. These stores will also hold a wide range of frozen foods, sometimes for quite a long while, and parts for the store may be let for public use. Wealthy residents of Hong Kong kept their winter fur coats in the Dairy Farm Cold Store during the hot humid summer to safeguard them from insect attack.

Vegetable and fruit stores are often kept either just above or just below 0°C according to the produce. Some vegetables and fruits freeze well, but others will only tolerate chilling without either physical or biochemical deterioration. Some fruits can very conveniently be picked green (unripe) and will then ripen satisfactorily in cool storage. Figures obtained from Holland (dated 1990) indicate that in general we (the UK and Europe) consume about twice as much chilled produce (218 kg per capita per year) as fresh, and seven times as much as deep frozen (31 kg per capita per year), and about the same as non-conditioned (i.e. non-refrigerated) food (228 kg). Thus, it is clear that refrigeration is playing an ever-increasing role in modern food production methods and logistics. This is summarized in a chart diagram (Figure 8.18) taken from Macrae, *et al.* (1993: p. 4616), where produce goes continuously from cold store to cold store until it is finally consumed.

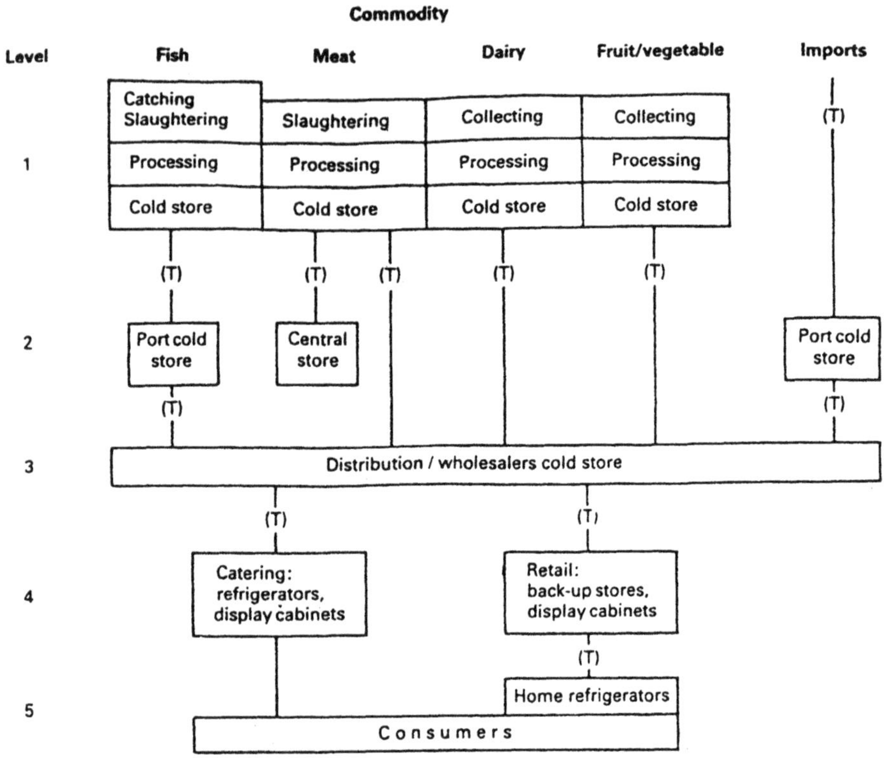

Figure 8.18. Distribution chains for refrigerated foodstuffs (T), transportation and transfers. Note that for products with very short quality life the distribution link is cut short by direct links between processing and retail/catering. (Adapted from Nowotny S. 1988.)

A good quality domestic refrigerator has a separate freezer compartment at around -20°C (similar to a domestic deep-freeze) for storage of deep-frozen items; the bulk of the refrigerator body is at about +3-5°C, and the vegetable compartment at the base some +5-8°C for fruit and vegetable storage. So it is important to realise that in the refrigerator body, life still goes on, albeit very slowly. Thus, a trapped bluebottle may not be flying around, but it is certainly able to walk (just), and will show little sign of dying. So, in a refrigerator, food is protected – but only for a time! We have a tendency to think that because the food is 'in the fridge' it is safe, but this misconception is one of the reasons for the alarming rise in incidents of food-poisoning.

8.6 NATIONAL FAMINE STORES (STRATEGIC STORES)

Most governments now have regionally distributed (but sometimes centralized) stores of large capacity in which essential foodstuffs are stored in case of any national emergency. Such an emergency could be earthquake, tidal wave, typhoon (hurricane), or war. In most countries, the precise locations of these stores are secret, and there is usually some non-informative notice board behind barbed wire – some such stores are underground, and some above. Most foodstuffs are stored for at least three years before being routinely replaced. The basic idea is that a country such as the UK should have at least one month's food supply in reserve in case of emergency. Obviously, such structures are well-constructed (it is to be hoped), and presumably designed for easy fumigation to keep pests at bay. Some publicity information has been released, and it appears that some stores are large underground caverns built of reinforced concrete and fully air-conditioned.

There are government grain stores which are less hidden and usually above ground, where animal grain feed is stored as a national reserve to safeguard the livestock industry in case of harvest failure or similar disaster.

But in general there is little information released about these stores, for reasons of national security.

CHAPTER 9

FOOD AND PRODUCE TRANSPORTATION

9.1 BACKGROUND

Traditionally, we have considered produce transportation in terms of shipment from the country of origin to the retail destination in Europe or the UK, and this involved ships, trains and road transport (lorries mostly), and more recently, aircraft. But in addition to this in the Western World, the recently developed supermarket system now demands a careful and precise control of the chain processes (logistics) to ensure a continuous flow of goods to the retail outlets. Each major Supermarket chain (there are about 12 in the UK now) has several hundred retail outlets throughout the country being supplied from the company headquarters in a central location. Because of the vast range of produce now retailed, there is little stock in reserve at each site, and the storage area is basically the display shelf area. This requires constant replenishment, at least on a weekly basis, and more often for some perishable items. The main question is how to control the flow of goods to reach pre-determined criteria of performance; namely, flow time, cost, quality, level of service, flexibility and added value. The application of logistic principles in the agrofood industry has revealed certain characteristics of the chain of activities which has resulted in the formulation of Agro (Food) Logistics; namely:

- The variation of basic material in quality and composition, within many primary products, over the seasons. The final product must still comply with firm specifications.
- The quality of intermediate and final products is strongly influenced by time and physical parameters (temperature, humidity and atmospheric composition) in the chain between producer and consumer. Hence, there is a special need for protection.
- Packaging technology plays an important role in the quality development, as well as protection, of foods. It influences also the design of processes, size of units, and equipment for physical transformation and preservation by heating, chilling or freezing.
- The agrorhythm of the primary production must be obeyed for biological and technical reasons. Nevertheless, a constant flow of final products of constant quality must be achieved.
- The production of linked products in the case of many intermediate and final products from one primary product.
- The high expenditure on logistic efforts to bring the primary product in a suitable state to the consumer, which can amount to more than 300% of the value of the primary product.

9.2 TRANSPORT AND STORAGE

The logistic flow of foodstuffs from producer to the supermarket retail outlets involves an alternation of transport and storage facilities. Storage facilities in a logistic chain are designed to buffer discontinuities of the product flow between supply and demand. They can be subdivided into types differing in task, size and operational conditions. (See Figure 1.1).

9.2.1 Primary (Production) Stores

Space for storage at the production site serves as a buffer between the discontinuities of the inflow from the production process and the outflow of products to the secondary stores. Residence time is short.

9.2.2 Secondary (Central) Stores

With seasonal produce, central stores are loaded quickly, but emptied slowly. The principle of first in, first out (FIFO) is often not realized. Non-seasonal products are supposed to pass central stores rapidly. Storage conditions are optimized within a narrow band of environmental factors. Residence times vary according to the type of product, but tend to be a substantial part of quality life, and FIFO cannot be guaranteed.

9.2.3 Tertiary (Distribution) Stores

Distribution stores are mere turnover facilities for rapid handling of flows of various products. Again, the FIFO principle cannot be applied, and mixing of products of different ages is usual. Environmental conditions are not optimal, but residence times are relatively short.

9.2.4 Quaternary Stores (Retail Cabinets)

At the retail outlet, display cabinets of fractions of a cubic metre shelf volume per metre present the product to the buyer (consumer). Residence times for produce here to be kept as short as possible, partly because of the cost of such display, and because of the danger of biodeterioration. For the reception of a discontinuous flow of goods at the retailer, a conditioned back-up store can form a buffer. This is kept as small as possible, and row storage is used. Conditioning is problematic owing to the high turnover rate, and residence has to be kept to a minimum, depending on consumers' buying practices.

9.3 SHIPS

In bulk grain carriers, the cargo area (hold) consists of several vast adjoining compartments, and the grain is pumped into the holds using suction or pumping equipment. The overseas journeys involved are seldom for more than a couple of weeks but could be for a month. If the hold is cleaned before use, and if, where there is any doubt, the space is fumigated, and if the grain is either clean or treated, there should be no pest problem. Ordinary cargo vessels which carry grain in sacks and produce in bags or boxes in the hold are the means by which so many pests have achieved their worldwide distributions. Such ships are not often fumigated and are seldom gas-tight, and the nooks and crevices in the hold are often filled with produce debris and detritus. The time periods involved will vary from a week to a month, usually, but some vessels go round the world, and could take two months or more to complete the journey. Produce may be delivered to, and collected from, a dozen or more ports *en route*. In many cases, there will be infested produce aboard, and there will be a strong likelihood of cross-infestation taking place. Most of the produce will have been kept in a godown or warehouse prior to shipment, and will often go into another on off-loading, so the possibilities for cross-infestation are almost endless.

This was the common state of affairs in the 1950s and 60s, but gradually the introduction of freight containers changed these practices. There are still bulk grain carriers for handling the largest consignments, but most of the produce is now carried in freight containers, and many cargo vessels can be seen with tightly packed containers, and most port facilities now have a container depot with the appropriate machinery for handling these structures.

9.4 FREIGHT CONTAINERS

This recent development in commercial shipping has considerable advantages when food materials and other perishable goods are being carried. The standard container is a rectangular box of approximately 12 or 20 ft x 6 ft x 6 ft, with one end removable. They are now used for every type of cargo that is small enough to fit inside the box. Fruit and vegetables in cartons, and other foodstuffs, can be packed inside the container right up to the roof. Most containers when new are more or less air-tight, and some of the best are completely gas-tight and have small gas taps built into the structure for fumigation purposes. On the back, there is a large red-bordered label which gives details of any fumigation treatment that has been carried out. This is a very effective method of transporting produce, and lends itself well to pest control measures.

There are a few aspects of the use of freight containers that are unexpected, all relating to the fact that the containers have to be returned after use. In Hong Kong, there is much export of electronic goods and clothing to the USA, especially California, and in the 1970s, rather than have the freight containers return empty, they were used for vegetables and fruit (celery, peppers, melons, citrus etc.) which could, apparently, be sold at prices competitive with local and regional produce.

The journey across the pacific took only about a week, so the produce arrived fresh. It did seem odd to see Californian celery on sale in the markets next to the excellent celery from the New Territories and South China. Such a practice could have many repercussions with regard to pest and disease distributions if carelessness is permitted.

The intermodal container has united the sea, air, land, road and rail modes of transportation, and has closed the protected-chain operation. On a regional scale, the removable body or swap-body serves the same purpose.

9.5 ROAD VEHICLES (TRUCKS)

Road vehicles account for the largest part of the agrofood products transportation. In the EEC on a national basis, up to 90% of the flow of foodstuffs is by road transport. Because of the need to conform to traffic regulations, there has been standardization of size and mass, and containers and removable bodies (swap-bodies) comply with stringent ISO standards for both external and internal dimensions. Conditioned (refrigerated) road vehicles and swap-bodies can comply better with ISO unit loads than do ISO containers. For both types of vehicles, performance is regulated. Residence times are close to transit times and almost uniform, except for the delivery runs which are mostly short. For transportation of deep-frozen meats, the vehicle has the internal body temperature of -20-30°C. For chilled produce, a higher temperature just above 0°C is maintained.

9.6 RAIL FREIGHT CARS

Some of the early testing of 'Phostoxin' tablets was conducted in East Africa. The treatment was used for grain shipments in sealed freight vans on the railway from Kampala to Mombasa. The journey to the port at Mombasa took several days, and the phosphine fumigation was generally very effective. The older and somewhat battered vans are, of course, no longer gas-tight, which could be a drawback in some tropical countries.

In some countries, particularly in the UK, the old railway system especially with regard to carrying freight has been severely curtailed and is now almost non-existent. This is particularly true for the carrying of foodstuffs.

In general, the transportation staff complain that railcars do not comply with unit load standards, and so are inconvenient as carriers. Also, because of the dependence on railway time schedules, waiting times can be considerable.

However, in many countries railways are still a major carrier for many types of produce in bulk, particularly consignments for export by sea.

9.7 AIRCRAFT

For high value fresh fruits and vegetables, air freight is becoming a regular means of transportation, especially from the tropics to cooler temperate regions. There is seldom chance for cross-infestation because of the short time involved, but pests are being transported also, and there may be quarantine problems. Travellers will have seen the flies and other insects regularly carried in modern airliners from country to country, and apparently rats are occasionally carried. Some air freight is carried in small, stoutly built wooden or metal containers, but this practice is not so common.

Aircraft are built largely to the builders' specifications, which include compliance with unit load, pallet and container regulations, but unfortunately there are no ISO standards available, so air cargoes are not part of the ISO modular system.

Chapter 24 in Gorham (1991), pp. 351-356, titled 'Prevention and management of pest problems associated with transportation of food', is particularly relevant.

CHAPTER 10

PESTS: CLASS MAMMALIA

Under the heading of the pest name in the following chapters is included data on the life history of the pest as available from the published literature, as well as some personal observations. Sometimes the data presented would seem to be rather odd in that a particular temperature or relative humidity is quoted, or a time such as 29 days with an implied high level of precision, and its selection could appear to have been somewhat arbitrary. Basically, this cannot be avoided, since the sources of most information have to be the published results of previous researchers.

Rate of development is generally governed by a combined effect of temperature, relative humidity (or moisture content of the food), and the nutritive value of the diet; the controlling factor can be any one of these three alone, or it can be a combined effect. Diet preferences are not easy to establish, for many pests can survive on a wide range of foodstuffs, but they may only thrive on a rather limited range, and successful breeding may only occur on these foods. Recorded lists of produce damaged seldom show any clear dietary preferences. On less suitable foods, reproduction may take place but larvae will develop at a very slow rate, and natural mortality levels will be high. Under any conditions, there will be natural (endogenous) mortality at egg, larval and pupal stages of insect pests, but under optimum conditions, the natural mortality levels will be low (only 1-2%). But under conditions of higher or lower temperature, higher or lower relative humidity (sometimes), or if the diet is inadequate (in either quantity or quality), mortality rates could be 50-70% at each instar.

Published data often do not give any indication as to the precise suitability of the environmental factors. Optimum conditions for a pest are often referred to rather superficially on the basis of a few laboratory experiments; clearly, such information is very useful to the non-specialist, but it should also be remembered that often it is only an approximation (or an extrapolation) since there are several variable factors that may not have been taken fully into account. Optimum conditions in terms of temperature and relative humidity have to be equated to the limits of tolerance. Some species are cosmopolitan, some tropical, some temperate, and in relation to temperature requirements, they may be eurythermal or stenothermal in any part of the overall tolerated temperature range. The end result can be somewhat confusing, for a species with a high temperature optimum might be most abundant in a cooler region; this could be because it is a eurythermal species that faces severe competition from another species under the more tropical conditions. Several major stored produce pests are recorded as being generally most abundant under conditions of temperature or humidity that are not regarded as being optimum for the species.

Distribution of a pest species is thus sometimes a combined effect of ambient climate (or microclimate), food availability and suitability, and natural competition levels. Thus, it should be remembered that a pest may be locally abundant under conditions that are not at all optimum. Each species will have originated somewhere in the world, but most species have been transported from region to region during trade activities over several centuries or more (some probably came to England with the invading Romans), and now many are quite cosmopolitan. With some of the truly cosmopolitan species that are very abundant, it is to be expected that they now occur as a series of geographical subspecies or races, and in some cases also as 'strains' locally. These different groups of insects are likely to show some basic differences in food preferences, climatic requirements and susceptibility to pesticides. With a few of the serious pests, it can be seen that their distribution is not completely worldwide (or pantropical), for they have not yet reached a few countries; and in some other countries, they arrived but were successfully eradicated, and have since been kept at bay. Thus, a literature search may not reveal the precise distribution of a pest at the present time. Legislation and quarantine regulations are very important for the countries where a particularly serious pest is not yet established; such as the major fruit flies in the USA; *Acanthoscelides obtectus* in Australia; *Trogoderma granarium* in East Africa etc.

Class Mammalia

Several species of mammals are regular pests of crops in the field, but the only regular serious pests of stored foodstuffs are rodents in the family Muridae – these are rats and mice. One point should perhaps be mentioned here, and that is the effect of predations by man, the most highly evolved of the Mammalia. A major problem nowadays with on-farm storage of produce is, regrettably, theft. In countries and locations where fifty years ago no property needed to be locked, declining standards of morality are such that the safety of food stores and the like cannot any longer be taken for granted.

10.1 ORDER RODENTIA (Rodents)

The synanthropic species of rodents are usually rats and mice, and they live in association with man in all parts of the world. They damage crops grown by smallholder farmers, and can be a very serious pest in food and produce stores. Rodents are especially damaging, for in addition to feeding, they also need to gnaw hard substances quite regularly in order to grind down their large incisor teeth that grow continuously throughout their life. This is why rats and mice are recorded gnawing through lead pipes, cables, wooden structures etc., sometimes causing extensive physical damage. They will, of course, gnaw through wooden structures in order to gain access to the interior of produce stores and to edible substances. All rodents possess very large, continuously growing incisors, and they also have a gap behind these teeth (called a diastema) and the premolars that follow them. Because of this gap, rodents are able to gnaw inedible materials, the fragments of which fall to the ground and do not pass into the mouth.

In rural parts of the world, other rodents do occasionally invade food stores and buildings – these include squirrels, dormice and others, and occasionally even porcupines.

Damage by rodents is primarily the eating of the produce, but contamination by faeces, urine, fur etc. is often the most serious. In the USA, through the Food and Drugs Act, there is very strict legislation concerning importation of flours etc. contaminated by rodents (as measured by the presence of rat and mouse hairs in the produce), so contamination of produce for export can be very serious economically, and often quite disproportionate to any actual damage.

10.1.1 Family Muridae (Rats and Mice)

Members of this family are the main rodent pests of stored products. Taxonomically, the differences between rats and mice are slight, and basically a matter of size. The urban species have adapted for life with man extremely well, and their diet is very similar to ours. The taxonomy of the genus *Mus* and of house mice generally (and some field mice) is reported to be in need of careful revision, and the present situation with regard to names in the literature is very confusing. Some of the Asian and African urban species of mice are often regarded as belonging to other genera, but for present purposes it is simpler to regard these mice as belonging to the genus *Mus*.

One of the side-effects of rodent infestation of human food (and water) is the transmission of disease organisms. It has long been known that *Pasturella pestis* (= *Yersinia pestis*), the causal organism of plague, is transmitted to man via the feeding bite of the tropical rat flea (*Xenopsylla cheopis*). But in recent years it has been established that several major human diseases, such as Weil's Disease, are transmitted via urine or faeces of rats.

Faecal droppings (scats) of urban rodents are quite distinctive in size and shape, and so the different species of rats and mice present in an area can be identified quite easily with a little practice.

Mus musculus (Linn.) (House mouse) (Figure 10.1)

Pest Status. Quite important urban pests, always present, worldwide in distribution; sometimes large populations cause serious problems.

Produce. House mice eat more or less the same foods as man, including all grains, farinaceous materials, cheese, meats, potatoes etc.; insects are also regularly eaten.

Damage. Direct eating of foodstuffs, combined with contamination by faeces, urine and hair; contamination is very important in produce for export as many consignments to the USA and parts of Europe are rejected after inspection. Damage by gnawing to containers, electricity cables etc. can also be important. The collecting of materials for nest building may damage fabrics and woollen materials. Food material is often carried away in cheek pouches to be deposited in nests and their own food stores hidden in crevices and corners. Mice consume less food than rats, but do a lot of nibbling and gnawing, and thus cause widespread slight damage. They have adapted so successfully to the human environment that races are known to live and breed in meat cold-stores – they live inside frozen beef carcasses, and feed entirely on fat and meat protein, at a temperature of -20°C!

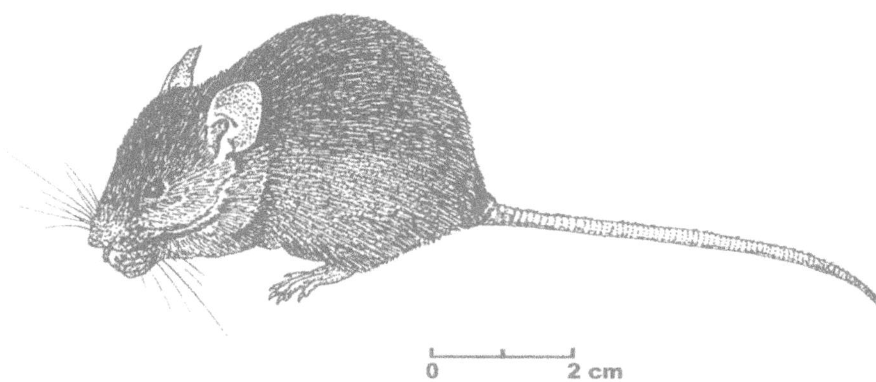

Figure 10.1. Mus musculus castaneus (Asian house mouse) (Rodentia; Muridae), adult mouse, from Hong Kong.

Life history. Mice become sexually mature at 2-3 months; the number of young per litter is typically 5-6, but 12-15 is recorded. Gestation period is 19-21 days. Some females in captivity have produced 100 offspring per year. Most mice prefer to feed little and often, and normally do not drink water much. The nest is made of shredded paper, fabrics, or other soft materials, and is often sited in the stored produce, but may be underground, in vegetation or in a crevice in the building. The adults are nocturnal usually, and very agile, climbing readily; they live for 1-3 years, but females seldom breed after about 15 months. Faecal dropping (scats) are rod-shaped and 3-6 mm long. In appearance, mice resemble young rats – body length is 70-90 mm, with tail 60-80 mm. Coloration varies somewhat, but most are grey-brown with belly varying, according to race, from grey, to buff, to white. Worldwide, several possible sub-species or races occur, but the taxonomy of *Mus* is at present in a state of confusion. When food is available and weather conditions mild, breeding is often continuous with many litters per year, and the young breed at 2-3 months of age. Thus, the potential for population increase is tremendous, and plagues of mice do occur from time to time in some rural areas, with dire consequences. There does not seem to be any regular irruption cycle (as with the four year cycle of field voles, *Microtis* spp., Cricetidae and some other rodents), but rather an irregular propensity for population increase when conditions are especially favourable. In Australia, there has been a series of very spectacular mouse plagues – the earliest in1916 and 1917, and the latest within the last few years, and damage to ripe cereal crops and in grain silos was extensive.

Distribution. *Mus musculus* occurs as a series of geographical sub-species and races, and *M. m. domesticus* is the common house mouse, now quite worldwide, but thought to have originated in the steppe zone of the southern Palaearctic region. Some of the races are quite urban, but others equally rural, and both types are found in produce stores. In many regions, the distinction between house mice and field mice is somewhat arbitrary.

Natural enemies. Cats, foxes, and various Mustellidae, and birds of prey (owls, kestrels etc.) are the natural predators of mice, as are some snakes. In most regions, the populations of natural predators have declined due to different human activities, and this does encourage mouse populations to increase more frequently.

Physical and cultural control. General sanitation methods as recommended for rat control (this Chapter) should be used, but mouse-proofing structures and buildings is more difficult owing to their small size. Mice can use very small holes for ingress to structures, and they are easily carried into stores in produce, crates and cartons.

Trapping. Mice are generally easier to trap than rats – they are less wary. Metal and wood-base snap traps are usually successful, and several traps can be spaced about 1 m apart. Other traps catch the mice alive; Longworth traps only catch one at a time, though in Victorian times in England, there were very elegant traps designed to catch a dozen or more mice, one at a time. Experiments have shown that mice will usually accept a new object, such as a trap, after only about ten minutes.

Tracking powders. These are poisonous powders applied to a floor surface, wall space and the like, which the mice pick up on their feet and later lick off. In the past, DDT (55% a.i.) was very effective (but not against rats), but in many countries the use of DDT is no longer permitted, and nowadays sodium fluorosilicate is often used, and the anticoagulant chlorophacinone (e.g. 'Rozol') is being used in the USA, as is valone.

Poisons. The same chemicals as recommended for rat control can be used against mice (with the exception of red squill). Because mice tend to nibble when feeding, rather than eating a large amount at any one time, the concentration of poison in the bait needs to be higher in order to ensure a toxic dose. Mice do not forage as widely as rats, so it is more effective to use many small baits rather than a few large ones, and frequent renewal is recommended, for it is recorded that mice are not attracted to old stale baits.

Apodemus spp. (Field Mice)

In Europe and Britain, the Common Field Mouse is *A. sylvaticus* Linn. (with several distinct races), and this rural species appears to be invading urban situations more often in recent years, and in some areas, it is as abundant as *Mus* in domestic buildings, and causing as much damage. The other species of *Apodemus* seem to be more rural, but can be found in on-farm stores.

Rattus rattus sspp. (Roof, ship or black rat) (Figure 10.2)

Pest Status. An important urban pest that is, in addition, a major pest of several tropical tree (palm) crops; it is also the main host of the tropical rat flea that is the vector of the plague organisms.

Produce. An omnivorous rat with a preference for fruits, nuts, seeds, and other vegetable matter (basically frugivorous). Rural *Rattus rattus* are arboreal in trees and palms, and usually climb to feed. Urban roof rats do not feed on garbage to the same extent as the common rat. In general, they eat much the same food as man.

Damage. Direct damage by eating the produce is usually outweighed by the contamination of foodstuffs with excreta, urine and hair.

Life history. Roof rats become sexually mature in 3-5 months, and may have 3-6 litters per year, with an average of 7 young (6-8 per litter). Nests are usually in buildings or trees, and seldom in burrows underground – this is essentially an arboreal rodent. Rats usually need water to drink, unless the food is moist vegetable material. These rats can climb vertical pipes up to 10 cm diameter, and along telephone wires, being very agile, as well as good swimmers. Roof rat infestations are often completely contained within the buildings and stores. Adults are moderately-sized rats, weighing about 0.25 kg, with a long tail and large ears; body length is 15-24 cm, and tail 100-130% of the head plus body length; males are generally larger than females. The tail is somewhat prehensile, and is semi-naked and uniform in colour. Droppings (scats) are spindle-shaped and about 12 mm long. Typical coloration is dark grey-brown, but sometimes black, with a pale grey belly. But several colour variations occur in different parts of the world – some are more brown, with either a white or grey belly. But it appears that these colour forms do not have any taxonomic status; however, it does seem that there are probably some distinct sub-species. Most activity is at night, but they are not completely nocturnal. Under suitable conditions of weather and food availability, roof rats may breed continuously.

Distribution. The centre of evolution of *Rattus rattus* appears to be S.E. Asia, and it has spread through the agency of man and commerce. Even on the most remote oceanic islands, the ship rat is to be found. Most of the dispersal has been made through shipping, as these arboreal rats easily climb along tie lines and ropes on to and off docked ships. In temperate regions, the distribution of *Rattus rattus* is largely confined to port areas and docklands, but in tropical and subtropical areas, the species is quite rural, and they are particularly associated with palms (especially when the old fronds are not removed annually); they are essentially arboreal, and prefer to nest aerially. In the tropics, they are more common at higher altitudes, at 2-3,000 m or more.

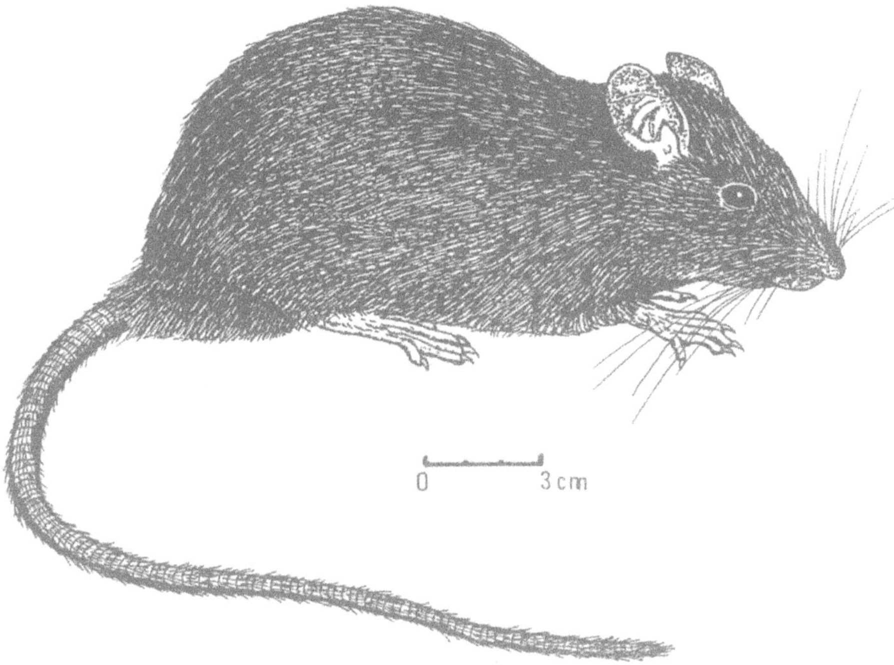

Figure 10.2. Rattus rattus rattus (Roof rat) (Rodentia; Muridae), adult rat, from Hong Kong

Control. Because it is an arboreal species, the usual practice for field baiting is to nail a solid bait block on to the side of a palm trunk.

Rattus norvegicus (Berkenhout) (Common Rat) (Figure 10.3)

Pest Status. A very serious urban pest and very damaging to stored produce, both on-farm and in city produce stores worldwide; also damaging to some field crops in the tropics.

Produce. Totally omnivorous, including almost anything edible in their diet; the diet is basically very similar to human diet, and includes quite a lot of meat – also, garbage is eaten. In some regions, it is known as the sewer rat.

Damage. As with other Muridae, the direct damage by eating is usually overshadowed by the very serious contamination by faeces, urine, and hair. With a rat so large, dead bodies following the use of poison baits can constitute a problem in food stores. But rat cadavers are usually located rapidly by blowflies, and the fly maggots devour the carcass quite quickly under warm conditions. The presence of blowflies in a store can be used as an indicator of rodent corpses. Gnawing of containers, pipes, cables etc. can be quite serious in some situations.

Life history. Common rats reach sexual maturity in 3-5 months, and they usually only live for about one year, but produce 4-7 litters in that year, each with 8-12 young. Thus, they have a higher reproductive potential than does *R. rattus*. They are basically ground dwellers, and nest in burrows underground – they swim well, but do not climb much. Their runs and tunnels are on the ground surface in grass and vegetation, and often clearly visible, being 5-7 cm in width. Nests are constructed of pieces of sacking, cloth, paper, straw, or similar soft material, and situated outside usually, under buildings or nearby in banks of earth. Common rats enter produce stores to forage and feed, and then return to their burrows and nests outside and underground. They often forage over a distance of up to 50 m daily. In towns, they inhabit the sewerage system, and often gain access to buildings via the underground pipes.

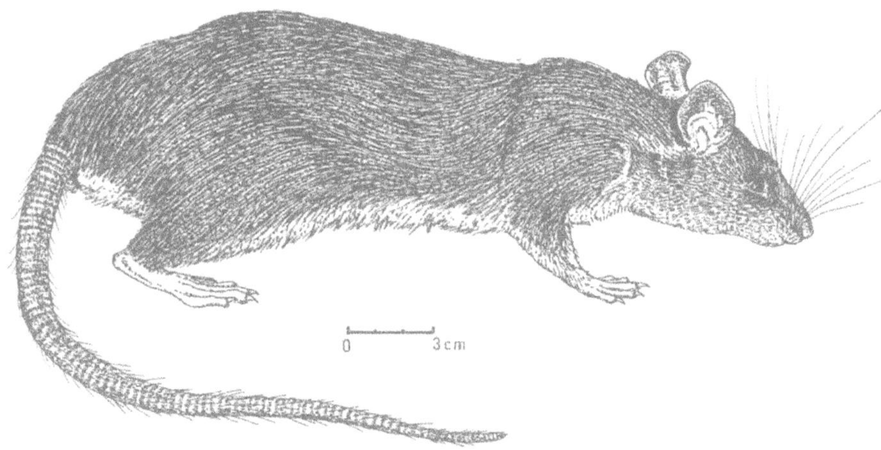

Figure 10.3. Rattus norvegicus (Common Rat) (Rodentia; Muridae), adult rat, from Hong Kong.

The adults are large, heavy rats (up to 0.5 kg) with body length 19-28 cm, and tail 80-100% of body length; ears are small; males usually larger than females. The tail is shortish, quite scaly, semi-naked and bi-coloured (paler underneath), and not at all prehensile. Body colour is typically grey-brown above and pale grey underneath, but melanistic and albino forms occur in most populations. Young rats can be distinguished from house mice by their relatively large head and large hind feet. Droppings are capsule-shaped and up to 20 mm long.

This species is essentially a ground dweller and is a lowland species. Under warm conditions and with plenty of food, breeding may be continuous – it has a tremendous capacity for population explosion. For example, in the UK in 1989, after two mild winters, the common rat population suddenly increased considerably, and damage was widespread. Because of their inhabiting sewerage systems and water generally, they are important vectors of some human parasites. They are nocturnal usually, but in some situations may be encountered in daylight. Water is usually required for drinking, and many colonies have runs to water sources. *R. norvegicus* often lives gregariously in a loose colony. With many food stores, there is often a rat colony nearby, either outside, or under the building, with a series of tunnels and runs into the store for foraging, and there are bolt holes at intervals for quick escape exits.

Distribution. Now completely worldwide, and on most oceanic islands. In the tropics, it is generally associated with port cities, but in temperate Europe and Asia, some truly rural populations occur many miles from the nearest human habitation, often in coastal regions or along river banks. In the tropics, they are usually not found in highland regions.

Natural enemies. For both *Rattus* species, the main predators in the tropics are snakes – ground snakes such as the cobras (*Naja* spp.), rat snakes (*Pytas* spp. etc.; Colubridae) for *R. norvegicus*, and the arboreal cat snakes (*Boiga* spp.) for *R. rattus* especially. But throughout the tropics, the local people invariably have a deep and unreasonable fear of all snakes, and kill them whenever possible. And in some countries, India for example, snakes are killed for their skins, and there are thriving local industries. The result is a serious depletion of urban snake populations. In addition, other predators such as civet cats (Viverridae), mongooses and other carnivores are killed by the local villagers. The end result is a greatly enlarged urban rat population. A recent FAO survey in India concluded that there were six urban rats for every human being in the country.

In Ethiopia, local cats killed the urban rats (*Rattus rattus* mostly) without hesitation, and the feral cats then ate them. The feline value in rat control was widely appreciated by the local villagers, and most huts had a hungry cat in residence; this practice had a major suppressive effect on the local rat population.

Control. A point that has been raised in several recent rodent control publications relates to the vast reproductive potential of these main pest species. In many pest control exercises, a kill of 90% rates a chemical pesticide as being effective commercially, and in some pest management programmes, an insect population reduction of 90% is adequate for control purposes. But with *Rattus norvegicus* and the other domestic rodents, their reproductive potential is so great that a survival rate of 10% would enable the population to re-attain its original size within a matter of only months. A population survival rate of 5% is basically too much to leave uncontrolled, and in many situations with isolated food stores, it is probably worthwhile to attempt a rodent eradication programme.

Physical and cultural control. General sanitation methods include removal of additional food sources, the packing of produce into containers, keeping the containers off the floor if possible, removal of garbage from the proximity of food stores. The building should be rat-proofed by constructing it of materials that rats cannot bite through (i.e. concrete, brick, sheet asbestos, sheet iron etc.). Fine mesh wire netting or metal gauze should be placed over ventilation holes. The sites where pipes pass through the walls and eaves of buildings are especially vulnerable, and access from nearby sewers and drains should be prevented. Buildings on stilts, or small on-farm containers on legs, should have rat guards fitted to the legs. But in the least developed countries, such recommendations are not easily followed because of the costs and the availability of materials. It should be remembered that rats have poor vision, but keen senses of smell, taste, touch and hearing.

Figure 10.4. Seed storage drawer with access hole gnawed by local rats (probably R. rattus; Forestry Department, Alemaya University of Agriculture, Ethiopia.

Trapping. Rats are not easy to trap, although if care is taken to disguise human smells (for example, in Ethiopia, by rubbing a cut onion bulb over a baited snap trap), such traps can be very effective. Traps should be tethered by a string in case the rat is not killed, and, because they forage so widely, traps should be spaced well apart. Snap traps are not expensive, and will work for years. Rats are nervous of new objects and new foods, and pre-baiting may be necessary for a few nights before the traps are set (this certainly applies to the use of poison baits).

Tracking powders. These rodents are not so successfully attacked using tracking powders as are mice, and DDT was never effective; but sodium fluorosilicate is said to be effective, and valone is used.

Figure 10.5. Typical poison bait dish for use with rat baits; the dish is brightly coloured yellow, clearly labeled 'Poison", and the bait grain dyed red

Poisons. The basic plan is to mix a poison with an attractive bait material to entice the rats to feed (Figure 10.5). Any fast-acting acute poison quickly causes the animals to become bait shy (same as trap shy). The earliest poisons were inorganic salts of arsenic and the like. But in the 1950s the anticoagulant poisons were discovered. Their action is insidious with no manifest symptoms of poisoning, and these chemicals were very successful. However, after some years of repeated use, many rat populations have built up resistance to many of the earlier anticoagulants, but new chemicals and formulations are being developed. Liquid baits are sometimes effective, especially in dry regions where water is scarce. Otherwise, food baits are most widely used, based on grain, oatmeal, meat, fish, fishmeal, fruit or other types of vegetable matter. Any type of human food can be used, and local preferences may be shown; for example, in Ethiopia onions are used very widely in local cooking, and it is said that *Rattus rattus* is attracted by this odour – and it appeared to be true. Pre-baiting for a few nights with the bait material alone is always recommended, especially if acute poisons are being used. Acceptance of new objects, such as a trap or a bait, with most rat populations typically takes several days, and for the shyest individuals it can take 10 days in experiments, even with a long-familiar food. Additives such as groundnut oil improve the acceptability of most baits. For tree baiting, the mixture is compressed into a block, often with wax added, and the block is nailed to the palm trunk where the rats run. For such exposed baits, a waterproofing additive is vital.

Poison grain baits are often coloured bright red as a safety measure. Acute poisons are one dose chemicals, and are still quite widely used in many tropical countries. Chemicals being used include strychnine, arsenic salts, zinc phosphide (sodium fluoroacetate), red squill, and norbormide.

The different rodenticides have different levels of general toxicity, and although often selective to *Rattus*, they do vary in toxicity according to the species of rat, and sometimes even to the race; in addition, there is the problem of the development of resistance to certain poisons by local rat populations.

Chronic, slow acting, multiple dose poisons are the anticoagulants, which cause the rats to bleed to death internally. Small, repeated doses are necessary over a period of time (a few days) to kill the rats, although some newer compounds can kill after a single feed – death occurs after a few days. The first chemical with anticoagulant properties developed as a rodenticide was warfarin. Now the list is quite extensive, and includes brodifacoum, bromodiolone, chlorophacinone, coumatetralyl, difenacoum, fumarin, pindone and others. Shell have recently publicised their new anticoagulant rodenticide, flocoumafen ('Storm'), and claim that it is very effective, giving control levels of 95 per cent in field trials in S.E. Asia; and they say that a single feed is sufficient since the active ingredient is highly potent.

Fumigation will, of course, kill any rodents present in the produce at the time, and if the underground nest system of *R. norvegicus* can be clearly established, it can be sealed and flooded with poison gas (methyl bromide, phosphine etc.).

Recent studies in S.E. Asia showed that for large rural food stores with a serious rat problem, the most effective approach was an integrated one whereby traps were used in runs outside the stores, the stores were carefully maintained physically, and poison baits were used both outside and inside the stores, and gas was used down the holes. Since rats are both intelligent and very nervous (shy), they are unlikely to succumb in large numbers to any one particular method of control.

For centuries, man has relied heavily on the use of poisons for rat control in domestic situations, but so far it is a moot point as to who is winning this war. The two main problems are the intelligence of the rats and their ability to learn, and the speed with which they can develop resistance to the poisons.

Rodenticides can be classified into three separate generations, as follows:

1st generation poisons (acute poisons)		
arsenic salts	}	
strychnine	}	80% kill the usual
zinc phosphide	}	maximum
red squill (scilliroside glycoside)	}	
norbormide	}	
cholecalciferol	}	
2nd generation (anticoagulant Rodenticides)		
warfarin (& prolin)	}	

chlorophacinone	}	90% kills achieved,
diphacinone	}	but resistance has
coumatetralyl	}	developed to all
coumaturyl	}	

3rd generation (2nd generation anticoagulants)
difenacoum
brodifacoum - single dose effective
bromadiolone
flocoumafen - Shell claims 95% kill

(See Gorham (1991); pp. 419-426; and Parshad (1999))

Although *Rattus rattus* and *R. norvegicus* are the most serious and most widespread international rat pests of stored foodstuffs, there are others. In each part of the world there are field rats of different species, and most of these do occasionally invade on-farm field stores, and sometimes regional stores, and serious damage has been recorded. These species include:

Arvicanthis spp. (Grass rats) Africa
Bandicota spp. (Bandicoot rats) India, S.E. Asia to China
Mastomys natalensis (Multimammate rat) Africa
Nesokia spp. India
Oryzomys spp. (Rice rats) Several important species are pests in Indonesia, Philippines and other parts of S.E. Asia.

10.1.2 Family Cricetidae (Voles)

Most of these rodents are small (size of mice), but a few are as large as rats; the body is more rounded and the tail is very short, and they are all rural species – none is synanthropic. But occasionally species of vole have been recorded entering rural food stores and causing damage. The main species concerned are:

Holochilus spp. (Marsh voles) South America
Meriones spp. (Jirds) North Africa and West Asia
Microtus agrestis (Field vole) Western Palaearctic; the genus is Holarctic and abundant – several species are involved.
Tatera spp. (Gerbils) Africa and South Asia.

CHAPTER 11

PESTS: CLASS AVES

As a group, the birds only occasionally cause concern as pests of stored products, but the open crib type of on-farm produce store is, of course, vulnerable to bird attack, for the stored grain is usually accessible. From the literature it appears that, in parts of S.E. Asia, grain stores of an open type of construction, built to allow free air circulation to help dry the crop, are regularly invaded by birds of several types, and grain losses can be of economic importance.

But the birds concerned are diurnal in habits, and large enough to be kept at bay by wire netting, although they can reach through the netting to eat grains that are accessible.

11.1 ORDER PASSERIFORMES (Perching birds)

These are the birds that live in trees to a large extent, and have feet adapted to gripping twigs, and most are quite small in size. Several species of perching birds are regular pests of ripening grain crops in the field, and they are biologically classed as being granivorous. Within this Order, the two main groups concerned are the families Fringillidae (Finches) and Ploceidae (Sparrows and Weavers). The bill is short and stout and capable of breaking open small cereal grains and seeds – ripe maize grains are generally too large and too hard for them to eat. The grain has to be crushed in the beak before the pieces are swallowed. Both families contain species that normally live in flocks, and these are the ones that cause most damage to ripening grain crops in Africa and parts of Asia.

11.1.1 Family Ploceidae (Sparrows and Weavers)

A large family, worldwide in distribution with many different species; many species are synanthropic and are also pests of cultivated plants.

Passer spp. (House sparrow etc.) (Figure 11.1)

Pest Status. These are mostly urban species and live in close association with man; but some are more rural; some live in definite colonies.

Produce. Grain in on-farm stores, particularly when still on the panicle, is at risk from bird attack, and flocks of sparrows can cause extensive damage. All small grains are vulnerable, but maize is usually not attacked, for the grains are too large and too hard. In food stores, various farinaceous products may also be eaten.

Figure 11.1. Passer montanus (Tree sparrow) (Aves; Passeriformes; Ploceidae), adult; specimen from Hong Kong where this is the common domestic sparrow

Damage. Direct eating of the grains is the obvious damage, and faecal contamination is secondary. Although this section is labeled '*Passer* spp.', the information given applies to most of the species within the family.

Life history. Some species nest singly, either in buildings or in adjacent trees, but a few species nest in small colonies. The nest is typically dome-shaped with a lateral or ventral entrance, and most nest in the vicinity of buildings. Several broods of 4-6 young are reared annually. The young birds after fledging typically form a flock, which may be joined by the adults in the winter period, when breeding has ceased. These flocks will attack on-farm stores where grain is housed. In parts of S.E. Asia, warehouse losses are reported to be serious at times; these rural stores have to be well ventilated in order to discourage mould development, and open eaves will allow birds to enter the buildings.

Distribution. The genus *Passer* is found worldwide as a number of different species – for example, six species are crop pests in Africa. The family contains 315 species; other species of economic importance are included in *Ploceus* and *Padda.*
Control: The easiest form of control is to ensure that access to produce stores is denied by using wire mesh over windows and ventilation sources. These birds are all diurnal in habits, and easily frightened by the presence of man and human activities, so it is seldom necessary to resort to the use of chemical avicides.

Ploceus spp. (Weavers etc.)

Many of these birds are more rural, but some are definitely to be regarded as domestic; they are often gregarious, and together with the rural *Quelea* spp. are most serious as field pests of ripening small-grain crops, but some are recorded causing damage in rural on-farm grain stores. They are abundant both in Africa and much of tropical Asia.

Padda oryzivora (Java sparrow)

This species is now widely distributed throughout S.E. Asia and many of the large islands, and is a serious pest of rice crops, but seldom recorded in grain stores.

10.2 ORDER COLUMBIFORMES (Pigeons and Doves)

10.2.1 Family Columbidae

In the world, there is a total of about 290 species; these are plump birds, many quite large, and they often live in flocks (for at least part of the year) and are basically seed eaters. Their beaks tend to be rather small and slender, and cannot be used to crack open seeds – they feed by swallowing the seeds which are then stored in the large crop, and later are crushed in the gizzard and then passed into the intestine for digestion. Normal feeding behaviour is to quickly fill the crop with intact grains and seeds, and then fly to a sheltered place for later digestion – thus, in a short time most pigeons can remove a large amount of grain from a store.

Most pigeons nest singly, but often form flocks later, especially in the winter in temperate regions. On field crops they also take leaf material, and they can be very damaging to crops of leaf vegetables. They are seldom pests of any importance in grain stores, but some records of appreciable damage have been made in parts of Indonesia and S.E. Asia.

The two main groups of birds recorded as grain pests are as follows:

Columba spp. (Pigeons)

Several species are worldwide, and important agricultural pests, and in some regions feral populations of *Columba livia* (Figure 11.2) are a nuisance. This is the native British rock pigeon that has been widely domesticated as the racing pigeon, and is also reared in large numbers in China as food.

Streptopelia spp. (Doves)

Generally smaller species; the group is more tropical, and at least half a dozen species are quite urban and regularly found on domestic premises in different parts of the world (Figure 11.3).

Figure 11.2. Columba livia (Feral pigeon) (Aves; Columbiformes; Columbidae), adult; specimen from Hong Kong

Figure 11.3. Oriental Turtle Dove (Streptopelia orientalis)

CHAPTER 12

PESTS: PHYLUM MOLLUSCA
(Class Gastropoda)

The Phylum Mollusca contains three main Classes – the Gastropoda (Snails), the Bivalvia (Bivalves), and the Cephalopoda (Squid and Octopus). The latter two groups are all aquatic and mostly marine, but the Gastropoda include many terrestrial forms as well as aquatic, and these terrestrial snails and slugs can be agricultural pests in that they feed on plant tissues, and some can be a nuisance on harvested vegetables and root crops.

Some of the bivalves, especially the marine oysters and mussels, can be of importance medically, because they are filter-feeders, and they can filter out pathogenic bacteria from the polluted sea water, and if eaten raw or insufficiently cooked, the infected mollusc flesh will cause food poisoning in the human who eats them.

12.1 CLASS GASTROPODA

The importance of terrestrial snails and slugs as pests of growing crops varies considerably in different parts of the world. In many tropical areas they are not very important, whereas in many temperate countries they are at times the cause of considerable crop losses.

In stored produce they are encountered from time to time, sometimes in grain but more usually in vegetables. In vegetables for canning and freezing, contamination by the body of a slug or the shell of a snail can be very serious, but production lines are generally designed so that such foreign bodies are recognised and removed.

Apart from the Giant African Snail in the tropics, there are no species of slugs and terrestrial snails that are very widely distributed; each area has its own endemic species. So two examples are given below that occur in the UK and Western Europe, but there are very similar species to be found in most other regions of the world, doing the same damage and causing similar problems.

12.1.1 Family Limacidae (Slugs)

There are three main families of slugs, the Limacidae, Arionidae and Milacidae. They are characterized by having an elongate foot and a small internal shell that in most species cannot be seen without dissecting the specimen. They produce great quantities of mucus on which the animal glides. Feeding in the Gastropoda involves the use of the strap-like radula bearing rows of pointed teeth with which the animal rasps the food plant. Since slugs have no shell large enough in which to shelter, they are greatly affected by environmental conditions, and are particularly susceptible to desiccation. The large species are soil litter and foliage inhabiting forms, and some climb high in plant foliage at night, when conditions are moist. The smaller species are usually soil dwellers, and sometimes quite subterranean in habits, having been recorded at depths of 1-2 m. They feed on plant roots, tubers and the like, but come to the surface at night and may be found in plant foliage when conditions are moist.

Deroceras reticulatum (Müller) (Field slug)

Pest Status. Rated as the most common slug in north western Europe; a major agricultural and garden pest damaging a very wide range of cultivated plants.

Produce. They bore narrow tunnels in potato tubers (Figure 12.1), and sometimes in vegetable root crops; in wet conditions they often climb into the foliage of vegetables, and may be found in celery and hearts of lettuce and cabbage etc., and also in the foliage of fresh peas.

Figure 12.1. Potato tuber (peeled) showing typical slug damage; Skegness, UK.

Damage. Feeding in potato tubers results in a deep narrow tunnel, and in leaf vegetables there may be feeding holes and scarification, but the damage is more often contamination by the body of the slug, and the presence of slime trails and sticky faecal matter on the foliage. Peas for freezing and canning are regularly contaminated by small dead field slugs.

Life history. Eggs are laid in batches in the soil in locations where they remain moist.

This species is rated as being of medium size, and measures 3.5-5 cm when extended; body colour ranges from cream, through brown to grey, often with a dense pattern of darker flecks, and the sole of the foot is pale. The mucus is colourless or white.

Smaller species are more likely to be found contaminating vegetables, whereas the largest species are seldom found thus.

Distribution. This species is purely European, but other similar species are found in temperate regions of the world, each region having its own particular species of small field slugs.

Control. In crop situations, pellets of methiocarb or metaldehyde are generally effective in killing slugs and snails, but in produce stores the usual problem is to remove the body of the slug from the produce – this often has to be done by hand.

12.1.2 Family Helicidae (Snails)

This is one of the largest families of terrestrial snails, and it also contains the largest of the European species. The single most important feature of the terrestrial snails is their coiled shell, into which the body can retreat when environmental conditions become unfavourable, the entrance to the shell (mouth) being sealed with a cap of dried mucus. From the point of view of contamination of produce, the shell is important because it is hard and calcareous and will persist after the death of the animal. The snails that are important in contaminating produce are usually the small species, as they are either small enough to escape casual produce inspection, or else their size is similar enough to the harvested produce (pea seeds, cereal grains) for them to be picked up by machinery. In all parts of the temperate world, there will be different species of small field snails that will be involved in produce contamination. Only the one species is mentioned here.

Cernuella virgata (da Costa) etc. (Small banded snail) (Figure 12.2)

Pest Status. Small banded snails are found as contaminants in a wide range of produce in most temperate regions of the world; occasionally, serious financial losses are incurred.

Produce. The best known contamination cases have involved barley in Australia, and peas for freezing and for canning in the UK.

Damage. In stored produce, the damage is invariably indirect in that it is the contamination of produce that is important. Their feeding damages the growing plants directly, and some plants will be taken into stores showing signs of snail feeding damage (which is often identical to that made by slugs).

Life history. Details are not really of relevance here since contamination is the main problem. This species is usually 6-12 mm in shell height and 8-16 mm in diameter, although it is said to grow larger; smaller individuals are the ones usually involved in produce contamination as they are closer to the size of fresh pea seeds. It is reported that in South Australia, barley crops are regularly infested with small field snails (*Theba* spp.; accidentally introduced from Europe) to such an extent that farmers rate the harvested barley as satisfactory if it contains no more than five snail shells per litre of grain. Needless to say, such a level of contamination of barley or wheat grain is not acceptable internationally, and in the past, shipments of grain from Australia have been rejected, costing millions of dollars in compensation payments.

Distribution. This species is confined to Western Europe and the Mediterranean region. But other species of small field snails occur in most other temperate regions of the world (Figure 12.2).

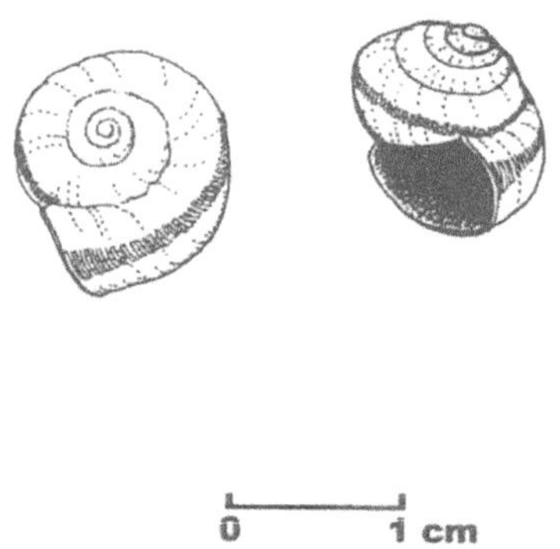

Figure 12.2. Small banded snails; these are Bradybaena similaris from Hong Kong, but there are similar snails to be found in most parts of the world belonging to different genera, causing similar pest problems.

Natural enemies. In an area such as South Australia, where a serious local snail problem exists, efforts at biological control are worthwhile. The main predators of small land snails are predacious beetles (Carabidae and Staphylinidae), and the main parasites are fly larvae (Sciomyzidae etc.); there are projects in progress to utilise such natural enemies.

CHAPTER 13

PESTS: CLASS ARACHNIDA

This is a large group of Arthropoda, mostly terrestrial, that are characterised by having the body divided into two main regions, the anterior prosoma or cephalothorax, and the posterior opisthosoma which resembles the insect abdomen. The adult arachnids have four pairs of legs. They do not have mandibles in the mouthparts for feeding, but have paired chelicerae instead.

13.1 ORDER ARANAEIDA (Spiders)

Spiders as insect predators in field crops have long been ignored in most parts of the world (with the exception of China), and only recently has their importance been recognised. Thus, in a produce store, spiders would be killing and eating insect pests; but, in general, the presence of many spiders' webs would be regarded as reflecting a low standard of store hygiene. It should be remembered that, apparently, many stored produce beetles, and others, evolved their life style from a beginning of eating dried insect remains in spiders' webs, under loose tree bark and the like.

A few other predacious arachnids will be found occasionally in produce stores, but their presence is not really beneficial, and they are not thought to be of any importance; the most likely species to be found are Chelifers (Order Pseudoscorpionida) and Harvestmen (Order Opiliones).

13.2 ORDER ACARINA (Mites)

Many species of Acarina are major veterinary and medical pests, and a large number of phytophagous forms are recognised pests of growing plants. Only a small number of mites are serious pests in produce stores, and there are some predacious species which attack stored produce insects and other mites, but the total number of species recorded attacking stored foodstuffs is large (Hughes, 1961); there are many minor pest species.

There is no generally recognised classification of the Acarina as yet, and several quite different schemes have been published in the past. Because of their microscopic size, and the relative lack of taxonomic research on the Order (there are many species as yet undescribed) and their specializations, mites are difficult for most entomologists to identify. In this book, only a few of the most important species are included.

From the egg hatches a larva which is 6-legged; this leads to usually two nymphal stages which are 8-legged. The mouthparts of most mites are based on a pair of tiny toothed chelate chelicerae with which a scraping motion is made. There have been reports in the literature of intestinal acariasis, citing the presence of mites or mite eggs in the stool samples, associated with attacks of diarrhoea. But now there is much doubt about the significance of mites in human faeces, and it is thought that the diarrhoea comes from micro-organisms found in the infested foodstuff. But dermatitis (grocer's itch etc.) is definitely associated with some mite infestations.

Sub-order Mesostigmata (= Parasitiformes)

This is the largest group of mites, with 65 or more families; they are mostly predacious or parasitic in habits, with well-armoured skeletal plates.

13.2.1 Family Phytoseiidae (Predatory mites)

These are the long-legged, active, large predatory mites naturally found on plant foliage, and they prey on phytophagous mites and insect eggs. Several species are used agriculturally for biocontrol of red spider mite in greenhouses, and are available commercially.

One genus is found in produce stores; but in some publications, it is placed in the family Aceosejidae.

Blattisocius tarsalis (Berlese)

A cosmopolitan species, but most abundant in the tropics, recorded as a predator of the eggs of *Sitotroga*, *Ephestia* and *Plodia*, and also some stored products beetles. It could possibly be used in biocontrol of moth populations. Several other species of *Blattisocius* are found in food stores on a wide range of produce, but all are predatory on eggs and larvae of a wide range of insect pests and other mites.

Sub-order Trombidiformes (= Prostigmata)

A diverse group of both animal and plant parasites, small to minute in size; a few are predators; in some groups there is morphological reduction, and the number of legs can be reduced to just two; some 35 families are placed here.

13.2.2 Family Eriophyidae (Gall, blister and rust mites)

These minute mites make galls on a wide range of host plants, or scarify the leaf epidermis or the skin of fruits. Leaf epidermal distortions are sometimes very distinctive and quite specific, and are termed 'erinea' – dorsally the leaf protrudes, and the ventral cavity is covered with wart-like outgrowths between which the microscopic mites live. The body of the mites is somewhat reduced – elongate and slightly worm-like with only two pairs of legs (the anterior pair).

Various harvested leaves and fruits could be taken into storage bearing eriophyiid mites, but this would be an occasional occurrence. The species that is quite often taken into fruit stores on infested oranges is:

Phyllocoptruta oleivora (Ashmead) (Citrus rust mite) (Figure 13.1)

Pantropical and quite common; it causes a thickening and brown scarification of the fruit skin of oranges and some other citrus fruits.

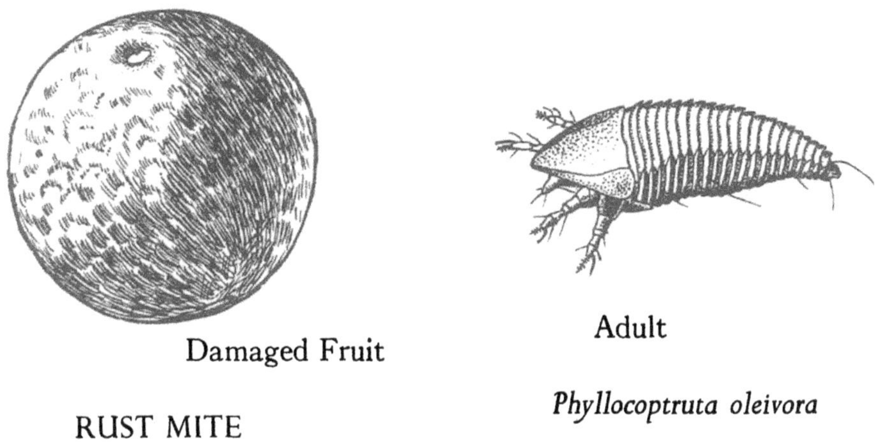

Figure 13.1. Phyllocoptruta oleivora (Citrus rust mite) (Acarina; Eriophyidae); adult mite and damaged orange; specimens from Uganda

13.2.3 Family Cheyletidae

Long-legged, active mites, often colourless but sometimes red in colour; the gnathosoma is freely moveable, and bears stylets with which the prey is attacked. Several species of *Cheyletus* spp. are known, and all live in association with tyroglyphid mites – they feed on the eggs and on the mites, and are regularly found in stored grain and other produce.

13.2.4 Family Tarsonemidae

These phytophagous mites are slightly elongate, with short legs (but not so reduced as in the Eriophyidae). They show some diversity of habits, but many are found on young leaves and shoots. Only a few are encountered in produce stores, so far as is known.

Stenotarsonemus laticeps (Halbert) (Bulb scale mite)

These tiny mites live between the bulb scales of *Narcissus* and other Amaryllidaceae, and are often a serious pest of bulbs in storage.

Tarsonemus granarius and spp.

These are sometimes found in grain stores worldwide, but it is thought that they feed on fungal mycelia in the grain. They are tolerant of very low temperatures.

13.2.5 Family Pyemotidae

Somewhat elongate mites, they are parasitic on insects, found on grasses and in the nests of small mammals.

Pyemotes ventricosus (Newport)

The females are ectoparasitic on larvae of Coleoptera and larvae and pupae of Lepidoptera, and they are regularly found in produce stores. The female mite when gravid has a greatly distended hysterosoma, almost globular, and reaching a diameter of 2 mm.

Sub-order Sarcoptiformes

A group showing great diversity of habits – most are free-living or parasitic; a somewhat confused group taxonomically.

13.2.6 Family Acaridae (= Tyroglyphidae; Astigmata *partim*)

A diverse group, most are free-living, but a few damage growing plants and bulbs, and some are stored produce pests. The gnathosoma is adapted for biting. The deutonymph is often adapted into a hypopus; an inert hypopus is developed as a resting stage (with reduced limbs), and the active hypopus is adapted for active dispersal, often using a mammal or an insect as a transporting host. Some authorities split this group into several families.

Acarus siro (Linn.) (Flour mite) (Figure 13.2)

Pest Status. A serious international pest of stored grain products and other foodstuffs; its minute size causes slight infestations to be completely overlooked.

Produce. All types of dry farinaceous produce, grains, hay, cheese, fishmeal, linseed etc.

Damage. Direct damage is by biting and eating the produce; they usually do not penetrate bulk flour to more than 5-10 cm. Only damaged grain is attacked, but once this pest is inside the seed coat, the plant embryo is eaten first, so that germination is impaired and nutritive value reduced. Heavily infested produce is tainted by a musty odour, and contamination of produce is often the main source of damage. A container of expensive tropical fish food was completely eaten out by flour mites, and, when opened, was found to contain only living and dead mites and frass to about one third the original volume. Flour mites in flour can cause bakers' itch, a dermal irritation.

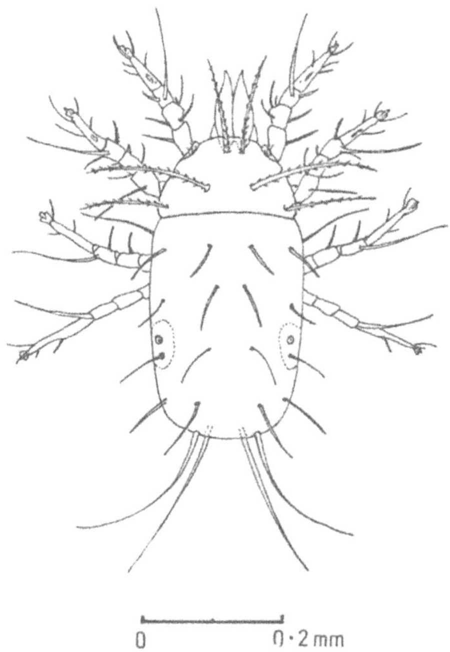

Figure 13.2. Acarus siro (Flour mite) (Acarina; Acaridae).

Life history. Eggs (200-300 per female) are laid into the produce; incubation takes 3-4 days.

They hatch into six-legged larvae, and later turn into the protonymph (with eight legs), and then the deutonymph before becoming adult. Optimum conditions for development appear to be about 20-25°C and 75-80% RH, and then the life cycle can be completed in 9-11 days, but they can survive very low temperatures (down to 3°C). Sometimes the deutonymph becomes a hypopus; there are two types of hypopus – the inert type, adapted for surviving unfavourable environmental conditions (dryness etc.), and which has reduced appendages and is capable only of feeble movement. The active hypopus has well-developed legs and is adapted for dispersal, either by its own efforts or by transportation on the fur of a rodent, or by attachment to an insect or an adult mite.

The adult mite is pearly white with pinkish legs, and has a distinctive enlarged basal segment on the foreleg; body length is 0.3-0.4 mm. To the unaided eye, the adults resemble dust specks. Under optimal conditions, females can live for 40-50 days.

Distribution. Worldwide; abundant in farm-stored grain in the UK.

A. farris is a field mite on moist vegetation or in bird and rodent nests, and comes into stores on plant remains, but populations tend to decline in stores.

Caloglyphus berlesi

Cosmopolitan, but is a major pest in Japan on stored foodstuffs of many types. Elsewhere found occasionally on damp, usually mouldy, produce – wheat, copra, linseed and peanuts, as well as mushroom beds and chicken litter. Time from oviposition to adult emergence averages 8 days at 27°C and 80% RH. Optimal temperature range is 22-30°C (minimum 16°C). The species has a very high reproductive potential, and a high feeding rate.

C. anomalus is found on onions, other vegetables and fruits, where they feed on semi-fluid materials associated with decomposition. Optimum conditions are temperature 20-25°C and humidity close to 100%, when a female lays an average 737 eggs, with a life-cycle of 9-10 days. Adults live for 30-50 days. Not a serious pest.

Glycophagus destructor (Schrank) (= *Lepidoglyphus destructor*)

Pest Status. A very common pest in stored products, but thought to be mycophagous; one of the commonest stored products pests worldwide, cosmopolitan in warmer regions, and often found in association with *Acarus* and *Cheyletus*; abundant on farm-stored grain in the UK.

Produce. Found in stored grain, cereals, cereal products, seeds, linseed, rape, dried fruits, medicinal herbs, sugar, cheese, hay and straw.

Damage. As a fungus-feeder, it actually does little direct damage to foodstuffs, feeding only on produce already mouldy. But contamination can be a serious problem.

Life history. This mite cannot develop at or below 60% RH, and the temperature range is 3°-34°C. Optimal conditions are 25°C and 90% RH or 15-17% m.c. in food; then, one generation is completed in 12 days. Each female lays an average of 150 eggs. Hypopal stages may develop, both resting stages under adverse conditions, and active stages for dispersal by birds and insects. In the wild these mites live in hay and straw, and in nests of rodents, bumble bees, and ants.

Distribution. Cosmopolitan; especially abundant in Canada, Russia and Japan in both open fields and produce stores.

Lepidoglyphus zachari is a pest of dried meat products in temperate regions.

Glycophagus domesticus (DeGeer) (House mite)

Cosmopolitan, but thought to be a temperate species; recorded feeding on a wide range of plant and animal materials, including flour, wheat, linseed, tobacco, hay and straw, sugar, medicinal herbs, ham, fish and cheese; commonly found in house dust, and causes dermatitis and asthma in man.

Lardoglyphus konoi (S. and A.)

Found in England, Japan, India; it feeds only on a high protein diet, and is recorded on dried fish, offal, bones, hides and sheepskins. A serious pest on dried and salted fish in India. Optimal conditions are 30°C and 83% RH, when the life cycle takes 8-10 days. Hypopi are carried by *Dermestes* and *Necrobia* beetles.

Rhizoglyphus echinopus (F. and R.) (Bulb mite)

Cosmopolitan in cooler regions; on bulbs, and all types of vegetables, also mushrooms, decaying grain, and roots of cereal plants; a primary pest of stored plant bulbs. This and *R. robini* tend to feed on fungi growing on stored products, but they can feed on a wide range of medicinal herbs and wheat germ. They can develop at temperatures between 3°C and 31°C if the relative humidity is greater than 86%; the life cycle at 30°C takes 8-14 days, 40-50 days at 10°C.

Suidasia spp.

Recorded from England, Europe, North Africa, and S.E. Asia; on wheat bran, rice bran, wheatgerm, groundnuts, and in bees' nests.

Tyrophagous casei (Oud.) (Cheese mite) (=*Tyrolichus casei*)

Cosmopolitan; feeds on cheeses, grain, damp flour, dogmeal, rodent nests, dried insect collections; it is essentially a fungivore, and has a life cycle of 15-18 days at 23°C and 87% RH.

Tyrophagous putrescentiae (Schrank) (Mould mite) (Figure 13.3)

Cosmopolitan; found on foods rich in fat and protein such as cheese, bacon, peanuts, copra, fishmeal, tobacco, linseed, dried eggs, flours and some cereals, dried fruits and medicinal herbs; they can feed on insect eggs. This species does not have a hypopial stage. Optimum conditions are thought to be about 23°C and 80-90% RH. Females can lay up to 500 eggs. Development time for one generation varies from 9.5 days to 130 days according to environmental conditions. Adults live for about 2-5 months, and can endure starvation for at least 2 months (longer at lower temperatures). It is recorded to cause grocers' itch and copra itch.

Other species of *Tyrophagous* are found in domestic premises and produce stores; some appear to be mycophagous.
(See Gorham, 1991; pp. 57-79)

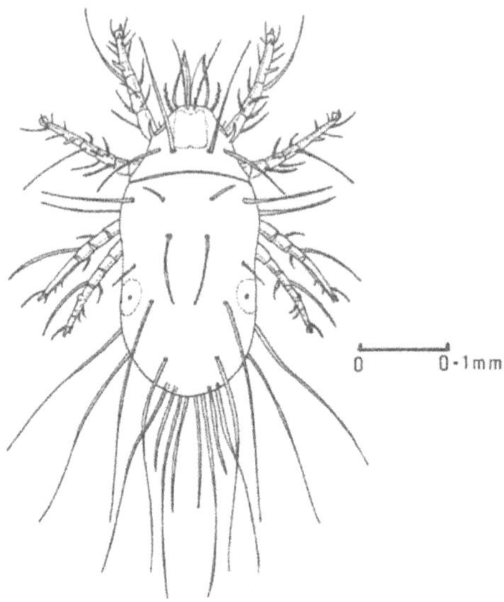

Figure 13.3. Tyrophagus sp. (probably putrescentiae) (Cheese or bacon mite etc.) (Acarina; Acaridae s.l.), adult mite.

13.2.7 Family Carpoglyphidae

Carpoglyphus lactis (Dried fruit mite)

Found on a wide range of foodstuffs, dried fruits, milk products, honey, beer, wine, animal feeds, and medicinal herbs. Yeast appears to be the best food, and on this females lay up to 280 eggs, and adults live for 20 days. With high humidity and low temperature, the mites can live for 200 days or more; adults can withstand starvation at high humidities.

13.2.8 Family Pyroglyphidae (House dust mites)

Dermatophagoides farinae (American house dust mite)

Cosmopolitan, and found in flour, animal feeds, insect cultures and house dust. In North America and Czechoslovakia, this species is more common than *D. pteronyssius* in house dust and mattress dust. Development takes 30 days from egg to egg at 70% RH and 25°C. Exposure to -18°C killed the entire population after 48 hours. It thrives on a high protein diet containing animal fats, and yeast is an important dietary supplement. Adults live about 2 months under optimal conditions.

Dermophagoides pteronyssinus (House dust mite)

Also a cosmopolitan species, but more common in Europe where it has become a major domestic pest, causing respiratory disorders such as asthma, and rhinitis, as well as eczema. Potent allergenic compounds have been obtained from both these species of House Dust Mite. In food stores, this is only a minor pest.

Table 13.1 Ecological parameters for some important stored product mites

	Physical limits				*Results under optimum conditions*		
	Temperature (°C)		*RH (%)*		*Fecundity (number of eggs)*	*Develop-mental time (days)*	*Longe-vity (days)*
Species	*Range*	*Opt.*	*Min.*	*Opt.*			
Acarus							
farris	3-31	20	74	89	137	10	19
siro	3-31	20	63	80	230	9	30
Caloglyphus							
anomalus	?-34	23	-	98	738	9.5	30
berlesei	16-35	27	-	95	588	8	24
Carpoglyphus lactisa	3-35	20	60	80	278	10	20
Cheyletus eruditus	8-31	20	56	80	71	17	35
Glycyphagus domesticus	-	22	-	-	135	22	31
Lardoglyphus konoi	-	30	75	82	-	8	-
Lepidoglyphus destructor	3-34	25	62	90	150	27	-
Rhizoglyphus echinopus	3-31	25	86	90	600	8	35
Tyrophagus putrescentiae	8.5-36	25	69	85	500	10	62

Duration of starvation (days) is 29 for Carpoglyphus lactis, 69 for Cheyletus eruditis, and 60 for Tyrophagus putrescentiae. (From: Gorham, 1991; p.73)

CONTROL OF STORED PRODUCTS MITES

Mites are not primary pests of most foods, except possibly for cheeses, mushrooms, and similar products associated with fungi and fermentation. There are only a few mites of importance to foods, but sometimes they cause very serious problems, with a great loss of product.

When serious mite infestation occurs, it is usually the result of poor storage conditions. They need a cool moist environment which is also conducive to the growth of fungi and yeasts. Usually, the mites feed on both the fungi and yeasts, and on the substrate which has been prepared by these micro-organisms. Typically, the micro-organisms break down the testa on the grain and legumes, and moisten the food for the mites to feed on.

The control of mites is largely one of correcting the environmental conditions so that they are unfavourable for the mites. Contact insecticides (acaricides) are effective on growing crops against phytophagous mites, but are generally not very effective on stored products, partly because the micro-organisms and the moisture degrade the chemicals. Fumigants can be effective, particularly methyl bromide, but phosphine gives a poor level of control.

Grain mites are sensitive to antibiotics and antimicrobial agents. Minute dietary doses of the preservatives potassium sorbate, calcium propionate and methyl paraben inhibit their development, and higher doses decrease fecundity and increase mortality. Antibiotics in high doses diminish egg production. But the other food mites are mostly unaffected by antibiotics.

Physical methods include humidity control; at 60% RH, or with a grain moisture content of less than 13.4%, development ceases and most mites die. Similarly, at -20°C death occurs to all known storage pest species after 48 hours.

Other methods of control include the use of gamma radiation, hermetic storage, and the use of modified atmosphere (CO_2 is especially effective at 100%).

CHAPTER 14

PESTS: CLASS INSECTA

As already mentioned, some of the data included in the following chapter would seem to be very specific and detailed, and would imply a very high level of precision, but this may not necessarily be so – most of the sources of information have to be the published works of previous workers, and so are used as given.

Most of the pest species are now quite cosmopolitan, especially in heated stores, although some are basically more tropical, and others more temperate. However, a few are still restricted to one part of the world, but can at any time extend their range if given the opportunity, as was the case with the Greater Grain Borer, which recently invaded East Africa from Central America (1970s).

In this chapter, the species are dealt with in systematic sequence; for the minor pests, a brief outline is given, but for the major or more common pests, more detail is included.

14.1 ORDER THYSANURA (Silverfish, Bristletails) (5 Families, 650 Species)

A primitive, wingless group, with no discernible metamorphosis; long antennae, and body scaly with two long terminal cerci and a median filament; compound eyes, and tiny abdominal appendages projecting laterally. Now regarded as consisting of five families and some 650 species, found worldwide; a few species are found in domestic situations, belonging to the family Lepismatidae.

Lepisma saccharina (Linn.) (Silverfish) (Figure 14.1)

Pest status. A minor pest in food stores, but abundant and widespread, and can be damaging in homes when paper may be eaten; it requires a high humidity; usually only found in domestic situations. Similar species found in woodland litter.

Produce. Apparently feeds mostly on carbohydrate material, including damp paper, starch used in paper, wallpaper paste, book bindings etc; also recorded damaging cotton and linen materials.

Damage. Holes are eaten in the paper etc., usually of irregular shape and small in size, by both nymphs and adults. On food materials and food debris, damage is generally negligible.

Life history. Each female lays about 100 eggs, singly or in small batches, in crevices; eggs are white and oval becoming brown and wrinkled, 1.5 x 1 mm. Incubation recorded as 19 days at 32°C to 43 days at 22°C.

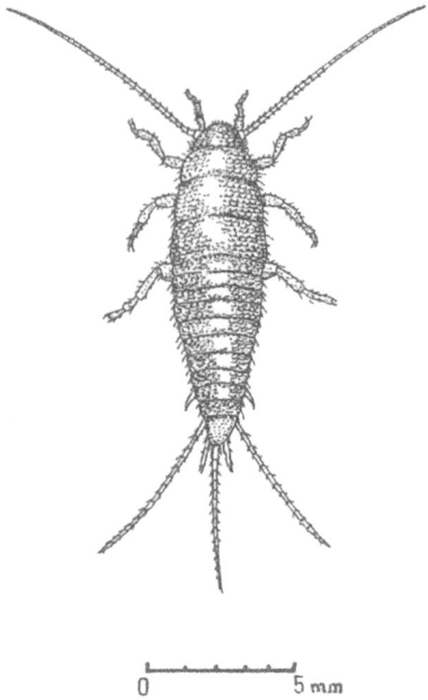

Figure 14.1. Lepisma saccharina (Silverfish) (Thysanura; Lepismatidae), adult.

Nymphs are whitish; there are many moults (40-50^{+}), and, in fact, moulting continues throughout the life of the insect. The silvery body scales appear at the third moult, and external genitalia at the eighth. At 27°C nymphal development takes 90-120 days; high humidities are preferred (80-100% RH).

Adults are wingless, with long antennae and cerci, body length 10-12 mm, silvery in colour. They are nocturnal in habits, and hide during daylight hours; they are long-lived, 1-3 years at 32°C to 27°C.

Breeding is often continuous, depending upon temperature and humidity, and all stages may be found at most times of the year.

Distribution. Widespread and cosmopolitan. Other species are to be found in leaf litter, caves etc., and some may invade domestic premises.

Control. Most produce stores are given general (blanket) pesticide treatments periodically (such as fumigation) which will kill these insects, but if specific treatment should be needed, sprays or dusts of 2% malathion, 1% HCH, or 0.5% diazinon are reputed to be effective. Formerly, DDT and dieldrin were widely used.

Thermobia domestica (Packard) (Firebrat)

A larger species (13 mm), darker in colour, only found inside heated buildings, often in hot situations such as kitchens, bakeries and by stoves. They usually eat more food particles than do silverfish. The species is cosmopolitan, in heated buildings. It prefers temperatures above 32°C, with an optimum of 37-41°C. Adults live 1-2 years usually; nymphs 50-300 days.

Ctenolepisma lineata pilifera (Fourlined Silverfish)

A common species in the USA, more adaptable regarding temperature and humidity than the previous two species.

14.2 ORDER COLLEMBOLA (Springtails)
(11 Families, 1,500 species)

A very widespread group of primitive wingless insects to be found mostly on the soil surface and in leaf litter; they are worldwide, but more abundant in temperate regions. They are small (1.5 mm) insects, with a forked springing organ under the abdomen with which they jump. They feed on fungi, pollen grains and dead or living plant material. They can destroy seedling plants and are often pests in glasshouses. A few species are regularly found in households and produce stores, where they, apparently, feed on food fragments and fungi. They are pests by virtue of their presence, as they cause no discernable damage. A common domestic species found in Hong Kong was *Homidea* (Figure 14.2).

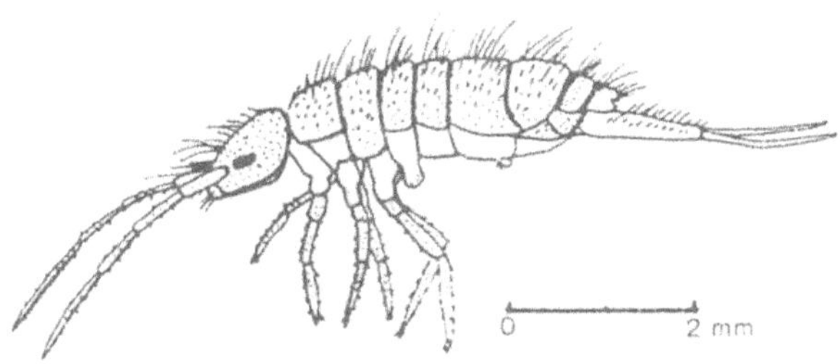

Figure 14.2. A common domestic species of springtail (Homidea sp.); Hong Kong.

Entomobrya is another genus found associated with man and stored food;
E. atrocinta is a pest of dried milk.

14.3 ORDER ORTHOPTERA (Crickets, Grasshoppers etc.)
(17 families, 17,000 species)

These are the grasshoppers and crickets, most of which live in either vegetation or the soil; an ancient group with biting and chewing mouthparts; forewing is modified into a hard narrow tegmen used to protect the large hind wings used for flying; most have stridulatory organs, and can be quite noisy. There is considerable diversity of form and habits within the group, but only a few crickets are of any importance in domestic situations.

14.3.1 Family Gryllidae (Crickets) (2,300 species)

Nocturnal insects, most living underground or in forest leaf litter; black or brown in colour, with long antennae and cerci and stout hind legs with which they jump; the female usually has a long straight ovipositor; they stridulate by rubbing the wingbases (tegmina) together. Many species have a nest underground; eggs are usually laid underground; many are scavenging in habit and omnivorous in diet, and they eat other insects.

Acheta domesticus (Linn.) (House cricket) (Figure 14.3)

Pest status. A minor pest, but common and often conspicuous at night because of the 'singing' (chirruping) of the males. In the warmer parts of the world, there may be several other species of domestic crickets.

Produce. Generally scavenging on food fragments; softer foods are preferred. Most crickets are omnivorous, and will eat meat of most types if available, including other insects.

Damage. Mouthparts are of the biting and chewing type, with well-developed mandibles. They often nibble material upon which they do not feed, and may damage wool and cotton textiles, leather, and even wood.

Life history. Eggs are laid in crevices or in the produce; each female lays 50-150 or more eggs, whitish in colour, elongate in shape (2 x 0.5 mm); incubation takes 1-10 weeks, according to temperature; eggs are very sensitive to atmospheric moisture, and desiccate under dry conditions. In northern regions, the eggs typically overwinter.

Nymphs are initially 2 mm long; the seven instars require 7-32 weeks for development. Out-of-doors, this cricket is often found in rubbish dumps, where the heat of fermentation provides warmth in cooler climates (northern Germany, for example).

Adults are stout-bodied, brown in colour, with long antennae, long cerci, and long ovipositor; body length 12-16 mm. They are nocturnal in habits, and the males 'sing' to attract mates; the chirruping sound is loud and penetrating, and in houses can be a nuisance. Locomotion is by crawling, jumping and flying. Adults live for several weeks or more, if warmth and food are available.

In North America and Europe, it is usually univoltine, but in the tropics, there may be several generations annually.

Distribution. Native to the Middle East and North-west Africa, but now widely distributed throughout the world by its synanthropic habits. Found outside in the cooler areas only in rubbish dumps, where it may be very abundant, for example, in Germany.

Control. If treatment is required, dust or sprays of HCH or chlordane have been the most widely used chemicals in the past; the chemicals listed in Chapter 21 are most effective against crickets.

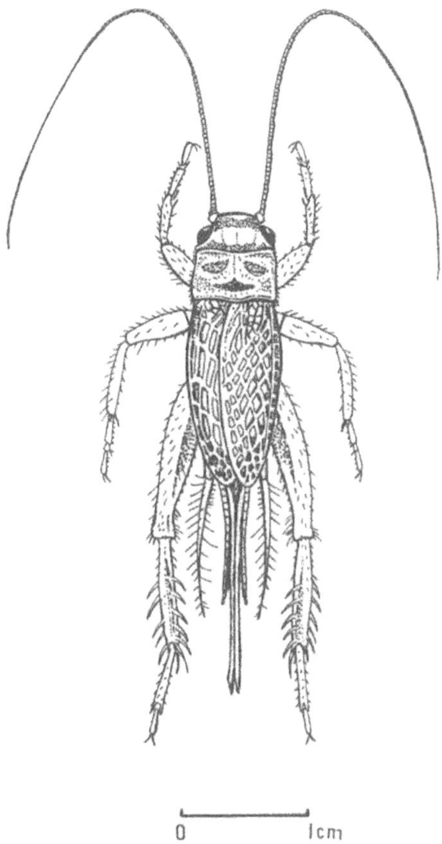

Figure 14.3. Acheta domesticus (House Cricket) (Orthoptera; Gryllidae), adult female.

14.4 ORDER DERMAPTERA (Earwigs)
(8 families, 1,200 species)

A small group, but worldwide and commonly found, although seldom very abundant. They are easily recognised by the large forceps at the tip of the abdomen and the elongate antennae; the winged species have the forewings shortened into leathery brown tegmina, under which the large ear-shaped hindwings are folded so that only a small part protrudes. The unsegmented, heavily sclerotized forceps are the cerci; in the male of some species, they are strongly curved, and in the female they are straighter. The mouthparts are not strong, but are of the biting and chewing type – they are omnivorous, and feed on a wide range of plant and animal materials, both living and dead, and a few species are to be found in produce stores and domestic premises, but they usually attack produce already damaged.

Euborellia spp.

A tropical genus, most about 10-14 mm long, and without tegmina; a wide range of stored plant materials have been recorded to be damaged, some in the field and some in storage.

Forficula auricularia (Linn.) (Common earwig) (Figure 14.4)

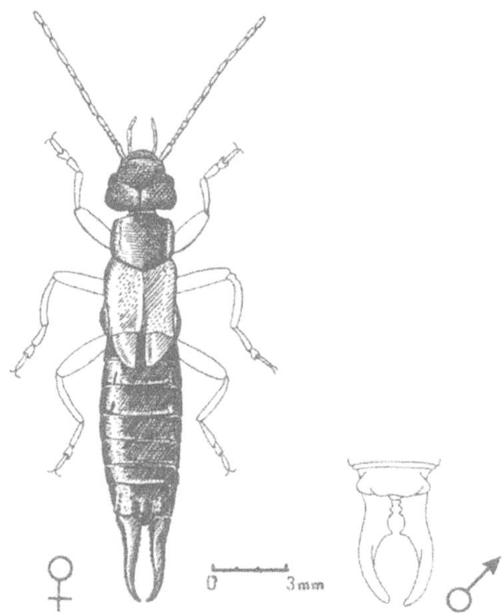

Figure 14.4. Forficula auricularia (Common earwig) (Dermaptera; Forficulidae), adult female, and 'forceps' of adult male.

A widely distributed species found throughout the Holarctic Region and also Australasia, some 12-20 mm; they are nocturnal and fly at night; the hindwings are large and well-developed; a very wide range of plant and animal material is recorded being damaged. Damage levels are, however, seldom more than slight.

Marava spp.

Another tropical genus, of smallish size (6-12 mm), adults have short tegmina but no hindwings, and thus cannot fly. One or two species are occasionally recorded from produce stores; mostly found in warmer regions, but sometimes introduced into the UK in food cargoes.

14.5 ORDER DICTYOPTERA (Cockroaches, Mantids) (9 families, 5,300 species)

A large, ancient and primitive group, widely distributed throughout the world, but most abundant in tropical countries. The forewings are hardened and thickened into tegmina to protect the large hindwings, but many wingless forms are known. They are mostly nocturnal, cryptozoic, with body flattened for easy concealment. Eggs are laid within a protective oötheca (a hard brown purse-like structure) stuck firmly into a crevice or a dark corner. Associated with the flattened body, the pronotum is often enlarged and anteriorly extended to protect the deflexed head. The predacious mantids are of no concern to produce stores.

Some 4,000 species of cockroaches are known; most are forest litter dwellers, with a few in caves, but some have become associated with man and his dwellings and food stores, and a few are found in temperate regions in warm buildings. A number of field and forest species are occasionally recorded invading domestic premises, but are usually of little importance. All have biting and chewing mouthparts, and an unspecialized diet mostly of vegetable matter and prepared foods; they are opportunist feeders. Adults are all long lived.

Despite the various common names, it now seems clear that the centre of evolution and origin of the group has been in the northern half of Africa. Some of the synanthropic species are recorded to have transmitted various pathogenic bacteria in domestic premises (hospitals etc.), and some produce allergens. It is commonly assumed that the regular stored foodstuffs insects (beetles and moths) are the most important pest problem for the food industry, but some leading consultants consider that the German Cockroach is the more important, partly because of its vast reproductive capacity, and its widespread distribution.

14.5.1 Family Blattidae

Taxonomic characters used now include male and female genitalia, shape and method of folding the hindwing, and the spines on the femora. Some taxonomists also use oviposition behaviour, structure of the gizzard and malpighian tubules, as well as wing venation, and they would prefer a greater number of families. Many genera are included in this large family.

Blatta orientalis (Linn.) (Oriental cockroach) (Figure 14.5)

Pest status. In parts of the USA, this is a serious domestic pest, but in the UK and Europe it is generally less abundant. In the USA it is becoming more abundant, and is spreading in some regions. In tall buildings, infestations are confined to the lower floors – adults do not fly at all. In the USA, it is generally associated with sewers; on ships, the infestations are usually in the cargo holds.

Produce. Feeds on a wide range of foodstuffs of vegetable origin. In some areas, summer infestations of rubbish dumps are increasingly being recorded.

Damage. Well-developed mandibles are used to bite off pieces of food which are then chewed. The smell associated with infestations is a major factor, and usually regarded with disgust; contamination of stored foodstuffs is often far more important than actual damage.

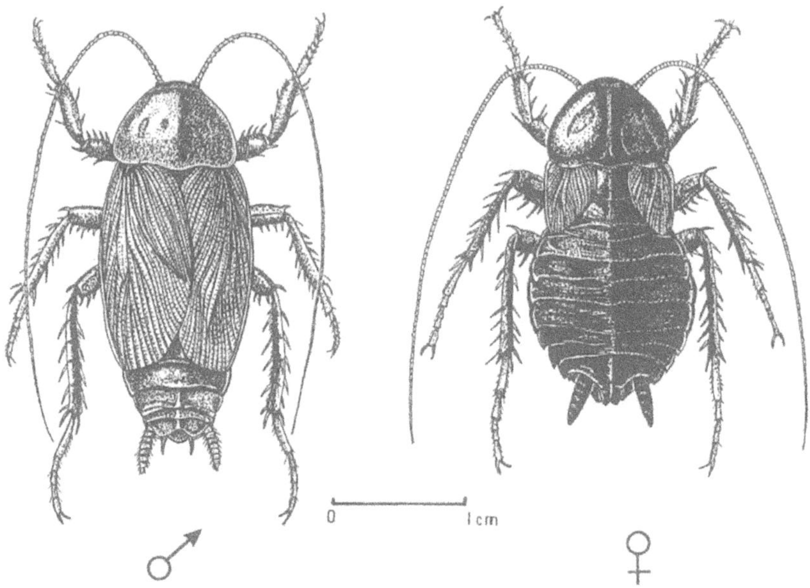

Figure 14.5. Blatta orientalis (Oriental cockroach) (Dictyoptera; Blattidae) male and female.

Life history. Oöthecae (10 x 5 mm) are laid in crevices, often underground in buildings in temperate regions, often near sources of water. About 15 eggs per oötheca; incubation takes 40 days at 30-36°C; but in cooler locations the eggs overwinter. Each female lays 5-10 oöthecae.

Nymphal development is a lengthy process; at 30-36°C, males have seven moults through 160 days, and females ten moults through 280 days.

Adults are large (20-24 mm), dark brown insects, and the females have very short tegmina – neither sex can fly. Both sexes are long-lived. Neither nymphs nor adults have a tarsal arolium, and they are unable to climb smooth surfaces. Adults sometimes show gregarious behaviour.

At 28°C, the life cycle is completed in less than a year, but at 25°C is recorded to take 530 days; availability of food and temperature are equally important in controlling rate of development. The preferred temperature range is reported to be 20-29°C, which is lower than for most other species.

Distribution. Now thought to have originated in North Africa, in a region with a hot summer and cool winter, and has been spread by human activity and commerce. Most abundant in Europe and North America, though now quite cosmopolitan, but it does not establish itself in the humid tropics. Many other species of *Blatta* are known.

Control. (See page 150).

Periplaneta americana (Linn.) (American cockroach) (Figure 14.6)

Pest status. A major domestic pest in tropical parts of Asia, Africa, the Pacific, the USA and Central and South America. It is becoming more abundant as urbanization increases; in Hong Kong, infestation of flats up to the twentieth floor were recorded within a month of habitation, some adults observed flying through open windows at night.

Produce. All types of foodstuffs are eaten, and rubbish dumps in the tropics, as well as sewers, are infested; they are quite catholic in their food choice, but prefer material of plant origin.

Damage. Direct eating of food materials is the main actual damage, but contamination and fouling can be very serious where people are fastidious. Bookbindings and papers may be defaced or destroyed in houses. Infestations are usually associated with a distinctive and unpleasant smell.

Life history. Oöthecae (8 x 5 mm) are laid at weekly intervals, over a period of a year or two; each female produces 20-40 oöthecae. After extrusion (when it is white), the oötheca is carried for a day or more prior to deposition, when it turns brown and then finally almost black. Each oötheca contains 12-20 (average 16) eggs, and the average hatch is 14 nymphs; incubation takes 35 days at 30°C (50 days at 26°C).

Nymphal development takes 4.5-5 months (20-30°C), and females usually have nine moults, but males up to 13; under cooler conditions, development takes a year or more, but much variation has been recorded in different parts of the world under different conditions.

Adults are long-lived, but males usually less than one year, and females 1-2 years; they are very active, and fly well at night. This species is large (28-44 mm), a shiny reddish-brown with a yellow edge to the pronotum. It prefers a moist warm environment, with the upper limit of its preferred temperature range at 33°C.

In the tropics, breeding is generally continuous, but the life cycle usually takes six months for completion.

Distribution. Native to tropical Africa, this species has been spread by man throughout the warmer regions of the world; it is most serious in India and tropical Asia, and in parts of the New World. In temperate regions, it is established in heated buildings. It is very abundant in 'subtropical' locations such as New York and Hong Kong, where it may be quiescent over the cold winter period.

Several other species of *Periplaneta* are regular domestic pests, and a number of other species occur in the wild in tropical Africa.

Natural enemies. Large numbers of cockroaches are preyed upon by synanthropic geckos and skinks; adults are killed by *Ampullex* wasps (see page 321), and the oöthecae parasitised by *Evania* wasps. Spiders can also be important predators, of all instars; *Heteropoda* spp. will kill the adults in S.E. Asia.

Control. Traditional control has been achieved by spraying the surfaces where they hide, or are active at night, with insecticides (chlordane, malathion, trichlorphon, HCH, diazinon, or dieldrin). Pesticide resistance has become a problem in most countries, and now the recommended chemicals include the pyrethrins, propoxur, bendiocarb and dioxacarb (see page 150).

Impregnated paints or varnishes have generally not been very successful; neither has been the use of poison baits. Generally, all species of cockroaches are susceptible to the same range of chemicals, but different species of insect develop resistance at different rates.

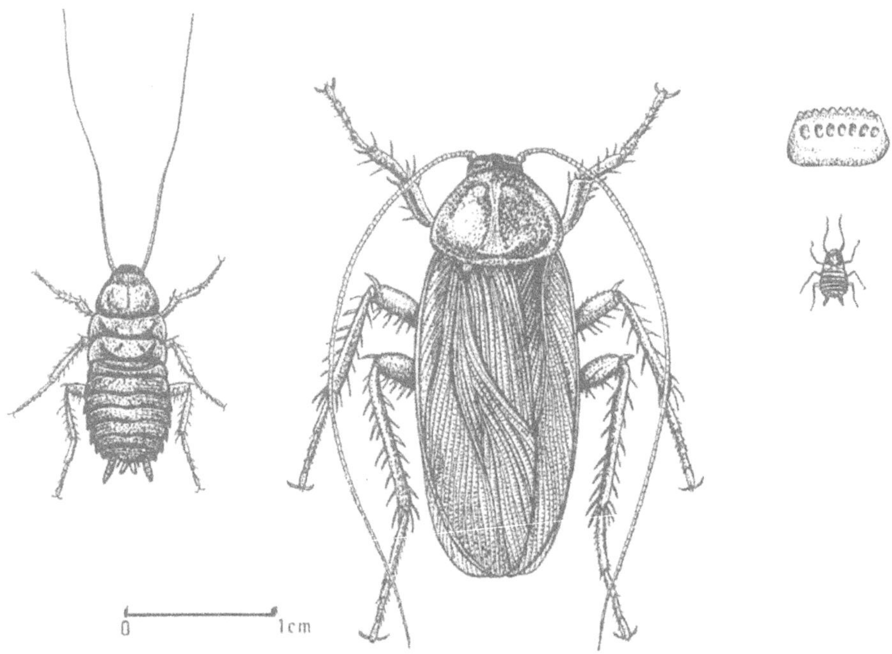

Figure 14.6. Periplaneta americana (American cockroach) (Dictyoptera; Blattidae), adult, two nymphs and oötheca.

Hygiene is an important aspect of population control, and the use of sealable food storage containers is recommended.

In large food stores, the routine fumigations usually keep cockroach populations in check, but in on-farm stores and houses, control may be difficult.

Volume 2 of *The Cockroach* (Cornwell, 1976) deals with insecticides and cockroach control.

Periplaneta australasiae (Fab.) (Australian cockroach)

Another tropical species, probably from Africa, which is quite important in tropical Asia and Australasia, as well as the southern USA and West Indies. It is similar to *P. americana* but less tolerant of cool conditions; it is a little smaller (30-35 mm), and with pale basal margins to the tegmina; not often recorded in sewers, but sometimes found in field crops; mostly found in domestic premises.

Periplaneta brunnea (Burmeister) (Brown cockroach)

Tropical in distribution; seldom abundant; thought to be native to Africa; it resembles *P. americana* but is smaller (31-37 mm) and browner. The biology of this species is little-known, but the oötheca is large (12-16 mm long , as opposed to about 8 mm), and it contains more eggs (average 24). Now a major pest in Florida and California.

Periplaneta fuliginosa (Serville) (Smoky-brown cockroach)

A subtropical species, found in the southern USA; quite abundant in some areas; similar in size to *P. americana* but very dark brown or even blackish; a common greenhouse pest, also a major pest in the Orient. It is sensitive to dehydration; mostly a commercial and structural pest.

14.5.2 Family Epilampridae (= Blattellidae *auct.*)

A large group, mostly of smaller species, Old World in origin; the Blattellinae contains 1,300 species, and the epicentre of evolution appears to be in North-east Africa, judged by the number of local species.

Blattella germanica (Linn.) (German cockroach) (Figure 14.7)

Pest status. This is the most widely distributed cockroach, originally from Ethiopia and Sudan and now worldwide after being distributed by man and his commerce. It is small in size, but often very abundant, totally polyphagous, and a serious international pest. Generally, there is some activity to be seen in daylight, but most active at night. It has the shortest life cycle of the domestic cockroaches, which gives it a tremendous reproductive potential. Under optimum conditions, one fertilized female German cockroach could, theoretically, produce over 10 million females within one year (3.5 generations), and over 10 billion females in just 11/2 years. The speed of development means that it develops resistance to insecticides rapidly.

Produce. A wide range of stored foodstuffs and produce is eaten, including fur, hair and leather.

Damage. A combination of direct damage by eating and of contamination and fouling of the stored produce. Adults and nymphs both produce allergens that cause asthma in humans.

Life history. Each female produces 4-8 oöthecae, each at three week intervals; the oötheca is small (8 x 3 mm) and contains 20-40 eggs; incubation takes 17 days at 30°C; hatching is recorded at temperatures from 15-35°C. Each oötheca is carried by the female for 2-4 weeks until the eggs hatch.

Nymphs go through 5-7 moults, and males take 38-40 days (females 40-63 days) at 30°C to develop; at 21°C, development takes 170 days.

Adults are small (10-15 mm), very active and fly freely; in colour, pale brownish with dark bands on the pronotum; males are slimmer than females. Basically, they are nocturnal, but some are quite active in broad daylight in heavily infested premises. Males live on average for up to 130 days, and females 150, but they do not live long without food. This is a tropical species, with the upper limit of preferred temperature at 33°C, but they may be quite active at lower temperatures. In Ethiopia, this is the most common species in the Highlands, which are really quite cool most of the time.

The development cycle is completed in 6 weeks under optimum conditions of food, temperature and humidity, and there may be many generations per year, but under cooler conditions, the species may be univoltine.

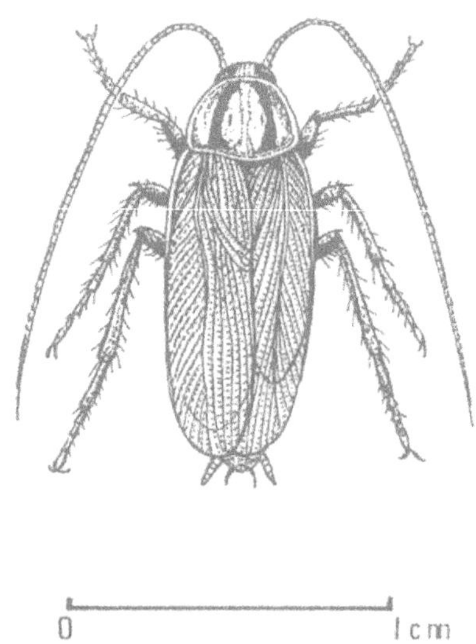

Figure 14.7. Blatella germanica (German cockroach) (Dictyoptera): Epilampridae), adult.

Distribution. Thought to be native to Ethiopia, but has spread since ancient times through commerce to most parts of the world, as far north as Canada and Scandinavia. To be found hiding in the internal structures of most ships. Severe infestations have been reported from Alaska in heated buildings.

Control. (See page 150).

Blattella asahinae (Mizukubo) (Asian cockroach)

Pest status. Described in 1981, this species closely resembles *B. germanica*, but is differentiated by the male genitalia and wing venation in the female. Needless to say, this species is not often identified in the field, or in the laboratory either. But, apparently, they differ considerably in habitat requirements, activity patterns, flight ability, and their reactions to light.

Produce. A wide range of fruits and vegetables both in the field as growing crops and in storage, and many different types of foodstuffs are eaten.

Damage. Direct damage involves the eating of the produce which is done readily, but it is a very serious contamination pest. It is very active in daylight just before sunset, and flies readily, so infestation spreads rapidly and very obviously, causing great concern in restaurants and food stores. They are especially attracted to lights, and in these evening forays may fly for several hundred metres. They are more dispersive than any other cockroach, most of which are relatively sedentary. They produce allergens that cause asthmatics to have serious reactions to their presence.

Life history. Basically, this species is similar to the German cockroach in many respects. But field populations in Florida often reach an estimated 100,000 per acre, and sometimes up to 200,000 per acre. Adults and nymphs hide in ground cover during the day, but at sunset they emerge and feed all night until about 90 minutes before sunrise.

Distribution. Records place the Asian cockroach throughout India, China and much of S.E. Asia; now firmly established in Florida in the USA, and spreading annually within the state.

Control. (See page 150).

For further information, see Gorham (1991; pp. 121-30), and Cornwell (1968, 1976).

Blattella vaga (Hebard) (Field cockroach)

First recorded in Arizona in 1933 as a pest in field crops but doing negligible damage. It is reported to be gradually spreading to domestic premises in parts of the southern USA.

Ectobius spp. (Temperate field cockroaches)

Several species are recorded in leaf litter in woodlands and various other outdoor habitats in the UK and Europe, and other species in Australia, now to be found in domestic premises. Relatives occur in the USA, but to date these are not recorded invading human habitations. As a group, these are essentially temperate species.

Supella supellectilium (Serville) (Brown-banded cockroach) (= *S. longipalpa*)

A domiciliary species from tropical Africa, now widespread in both Old and New Worlds in the tropics and subtropics; the distribution appears to be spreading further. Its entry into the USA is recorded for 1903 in Florida. It is a small species (10-14 mm), and variable in coloration but with two pale bands across the tegmina.

14.5.3 Family Blaberidae

A large and diverse group, some are spectacular insects, but they are mostly of little real importance as stored produce pests, although they may be regularly encountered.

Pycnoscelis surinamensis (Linn.) (Surinam cockroach) (Figure 14.8)

An oriental species, usually found out-of-doors in tropical Asia as a litter dweller, and sometimes damaging field crops. Now, it is widespread throughout the warmer regions of the world, and in heated greenhouses in North America and Germany, and is becoming a domestic pest in parts of Asia, Africa, the southern USA and the West Indies.

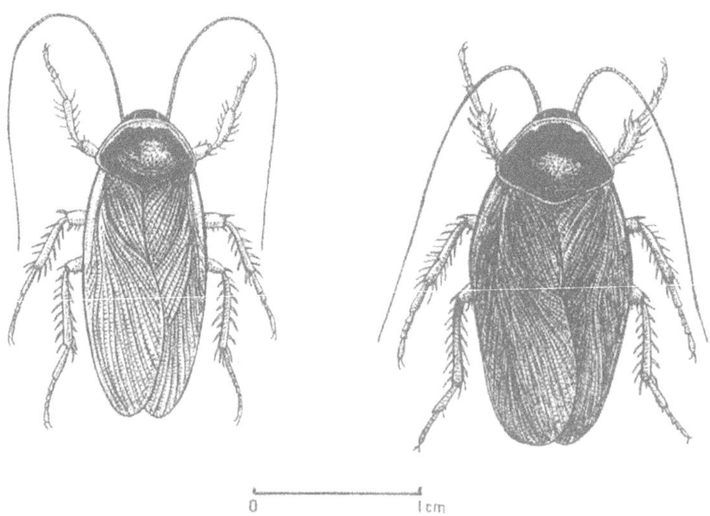

Figure 14.8. Pycnoscelis surinamensis (Surinam cockroach) (Dictyoptera: Blaberidae), adult male and female.

It is a medium-sized cockroach, 18-24 mm long, and dark in colour, with long tegmina. In the New World, males are not recorded, and reproduction is by parthenogenesis. They practise false ovoviviparity, and produce an oötheca which is partly extruded and then withdrawn into the internal brood sac, where the eggs develop.

Blaberus spp. (Giant cockroaches)

These species are known in South America, usually in rotting vegetation and caves; they are up to 80 mm in length, brownish in colour, and often associated with fruit (bananas etc.), and sometimes found in human dwellings.

Leucophaea maderae (Fab.) (Madeira cockroach) (Figure 14.9)

A West African species now firmly established in the West Indies as a domestic pest, and spreading into South America and the USA. A large species (40-55 mm, with tegmina extending a further 10 mm) of mottled pale brown coloration, and gregarious in habits. In the tropics, it is often also associated with certain field crops (bananas, sugarcane, palms), and is often transported in banana cargoes.

Figure 14.9. Leucophaea maderae (Madeira cockroach) (Dictyoptera; Blaberidae), adult; body length 55 mm; Kampala, Uganda.

Nauphoeta cinerea (Oliver) (Lobster or cinereous cockroach)

This species is so named because of a lobster-like mark on the pronotum. Size is moderate (25-29 mm), colour ashy, with wings that do not cover the whole abdomen. a tropical urban species, probably from East Africa, now widespread throughout the tropics, and carried by commerce into many temperate countries. It is quite omnivorous in diet, and is now recorded as a pest in mills and stores of animal feeds containing fish oil; it is recorded killing and eating other cockroaches. It might well become a serious pest in more countries in the future.

CONTROL OF COCKROACHES

The areas most frequented by cockroaches have to offer them physical shelter in dark locations near to sources of food and water. They are cryptozoic by nature, and need crevices in which to lurk. This tendency has been exploited by pest control operators in what is called the 'crack and crevice' treatment, whereby small amounts of insecticide are injected into these small cracks and crevices (hiding places) in the infested building and its fitments. This is a useful technique for food-handling establishments.

The blatticides commonly used for 'crack and crevice' treatments in liquid formulation are acephate, bendiocarb, chlorpyrifos, diazinon, propoxur and cypermethrin, often with synergized pyrethrins or dichlorvos added for their flushing value. Application of the fine needle (pin-stream) spray takes longer than a fan-spray application but gives far greater control. (See Chapter 21.8 for more details on pesticide use.)

Dust or baits may also be used. Microencapsulated chemicals gave excellent persistence in several recorded trials.

Traps of various types are useful for monitoring purposes.

Several parasitic wasps are regularly found in food stores – these include *Ampullex* (p. 313) which prey on the cockroaches, and *Evania* (p. 305) which parasitize the oöthecae.

14.6 ORDER ISOPTERA (Termites)
(7 families, 1,700 species)

These are social insects that live in large communities, sometimes very large colonies; they are soft-bodied, pale in colour, with mouthparts for biting and chewing. A tropical group, in some respects regarded as primitive; winged forms have two pairs of large equally developed wings, easily shed; polymorphism is usual, with workers apterous and usually eyeless, and soldiers with large mandibles and rudimentary eyes; egg-laying is done by the queen, which has an enormous distended abdomen filled by the huge ovaries and fat body.

They are vegetarian; some species have micro-organisms in the intestine that break down plant celluloses; others collect the plant material and construct fungus gardens in the nest on which fungi are cultivated and then eaten as food. Most species can thus digest wood as well as leaves, seeds and other plant materials.

Termites are of some importance to stored products entomologists on two counts; firstly, they can destroy wooden structures so that stores can literally collapse, and, secondly, subterranean species may invade food stores, at or below ground level, and foraging workers will remove the stored food materials. The food material is carried back to the nest in all cases, although a little might be eaten in the store. Termites are often most serious to small on-farm stores or small wooden regional stores in rural areas.

Many different species of termites can be involved with produce stores, and the precise species will vary from country to country, and even within the country. Three groups will be mentioned here, somewhat briefly, as being representative.

14.6.1 Family Termitidae

These are the mound-building or subterranean termites, sometimes also called bark-eating termites. The nest is underground, often under a huge mound, and the colony may be very large. They are fungus growers, and build fungus gardens deep in the nest from a mixture of chopped plant material and saliva; parts of the fungal mycelium are eaten as food. They are pests of growing and dead plants, but also attack underground wooden structures, and may enter buildings; buildings are usually invaded following attack of adjacent tree stumps and roots.

Macrotermes spp. (Mound-building termites) (Figures 14.10, 14.11)

Figure 14.10. Winged adult Macrotermes (Isoptera); Termitidae), wingspan 52 mm; Hong Kong.

Pest status. Major pests of agriculture and forest trees; the colony is indicated by the presence of the large nest mound, which may be up to 5 m or more in the tropics, but is usually 2-3 m, although in South China and the Ethiopian Highlands under cooler conditions the local *Macrotermes* species makes no mound at all. Colonies are large, there may be a million individuals or even more in the largest nests in the hot tropics. In produce stores, they are only occasional pests, but they can be quite a nuisance; wooden structures are very vulnerable, but more often the termites that attack them are wet-wood termites. Concrete and brick structures with a crack at ground level, or small gaps, are liable to be invaded by foraging termite workers. When a nest is located near a building (Figure 14.11), the underground tunnels usually run right under the building, and the foraging workers will seek any cracks or crevices leading inside.

Produce. Grain and a wide range of materials of plant origin are damaged or removed by the termite workers, and wooden structures may be gnawed. On-farm stores constructed of wood have been destroyed, particularly the parts that are underground. Bones and animal skulls have been eaten.

Damage. Produce such as grain is removed from the store by the foraging workers; anything too large to be removed is gnawed, and the fragments removed. Most workers will follow a definite trail into the building, and the trail is sometimes covered by a roof of fragments of earth. The nest will be outside and underground. Wooden structures are gnawed away, and the fragments taken down into the nest for construction of the fungus garden.

Life history. Eggs are laid by the queen inside the royal cell in the base of the nest mound, and they are removed by attendant workers who nurse and feed the young nymphs.

Workers are usually about 4-6 mm in length, and adults 7-15 mm. Adults initially have a pair of long, usually hyaline, wings which are shed after the dispersal and mating flight. The queen has an abdomen that progressively enlarges as the ovaries develop, and she may reach a size of 70 mm. The main food source is the bromatia produced on the fungal mycelium in the fungus gardens made of chewed pieces of wood and plant materials. Most colonies live for about 3-8 years, and so can constitute a threat to a food store for a long time. New colonies are established after an annual dispersal and mating flight of young winged adults usually timed for the start of the rainy season.

Distribution. Up to about ten species are known throughout tropical Asia and Africa, occurring as far north as South China. Similar species are found in Australia and in the New World.

Control. There are two basic approaches to termite problems; the simplest is to avoid attack by using concrete and metal structures where possible, and to protect the wooden structures at risk by treatment with creosote, tar oils, various zinc compounds, or persistent insecticides. The more difficult task is to destroy the colony; killing a few workers is just a waste of time - the queen in the centre of the nest must be killed. It is necessary to locate the nest and to pour the pesticide down the entrance/ventilation funnels into the interior. In S.E. Asia, standard building practice incorporates a layer of dieldrin into building foundations to deter termites. The insecticides most effective have for long been aldrin and dieldrin, and these chemicals are still widely used in the tropics, although production in other countries has officially ceased now. Other chemicals used were chlordane, heptachlor, and HCH. But in the search for alternatives, malathion, dichlorvos, and the carbamates have been tried, and in the USA attention is now focused on the synthetic pyrethroid, cyhalothrin, and chlorpyrifos. With large termite mounds, the use of fumigants may be feasible.

Figure 14.11. Nest mound of Macrotermes bellicosus in urban location; Kampala, Uganda.

Odontotermes spp. (Subterranean termites) (Figure 14.12)

Pest status. This is probably the dominant genus of soil-dwelling termite in the tropics, but the colonies are smaller and there is usually no nest mound, and so they are less conspicuous; if a mound is present, it is typically quite small. Occasionally recorded invading produce stores, but more threatening to farmers than to large regional stores.

Produce. As with previous species, these will collect many different types of dead and living plant material, as well as gnawing wooden structures.

Damage. Food material is removed from the store and taken back to the termite nest underground; the colony is often quite small, so damage is often slight, but damage to wooden structures (especially underground) can be extensive (Figure 14.13).

Figure 14.12. Adult winged Odontotermes formosanus (Subterranean termite) (Isoptera; Termitidae), wingspan 50 mm; Hong Kong.

Life history. As fungus feeders, members of this genus take the food materials back to the nest to construct fungus gardens. Colonies are often quite small, and usually there is no nest mound evident. Adults are slightly smaller than *Macrotermes* spp., usually 7-13 mm in body length, with wings 15-28 mm long, and most species have wings dark brown in colour.

Distribution. In the book *Termites* (Harris, 1971), a total of 23 species are recorded as crop pests throughout tropical Africa and tropical Asia, most of which could attack produce stores, especially on farms.

Control. The same as for *Macrotermes* (page 152).

Figure 14.13. Fence post destroyed by subterranean termites underground; Alemaya, Ethiopia.

14.6.2 Family Rhinotermitidae (Wet-wood termites)

These are small termites, subterranean in habit, and wood eating; they all have mutualistic micro-organisms in the intestine which digest cellulose materials. Generally, they only attack wood that is moist or wet, hence the common name. Small colonies can be found inside domestic premises where water seepage or condensation wets a small area of wooden structure or flooring. In this family, the frontal gland on the head is well developed, and in some soldiers it is used as a defensive organ.

Coptotermes spp. (Wet-wood termites) (Figure 14.14)

Pest status. Very damaging to wooden structures in the tropics, these are important domestic pests. They are the main species that destroy posts and wooden structures in the tropics. They are subterranean, and live naturally in wet dead tree stumps, and some are recorded as pests of field crops and forest trees.

Produce. Most recorded damage concerns fence posts, structural timbers, and other wooden structures in the presence of moisture, but stored produce of different types has been attacked; all types of cellulose can be eaten.

Damage. Direct eating and removal of tissues is the usual damage. These termites have symbiotic micro-organisms in the intestine and so can digest the cellulose material ingested; they usually eat the material *in situ* rather than carry it back to the nest. Typically, they damage the structure of the store more often than the food materials inside. Small wooden on-farm stores may collapse after their supports have been destroyed by these termites underground. In many parts of S.E. Asia, houses are built on small concrete pilings so as to make them less vulnerable to termite attack (Figure 8.1). In 1979 in Hong Kong (Kowloon), *Coptotermes formosanus* caused an extensive electricity failure (for two days) by gnawing through underground power cable coverings, both rubber and plastic, in order to eat a Hessian wrapping; the gnawing termites exposed part of the copper core, and the moisture caused a short circuit (Figure 14.15).

Life history. The colonies are small with only a few thousand individuals, living either in a tree stump or wet (damp) structural timbers underground, or above ground in a damp situation; there are no fungal gardens.

Workers are small (4 mm), with a soft white body and yellow head; soldiers are larger (5 mm) with a brown head capsule and black mandibles.

Adults swarm annually in warm wet weather, and wings are shed after the mating and dispersal flight. The queen is large when fully grown, 30-40 mm. Each colony usually lives for several years.

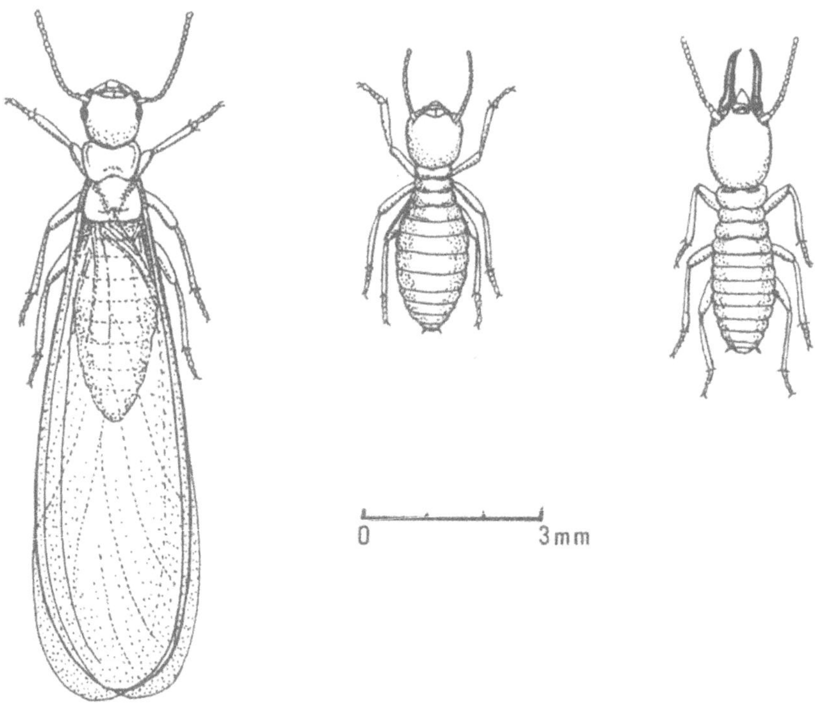

Figure 14.14. Coptotermes formosanus (Wet-wood termite) (Isoptera: Rhinotermitidae), winged adult, worker and soldier forms; Hong Kong.

Distribution. *Coptotermes* is a tropical genus with some 45 species worldwide, but best represented in S.E. Asia and Australia.

The temperate wet-wood termites belong to the genus *Reticulitermes*; they can tolerate cooler conditions and are found as far north as France, China and Japan, and several species are important subterranean pests in the USA. Recently, one species invaded southern England.

Control: (See page 152).

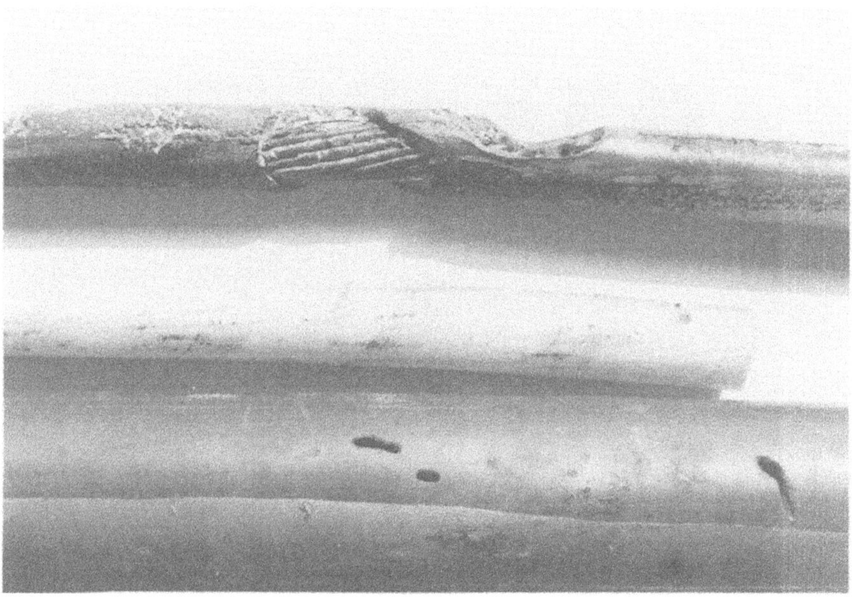

Figure 14.15. Power cable gnawed by Coptotermes formosanus; outer sheath (lower) made of thick rubberized plastic, and inner plastic covering to the copper core (upper) bitten through while the insulating layer of Hessian was being eaten; when the copper core was exposed, water caused a short circuit which melted part of the core (top right); Kowloon, S. China, 1979.

14.6.3 Family Kalotermitidae (Dry-wood termites) (250 species)

A small family of small termites, pantropical in distribution, and of some importance in domestic situations because of their ability to live in dry wood and structural timbers. They all have symbiotic protozoa in the intestine, typically in their large rectal pouch, which can break down celluloses so the termites can directly eat dry wood, and they are able to utilise the metabolic water produced. In most countries, these termites are not abundant, but it is reported that they are very serious domestic pests in parts of the West Indies.

Cryptotermes brevis (Dry-wood termite) (Figure 14.16)

This is a domestic species in S.E. Asia, occurring up into South China; the colony is usually quite small, and often measured in dozens rather than hundreds. In produce stores and other domestic buildings, they can be damaging because they can hollow out dry structural timbers and so weaken the structure, but they are generally not common.

Other species are found in Asia, the Pacific, and in the New World, and another in parts of Africa.

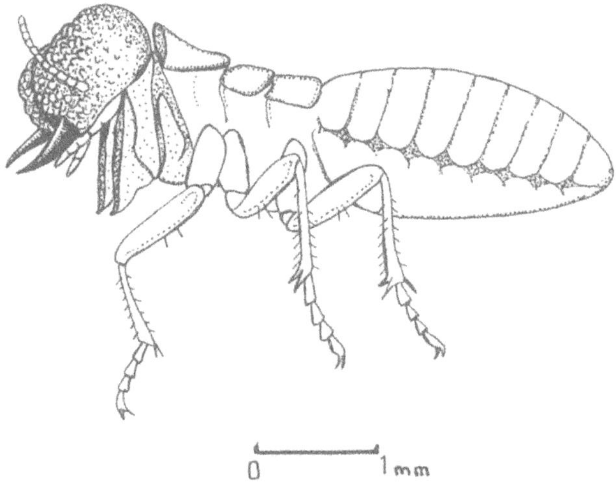

Figure 14.16. Cryptotermes brevis (Dry-wood termite) (Isoptera; Kalotermitidae), soldier; South China.

The other genus of urban and domestic dry wood termites is *Kalotermes* – several species could be encountered in the woodwork of produce stores.

14.7 ORDER PSOCOPTERA (Booklice, Barklice) (26 families, 1,000 species)

A group of small to minute, rather generalized or primitive insects; some are winged, some wingless; antennae long and filiform (13-50 segments); body rounded and soft, and eyes protruding. The biting mouthparts are modified in that the maxillary laciniae have been developed into elongate 'picks'. Most feed on fungi or lichens, or epiphytic algae, but a few eat dried animal material and stored flours and cereal products. Occasionally, psocids can be serious pests of stored foodstuffs when conditions are humid – large populations can develop. Colonies found on the walls of domestic buildings are usually feeding on moulds growing on damp plaster. The group is worldwide, with a number of species that are truly tropical, and others restricted to temperate regions. The family designations are based on rather esoteric criteria that are usually too subtle for the non-specialist to use. In the past, there has not been much study of Psocoptera in food stores, and many infestations have been recorded but seldom identified. However, in the UK, Broadhead in 1954 (Mound, 1989) recorded some 30 species in produce stores and ships' holds – but only eight species were commonly found. The Psocoptera in food stores in the UK are now being studied quite intensively in relation to the food industry, and research is being funded by the Agricultural (and Food) Research Council.

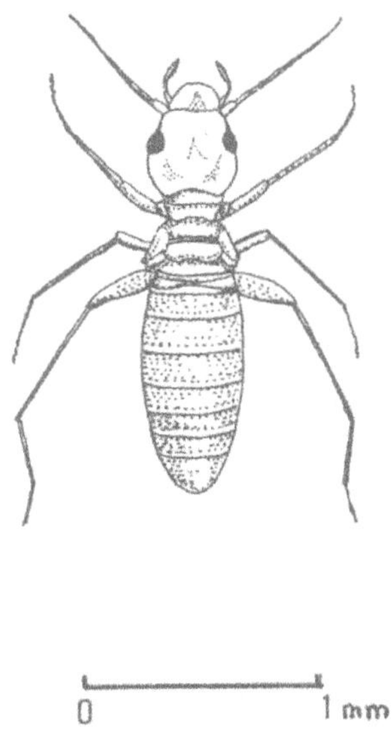

Figure 14.17. Typical wingless psocid (Psocoptera) to be found in domestic premises.

In museums and insect collections, the major source of damage to dried preserved specimens is often psocid populations – because of their tiny size, when numbers are low their presence is often overlooked, and they eat out the inside of specimens until the body shell collapses. They are a very serious threat to museum collections.

Control of Psocoptera in past years has often relied upon the use of synthetic pyrethroids, but recent work has suggested that *Liposcelis bostrychophilus* is not particularly sensitive to permethrin. This might in part account for the general increase in the importance of this group as stored foodstuff pests.

Most species cannot survive more than 2-3 weeks at less than 58%RH.

Lepinotus spp.

There are three temperate species recorded in the UK in stored products, often in grain stores; *L. inquilinus* Heyden, *L. patruelis* Pearman – a large (2 mm) dark brown species and *L. reticulatus* Enderlein.

Liposcelis spp. (Tropical booklice) (Figure 14.17)

Several species are recorded, mostly with swollen hind femora and a small dorsal tubercle by the leg base. Although generally regarded as a tropical group, there are 16 species recorded in England, and most in association with stored products; the most publicized species is probably *Liposcelis bostrychophilus* Badonnel, now quite cosmopolitan, but thought to be of African origin. It is tiny, about 1 mm long, pale brown, and in temperate regions only to be found in heated premises. This species is parthenogenetic, and all adults are female; the long-lived females lay about 200 eggs each, and, clearly, they have great potential as stored foodstuffs pests. Being tropical, they do not flourish at temperatures below about 20°C; they are normally wingless, and rely upon human agency for dispersal. This species has been recorded surviving 2 months of starvation, and then resuming normal fecundity when food became available.

The other commonly recorded species in the UK are *L. entomophilus* (Enderlein) and *L. paetus* (Pearman); *L. divinatorius* (Muller) is the cereal psocid of North America.

Trogium pulsatorium (Linn.) (Common temperate booklouse)

A larger (2 mm) but pale species; usually wingless; cosmopolitan but most abundant under cooler temperate conditions in stores and outhouses. This species produces a noise rather like the sound of a ticking watch (*Lepinotus* species make a faint chirping sound).

Lachesilla pedicularia is the cosmopolitan Grain Psocid.

Psyllipsocus ramburii Selys

This species is typically found in produce stores and on domestic premises both in Europe and North America; wings are often present in this species but are reduced in size.

Several other species are recorded by Ebeling (1975) as being common in stores and human dwellings in the USA.

14.8 ORDER HEMIPTERA (Bugs)
(38 families, 56,000 species)

A very large group of insects, abundant and widespread, and many species are agricultural pests of considerable importance. The group is characterised by having mouthparts modified into a piercing and sucking proboscis – most species feed on plant sap and fruit juices, but some are predacious and blood suckers. The sub-order Homoptera are all sap-sucking, plant-feeding bugs, but the Heteroptera are a mixture of plant-feeding bugs and predacious blood suckers; sometimes, the two sub-orders are treated as separate orders for convenience.

A few species will be encountered in fruit and vegetable stores, and they are sometimes of great potential importance if imported into a new country. In grain stores, a few predacious Heteroptera will be found – these are predators on some of the pests found in the stores, and thus to be regarded as beneficial species.

14.8.1 Family Aphididae (Aphids, Greenfly) (3,500 species)

On herbaceous plants and trees, this is a dominant group of insects in temperate regions, but some species also occur worldwide, and there is a small number of tropical species. Although very serious on growing plants, they are only found occasionally in produce stores, and then usually on vegetables. Leafy vegetables such as cabbage are likely to house a colony of cabbage aphid. In temperate regions, most aphids overwinter as dormant eggs stuck into protected crevices on the host.

Another characteristic is that aphids reproduce most of the time parthenogenetically and viviparously, and they have a tremendous reproductive potential, and massive infestation can arise from a single egg. On potato and some other root crops, overwintering aphids can survive and can be serious pests in chitting houses, where they can kill shoots on sprouting tubers. At least four species are involved in potato stores in the UK. Because of their small size, they are easily introduced into food-processing operations and incorporated into the finished product.

Brevicoryne brassicae (Linn.) (Cabbage aphid)

This mealy waxy aphid lives in small colonies on the leaves of cabbage and other Crucuferae; growth distortion of young leaves may occur, and older leaves may be cupped, but the major economic consequence of infestation of cabbage in store is a quality reduction.

Washing the cabbage heart with soapy water will remove the aphids.

Aulacorthum solani (Kalt.) (Potato and glasshouse aphid) (Figure 14.18)

Myzus persicae (Sulzer) (Peach-potato aphid) (Figure 14.19)

Rhopalosiphoninus latysiphon (Bulb and potato aphid)

These are the main species of aphid likely to occur in potato stores and later in chitting houses, and they are all capable of causing serious damage to the sprouting tubers; they are also virus vectors.

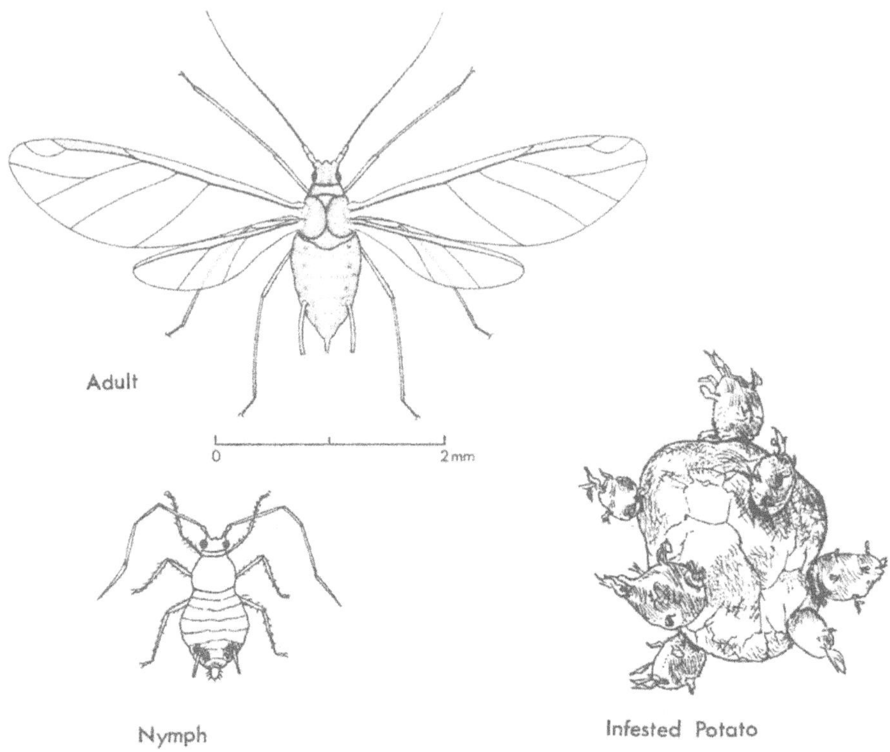

Figure 14.18. Aulacorthum solani (Potato and glasshouse aphid) (Homoptera; Aphididae), nymphs and alate female; to be found on potato tubers in store.

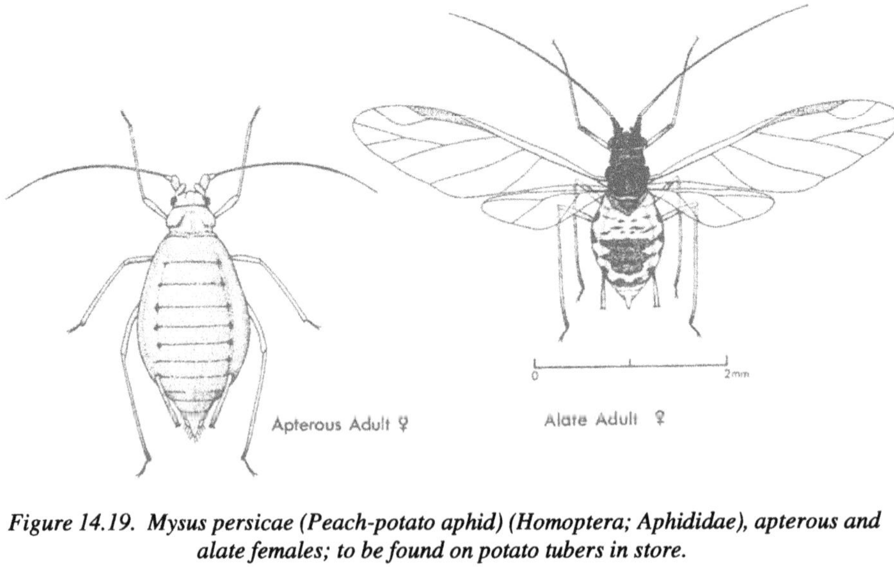

Figure 14.19. Mysus persicae (Peach-potato aphid) (Homoptera; Aphididae), apterous and alate females; to be found on potato tubers in store.

14.8.2 Family Pseudococcidae (Mealybugs)

A distinctive group of wingless (except for adult males), semi-sessile, somewhat degenerate insects with a segmented body covered by a white (or grey) waxy secretion, often with conspicuous lateral and terminal filaments. The body under the wax is often red or orange; legs are well-developed, but antennae less so; young stages are quite mobile, but the adult females are less inclined to move. Mostly a tropical group, but a few species occur in temperate regions. They are found mostly on fruit trees, various bushes, palms and sugarcane, seldom on vegetables. The bugs feed by sucking sap, and excess sugars are excreted quickly as sticky honey-dew, and on these sugars there can often be an extensive growth of sooty moulds that are very conspicuous and disfiguring. Mealybugs have no special methods of attachment to the host plant, and they often seek crevices in which to sit, where they survive casual dislodgement and may then be transported into fruit stores. Smooth fruits such as apples and citrus may carry a few mealybugs either in the eye of the fruit or at the top of the stalk, but not on the general surface which does not offer sufficient protection. The two tropical fruits most regularly and most heavily infested with mealybugs are custard apple (*Annona* spp.) and pineapple – the multi-segmented surface offers ideal shelter niches.

Some of the species of Pseudococcidae which have been observed on fruits in fruit stores and markets include the following:

Dysmicoccus brevipes (Cockerell) (Pineapple mealybug) (Figure 14.20)

Pantropical and polyphagous, but found on pineapples, lodged in the basal segments of the fruits.

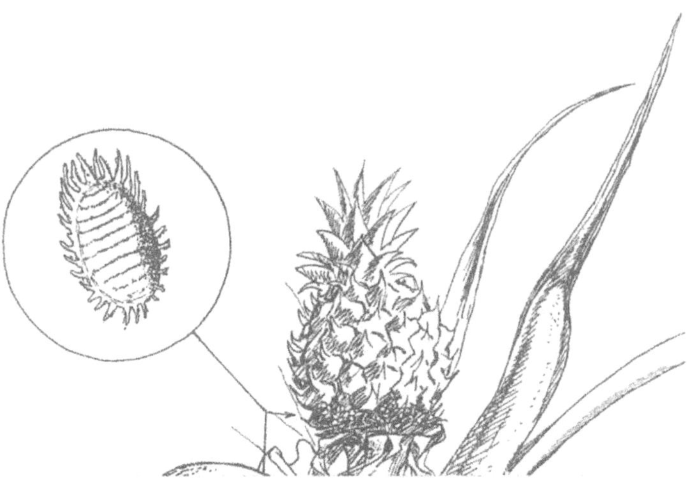

Figure 14.20. Dysmicoccus brevipes (Pineapple mealybug) (Homoptera; Pseudococcidae); to be found at the base of the fruit.

Planococcus spp. (Citrus mealybug, etc.) (Figure 14.21)

Several species are cosmopolitan, and recorded from many different hosts; heavy infestations of custard apple fruits are common, and the bugs are difficult to remove from between the protruding segments.

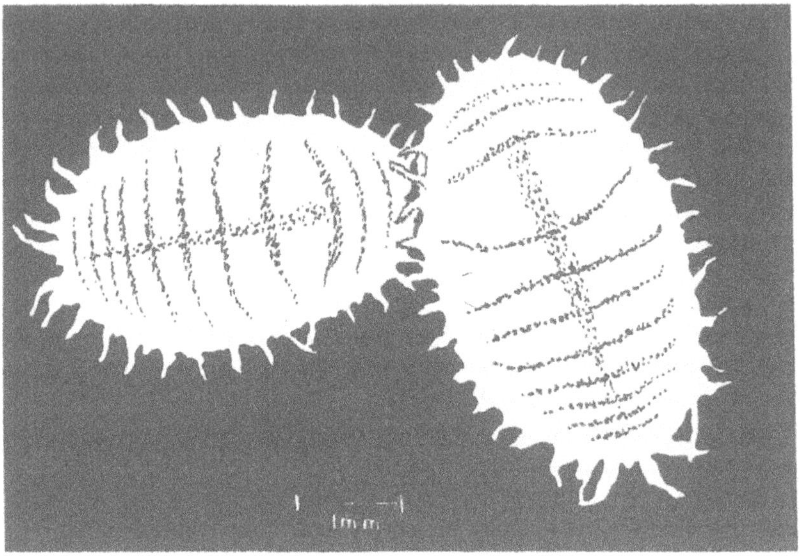

Figure 14.21. Planococcus citri (Citrus mealybug) (Homoptera; Pseudococcidae); often found on citrus and custard apple fruits.

Figure 14.22. Pseudococcus sp. (Long-tailed mealybug) (Homoptera; Pseudococcidae), on kumquat fruit, together with some sooty mould; South China.

Pseudococcus spp. (Long-tailed mealybug, etc.) (Figure 14.22)

Several species have the distinctive long waxy 'tails'; most are cosmopolitan; several are polyphagous, and some are regularly found on citrus fruits either by the eye or the stem base (as illustrated on the previous page).

14.8.3 Family Coccidae (Soft scales)

In this family there is great diversity of form, although all are wingless (except adult males), and the female body is degenerate, with obscure segmentation; antennae reduced, and legs somewhat reduced. The dorsal surface of the body is covered with a shield-like 'scale' which is part of the insect body integument. The first instar larvae are called 'crawlers', and are very active – this is the non-feeding dispersal stage. As they develop, they become more sessile, and the adult females normally do not move – usually they sit with their proboscis firmly embedded into the host plant tissues. The edges of the scale are firmly pressed on to the plant surface; the more flattened species are not easily dislodged, especially since they most typically sit on the undersurface of the leaves close to the edges of veins whose projection affords them some physical protection.

Soft scales are important pests of cultivated plants in the warmer regions of the world, but are not likely to be encountered in produce stores because of their preference for the leaves of shrubs. But two species have been observed on harvested bay leaves.

Ceroplastes sinensis (Pink waxy scale)

In Italy, it was noticed that local bay trees were very heavily infested with these scales on the leaves.

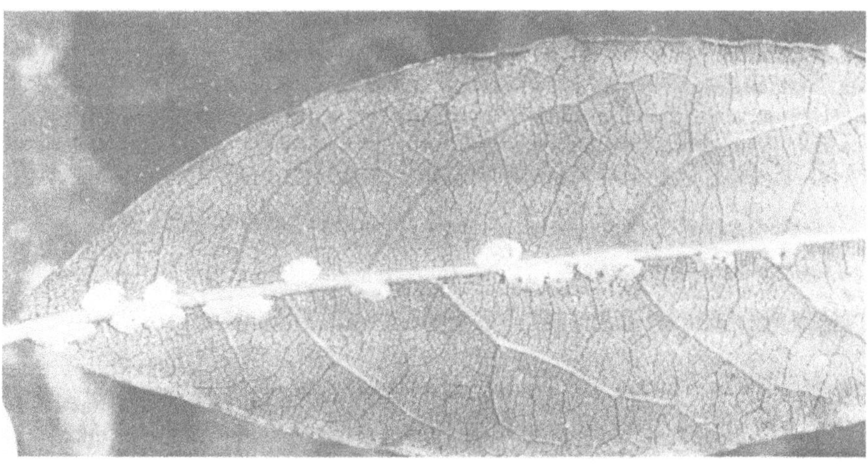

Figure 14.23. Coccus hesperidum (Soft brown scale) (Homoptera; Coccidae), polyphagous but commonly on bay leaves in the UK; Skegness, UK

Coccus hesperidum (Linn.) (Soft brown scale) (Figure 14.23)

An elongate greenish scale with brown markings, quite flat, on the soft twigs and younger leaves; the species is cosmopolitan in warmer regions, and polyphagous on trees and shrubs, and it is a regular pest of bay in the UK and parts of Europe. When the bay leaves are harvested, the scale may be included, particularly if younger leaves are also taken.

14.8.4 Family Diaspididae (Hard or armoured scales)

These are also wingless (except for adult males), but more degenerate physically than the other Coccoidea. Adult females are fixed permanently in position on the host, with the very elongate stylets (proboscis) fixed into the host tissues. An important family characteristic is that the 'scale' is separate from the insect body in the adult females. The larger scales (later instars) are very firmly fixed in position, and will resist considerable friction – dead scales, however, are often more easily dislodged by handling. Eggs are deposited under the old scale of the female, and eventually 'crawlers' will emerge from under the old scale and disperse.

This group is most important throughout the tropics and subtropics on trees and shrubs, but some species occur in temperate regions on fruit and other trees, and they can be very damaging. Favoured location sites are woody twigs, fruits, and on the upper surface of leaves. Most hard scales are quite small, only 1-2 mm in diameter, and thus light infestations on fruits may be unnoticed and overlooked, although heavy infestations are usually conspicuous.

Some species are particularly damaging to the host trees, and there is international phytosanitary legislation to restrict the distribution of these species and to prevent their introduction into countries where they are not already established. The most serious of these international pests is the San José scale, native to North China; in 1880 it was found in the USA (California), and it inflicted tremendous damage on the deciduous fruit industry until it was eventually brought under a measure of control. Throughout the world, there are about a dozen or more Diaspididae that are regularly found on *Citrus* and various deciduous fruits. The heavier infestations are usually noticed during quality inspection and packing, but because of their very small size and inconspicuous nature, light infestations may pass unnoticed. A few of the more widespread and important species are mentioned below.

Aonidiella aurantii (Maskell) (California red scale) (Figure 14.24)

Pantropical; on twigs and fruits, most serious on citrus fruits.

Chrysomphalus aonidum (Linn.) (Purple scale) (Figure 14.25)

Pantropical; on citrus and a wide range of other hosts, usually on leaves and fruit.

Figure 14.24. Aonidiella aurantii (California red scale) (Homoptera; Diaspididae), scales on a small orange; Hong Kong.

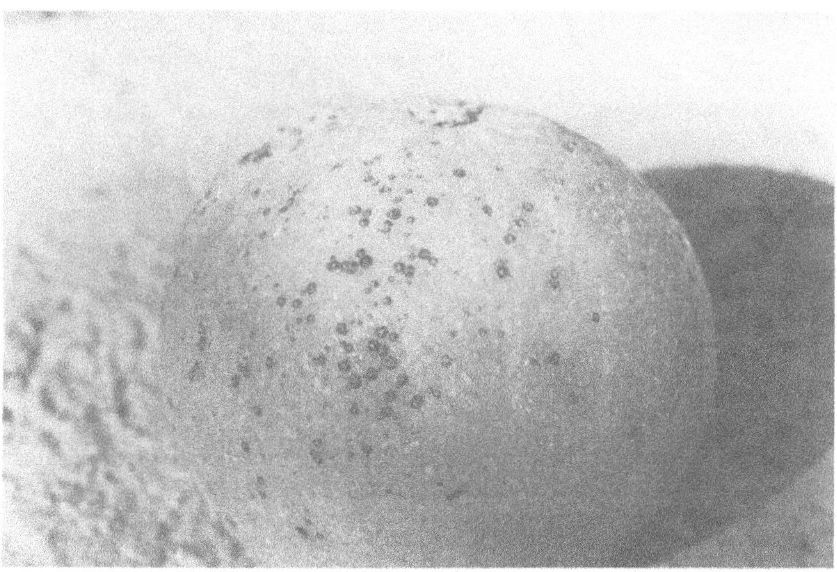

Figure 14.25. Chrysomphalus aonidum (Purple scale) (Homoptera; Diaspididae), scales on ripe orange; Dire Dawa, Ethiopia.

Ischnaspis longirostris (Sign.) (Black line scale) (Figure 14.26)

Probably pantropical in distribution; on citrus, bananas, mango and other plants, on both fruits and foliage.

Lepidosaphes spp. (Mussel scales)

Several species, both tropical and temperate, occur on a wide range of fruit trees, more often on twigs but sometimes on fruits.

Quadraspidiotus perniciosus (Comstock) (San José scale) (Figure 14.27)

Distribution is broadly cosmopolitan, but not recorded from some countries; polyphagous on deciduous trees and shrubs, and very serious on deciduous fruits.

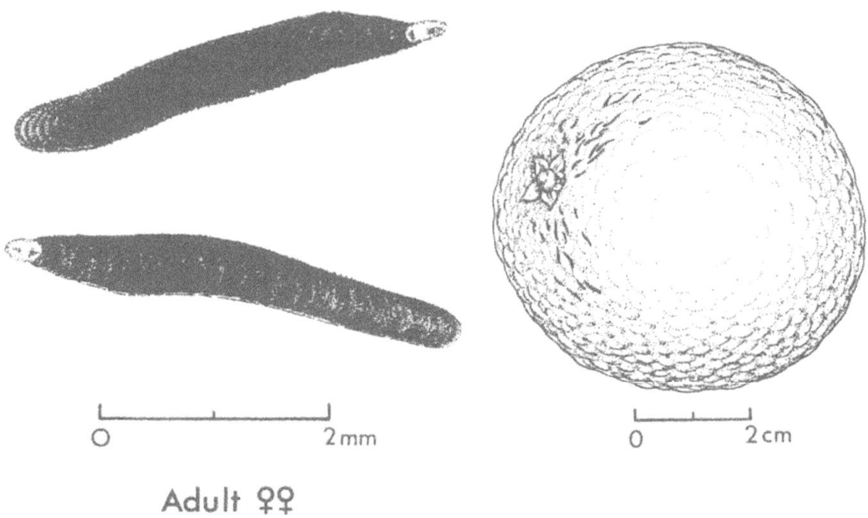

Figure 14.26. Ischnaspis longirostris (Black line scale) (Homoptera; Diaspididae), a distinctive scale drawn infesting an orange.

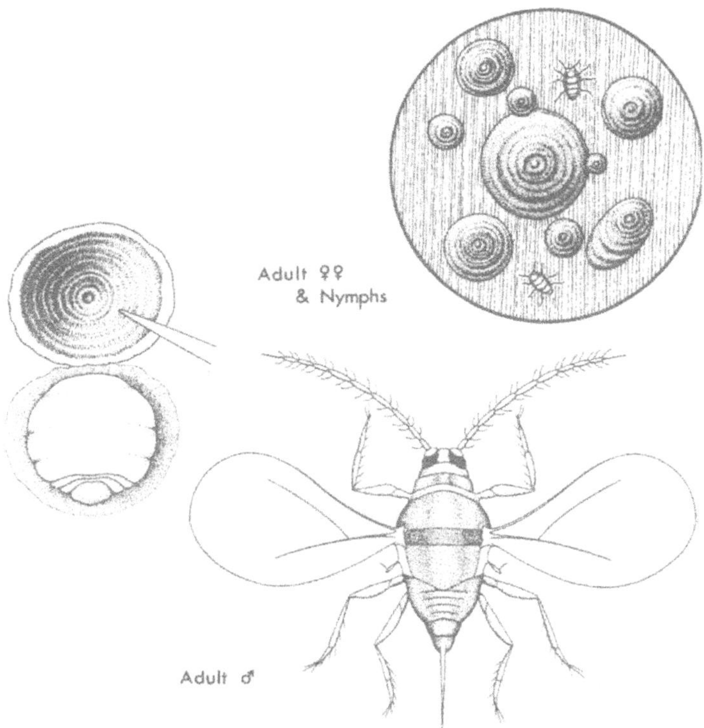

Figure 14.27. Quadraspidiotus perniciosus (San José scale) (Homoptera; Diaspididae), a serious pest of deciduous fruits, almost worldwide; often found in produce stores, although some other species might be recorded.

PREDACIOUS HETEROPTERA (Animal and plant bugs)

These predacious Heteroptera are very varied in size, shape, habitat etc.; some are plant sap feeders, and others are predacious and blood-sucking. The few that are likely to be encountered in produce stores are predacious forms that prey on other insects or that take blood from the rodents.

They are characterized by having the head porrect, forewings corneus at the base (hemi-elytra) and membraneous distally, with wings lying flat and overlapping over the abdomen at rest. The proboscis (beak or rostrum) arises from the front of the head, and is often short and curved.

Two families are often found in produce stores, although some others might also be recorded.

14.8.5 Family Reduviidae (Assassin bugs) (3,000 species)

The well-known domestic species are large and conspicuous, although seldom seen being nocturnal and secretive. Their normal prey are domestic rats and other rodents, from which they suck blood, but they also attack man (at night), snails and other insects. There are many small and inconspicuous species that will typically feed on other insects in stores, killing them in the process. The rostrum is characteristically short and curved, as shown in Figure 14.28.

Three species are sometimes found in stored produce in the British Isles; *Amphibolus venator* (Klug) is a dark brown species, 9-10 mm long, and is apparently frequently found in cargoes of groundnuts from Africa (Mound, 1989); the two less abundant species are *Peregrinator biannulipes* (Montrouzier), a small species 6-7 mm long, and the larger (16-17 mm) black *Reduvius personatus* (Linn.) whose nymphs cover their bodies with detritus.

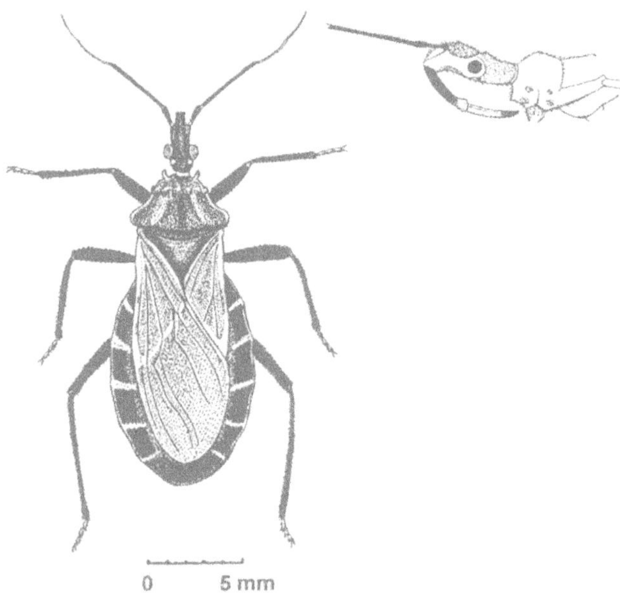

Figure 14.28. Triatoma rubrofasciata (Domestic assassin bug) (Heteroptera; Reduviidae), an example of this family of predacious bugs, side view of head shows the characteristic short curved proboscis.

14.8.6 Family Anthocoridae (Flower bugs, etc.) (300 species)

A small but widespread group, often encountered; the species are small flattened bugs, often red and black; they are predatory on small arthropods on flowers or in plant foliage, or under tree bark. A few species regularly occur in grain and produce stores where they are important predators of stored products pests. The last two antennal segments (as also with Reduviidae) are distinctly more slender than the basal two, and the rostrum is straight and three-segmented. The two genera quite frequently found in produce stores are *Lyctocoris* and *Xylocoris*. *Lyctocoris* bugs are quite small at about 4 mm body length, dark brown in colour, with legs and base of forewings yellow. *Xylocoris* are even smaller, being about 2-3 mm long, brown in colour, and some species have shortened wings.

PHYTOPHAGOUS HETEROPTERA (Seed bugs etc.)

A few phytophagous forms are sometimes found in produce stores, but since these are sap-suckers, their presence is not to be expected.

14.8.7 Family Lygaeidae (Seed bugs etc.)

These bugs are able to feed on oily seeds in storage, and with the aid of salivary enzymes they can suck out part of the seed contents. The species most commonly found in stored produce is *Elasmolomus sordidus* (Fab.), a brown coloured bug, about 10 mm long, with a straight four-segmented rostrum, and all four antennal segments of equal thickness and length. It is usually associated with groundnuts, cotton seeds and copra, and its feeding can cause shriveling of groundnuts. *Oxycarenus* spp. (Cotton seed bugs) may be found in stores with cotton; they are small (4 mm) and dark, and feed on the cotton seeds, but seldom survive in the stores.

14.9 ORDER THYSANOPTERA (Thrips) (6 families, 5,000 species)

These are very small (0.5-5 mm) slender insects with strap-like wings fringed with long setae. Body colour is usually black or brown, but can be red or yellow, especially as nymphs. Most feed on plant sap obtained by their piercing-rasping mouthparts, but some are fungus-feeding and others are predacious.

Figure 14.29. Adult Frankliniella sp. (Thysanoptera).

At certain times of the year, they swarm (dispersal flights in vast numbers, and may be known then as 'thunder flies'). *Limothrips cerealium* (Grain Thrips) is actually attracted to grain of all types, both in the field and in storage. The other species, by virtue of their tiny size and vast numbers when swarming, invade buildings of all types, and may end up in packaged foods where they die, but can be quite conspicuous contaminants.

14.10 ORDER COLEOPTERA (Beetles) (95 families, 330,000 species)

This is the largest order of insects, and there are many new species of beetles being described each year. It is also by far the largest group of stored produce pests – more than 600 species of beetle have been recorded feeding upon stored foods and food products worldwide.

Beetles are characterised by having the forewings modified into hard protective elytra under which are folded the hind pair used for flying. The elytra also cover the mesothorax and metathorax, except for the scutellum, and usually all the abdominal segments, but in a few cases, the terminal segments are left exposed. The prothorax is typically large, and its shape, colour, sculpturing and bristles are usually characteristic of the genus or species. They range in size from minute (0.5 mm long) to large (up to 100 mm), but the stored produce beetles are mostly small and some 2-4 mm in length. Adult mouthparts are typically unspecialized biting and chewing types, with well-developed toothed mandibles. Both adults and larvae show great variations in body form and habits. Larvae of course all feed, and thus may be pests, but adults are either short- or long-lived and may or may not feed, and thus may not cause any damage. Some adults feed, but not upon the stored produce; some take nectar from flowers, some feed on fungi etc. Predatory larvae have well-developed legs, and are active and agile, but many of the phytophagous species have reduced legs and are sluggish. Weevil larvae (usually regarded as advanced species) are mostly quite legless. Larvae that tunnel in plant tissues, or make galls, are usually apodous. Mouthparts have biting mandibles – some predatory forms have them modified for piercing prey tissues, but typically biting and chewing is the primary form of damage.

The arrangement of families below is according to the latest edition of Imms' *General textbook of entomology* (Richards & Davies, 1977). Groups which are predatory on other invertebrates, or which are fungivorous but commonly encountered in stored foodstuffs, are mentioned briefly – for further information, *Insect and arachnids of tropical stored products* (TDRI, 1984) should be consulted. Further information on the pest species involved in agriculture can be found in Hill (1987, 1997).

14.10.1 Family Histeridae (3,200 species)

Some 16 species are recorded in stores; they are small (2-5 mm), oval and compact; heavily sclerotised and glossy black, brown or metallic; the abdomen tip is exposed; antennae are elbowed and clubbed. Larvae are elongate, active, and have large mandibles.

Both adults and larvae are predacious on other insects and mites. A few species are typical of birds' nests, carrion or dung, and their presence indicates a poor standard of store hygiene. Other species prey on timber beetles (which may infest timbers in the stores), and a few attack Bostrychidae that occur in foodstuffs.

Notable species include *Teretriosoma nigrescens*, recorded on stored maize with *Prostephanus* in Central America, and *Teretrius punctulatus* found on dried cassava in East Africa.

14.10.2 Family Scarabaeidae (Cockchafers etc.) (17,000 species)

A large group, found worldwide; they are stout-bodied, robust beetles, some brightly coloured, others black or brown, with distinctive and characteristic larvae. Adults have short antennae which end in a club made of flattened plate-like segments. Some adults are pests of growing plants, but others have very weak mouthparts, and only eat pollen and very soft materials. The larvae are C-shaped, with a swollen abdomen and very limited mobility, well-developed thoracic legs, and large powerful biting and chewing mandibles. Most feed on rotting vegetable matter, but some on plant roots, and they bite large holes in tubers and root crops – these larvae are called 'chafer grubs' or 'white grubs'. Damage to potato tubers and root vegetables is common, but only occasionally are the large white fleshy chafer grubs taken into stores – they tend to make a shallow excavation in the tubers rather than an actual tunnel.

The species most likely to be encountered in produce stores are probably:

Prionoryctes etc. (Yam beetles)

A number of species in several different genera are recorded as yam beetles in different parts of the world, but the damage to the tubers is essentially the same. Tuber damage may include deep tunnels, and sometimes both adults and larvae may be found inside the tunnels.

Melolontha spp. (Cockchafers, Chafer grubs) (Figures 14.30, 14.31)

Different species are found in different parts of Europe and Asia, and the larvae may sometimes be found in vegetable stores in potato tubers.

Phyllophaga spp. (New World cockchafers)

Many species are recorded from Canada and the USA, and some are quite polyphagous and feed on potato tubers, various root crops, and other crop plants, and both larvae and adults are occasionally found in produce in storage.

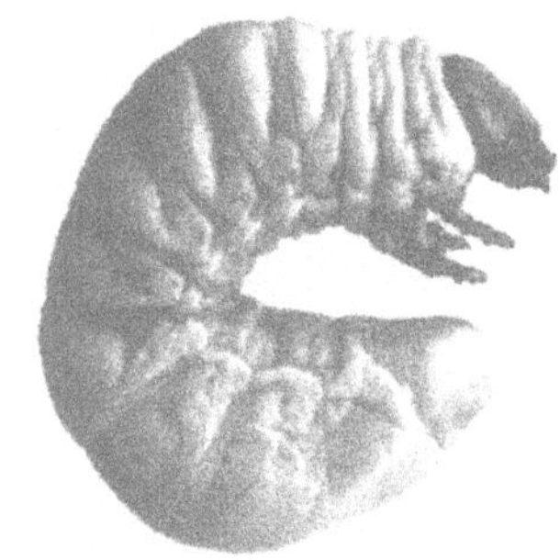

Figure 14.30. Chafer grub (White grub) (Coleoptera; Scarabaeidae), larva typical of all members of the family; Hong Kong.

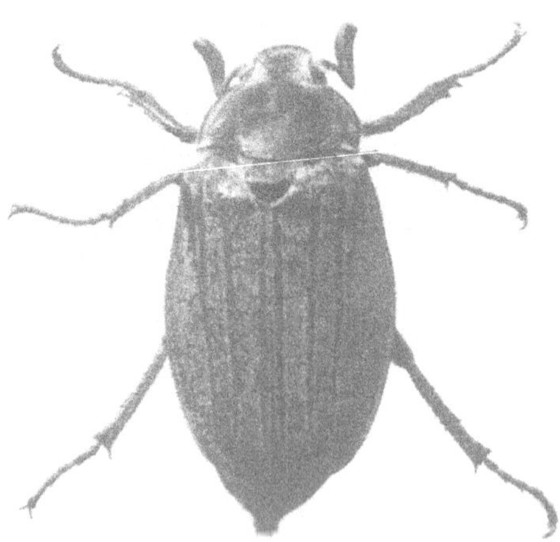

Figure 14.31. Melolontha melolontha (European cockchafer) (Coleoptera; Scarabaeidae), adult beetle, length 24 mm; Cambridge, UK.

14.10.3 Family Buprestidae (Jewel beetles, Flat-headed borers) (11,500 species)

A distinctive group of beetles, elongate, flattened, and often metallic green or bronze. The larvae are legless borers of wood, and have a very distinctive flattened and broad thorax; they live in their tunnels for a year or more, and adults may emerge from building timbers, bamboos and packing cases months after construction – occasionally emergence may occur after 1-2 years.

Chalcophora japonica (Pine jewel beetle) (Figure 14.32)

This is one of the larger species (30-35 mm) found in pine wood in the Far East. The specimen photographed emerged from a door (filled with blocks of pine under veneer sheets) that had been made more than a year previously (Figure 14.33). Other species of *Chalcophora* occur in Europe and North America, also in *Pinus.*

Figure 14.32. Chalcophora japonica (Pine jewel beetle) (Coleoptera Buprestidae), an Oriental species, length 35 mm; Hong Kong.

Figure 14.33. Piece of pine timber block from inside a veneered door, showing larval tunnel with most of the packed frass removed.

14.10.4 Family Elateridae (Click beetles, Wireworms) (7,000 species)

Another large, worldwide group of beetles with a characteristic appearance. Adults are elongate, flattened, with small fore-coxae, and the hind angles to the pronotum are acute and projecting; they have a special ventral mechanism whereby they can leap into the air with a 'click' if placed on their backs; they are often brown or greenish in colour;

The larvae are elongate and cylindrical in shape, shiny reddish brown, with well-developed short thoracic legs – they live in the soil and eat plant roots, or else in rotten wood, and are called wireworms. Some species, such as the temperate *Agriotes* (Figure 14.34) are very damaging to potato tubers – they excavate deep narrow tunnels into the tubers, and may be carried into potato stores at harvest. The tunnels are often secondarily infected by fungi and bacteria, and the tubers may rot, causing extensive damage to the stored crop. Other root crops are also attacked by wireworms.

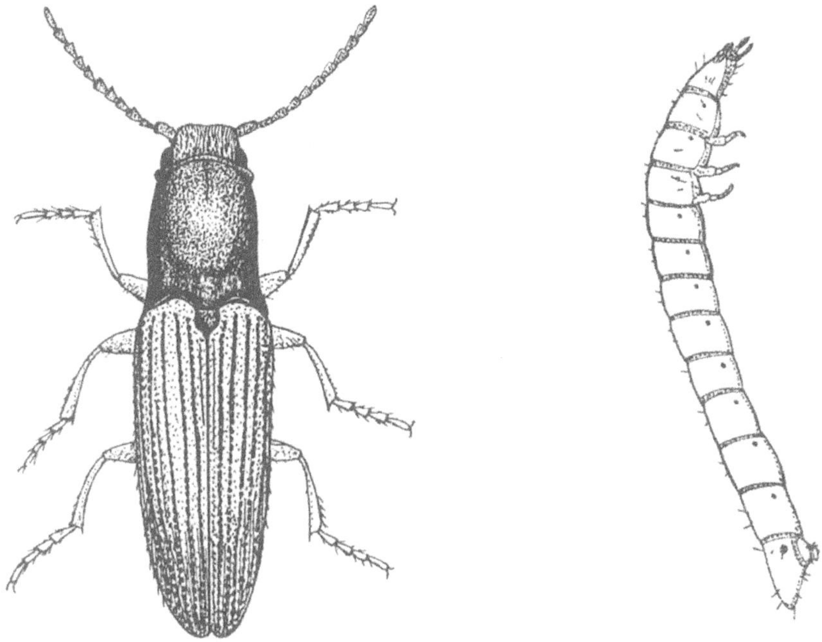

Figure 14.34. Agriotes sp. (Wireworm) (Coleoptera; Elateridae), adult beetle and larval wireworm, a common pest of potato tubers; length 2 cm; Cambridge, UK.

14.10.5 Family Dermestidae (700 species)

A small but important group of storage and domestic pests, showing great diversity of habits. The adults are rather similar, being small (mostly 2-4 mm), ovoid and convex, and covered with small scales; antennae are short and sometimes clubbed; there is a characteristic single ocellus between the two compound eyes (except in *Dermestes*). Larvae are very hairy, with setae of different and characteristic shapes. In the wild, these beetles are important scavengers which eat carrion and dried animal cadavers. Some 55 species are recorded from stored products. Some adults do not feed at all, but some do.

Dermestes lardarius Linn. (Larder beetle) (Figure 14.35)

Pest status. An important pest on a wide range of animal materials in warm regions.

Produce. Animal products of all types, including dried fish, bacon, cheese, bones, sausage, hair, rye bread; it also scavenges on dead insect bodies, and has been reared in the laboratory on wheatgerm.

Damage. The feeding insects bite holes in skins and furs etc., and the larval exuviae and frass contaminate foodstuffs. Damage is done by both adults and larvae, as they eat the same range of foodstuffs, but larvae do the most damage.

Life history. Eggs are usually not laid until about 10 days after emergence, and the oviposition period lasts about 90 days; but at 25°C, oviposition may start soon after emergence. Temperature of 30°C inhibits oviposition. Eggs (2 mm) are laid in crevices, and each female can lay 40-80 eggs. Incubation takes 3-9 days.

There are 5-6 larval instars, but more have been recorded. Larvae are very bristly, and have two posteriorly-directed terminal urogomphi; fully grown larvae are 10-15 mm in length. Temperature requirement for development is between 15 and 25°C; at 25°C and 65% RH, development takes 48 days.

Pupation takes place at a site away from the infestation, often in a crevice or a tunnel in wood. The larval exuvium acts as a tunnel plug to protect the naked pupa. The pupal period is usually 8-15 days.

Adults are 6-10 mm long, and characterised by having the basal part of the elytra grey-brown with three small dark patches, and the distal half is dark brown (the colour is imparted by the presence of coloured setae).

Adults feed on the same materials as the larvae; they are long-lived (3-5 months), and may hibernate in unheated premises in the winter period.

The life cycle can be completed in 2-3 months, but often takes longer.

Distribution. Cosmopolitan, but not occurring in the hot tropics.

Control. The general methods of population control described in Chapter 21 apply.

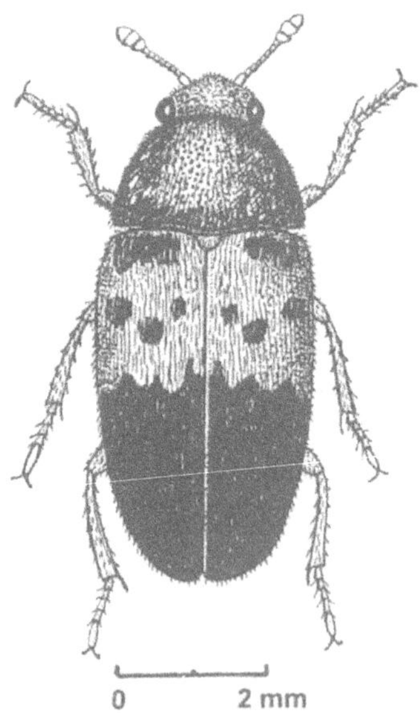

Figure 14.35. Dermestes lardarius (Larder beetle) (Coleoptera; Dermestidae), adult.

Dermestes maculatus Degeer (Hide or Leather beetle) (Figure 14.36)

Pest status. A serious pest of stored untreated hides and skins, and of air-dried fish in the tropics, especially in parts of Africa and Asia. Damage can be very extensive, and the rate of development is rapid in the tropics; an important penetrator species.

Produce. A wide range of animal materials is eaten, but the most important produce damaged is stored untreated (dried) hides, sun-dried fish, and fishmeal. Cheese, dog biscuits, ham, bacon and cacao beans are also eaten. Adults will fly to start new infestations.

Damage. Both adults and larvae eat the produce and bite holes in the skins, and the frass (especially larval exuviae) is a troublesome contamination of foodstuffs. The setae dislodge from the larval cuticle and may be ingested or inhaled by workers, causing considerable discomfort. Adults cause less damage than the larvae, and they are not long-lived (2-3 weeks only). Infestation of fish starts early in the drying process, as soon as the surface has dried; freshwater fish are more heavily attacked – marine fish have a higher salt content. Fish and fishmeal with a salt content of 10% are, apparently, immune to infestation. Pupation tunnels may damage timbers, and by cross-infestation may hole stored fabrics, materials, tobacco leaf, and the like.

Life history. Eggs (1.3 mm) are laid starting a day after copulation; on average, 17 eggs per day; the female needs water to drink, and lives for about 14 days, laying 200-800 eggs. At 30°C, eggs hatch in 2 days, with a survival rate of about 50%.

Larvae develop through 6-7 (up to 11) instars, taking 20 days at 35°C and 75% RH, or 28 days at 27°C; at lower humidities, development is prolonged. Larvae are cannibalistic, and will eat both eggs and pupae. The last larval instar tunnels into a solid substrate to make a pupation chamber. Fully grown larvae measure 15 mm, and are covered with long setae, and there are two terminal curved urogomphi.

In ships and stores, wooden structures can be weakened by the pupation tunnels, and in mixed stores and cargoes, cross-infestation can lead to damage and holes in fabrics, containers, dunnage, stored tobacco leaf, cigarettes etc. Pupation at 30°C takes about 6 days.

Adults are elongate to oval, and more or less parallel-sided, 6-10 mm long, with a dark cuticle clothed in black and grey setae dorsally. The antennae are 11-segmented, with a distinct club. The inner apex of each elytron ends in an acute spine. They have wings and regularly fly, and are thus able to infest sun-drying fish in the tropics. They are not long-lived (2-3 weeks), and although they do feed they do not cause much damage.

The life cycle in the tropics (30°C) is usually completed in about 30 days, and there can be many successive generations per year. This rapid rate of development makes this insect a very serious pest in some situations, particularly on sun-drying fish.

Distribution. Cosmopolitan throughout the tropics, and carried into heated premises and mills in temperate regions.

Natural enemies. The larvae are cannibalistic and will eat eggs and pupae; *Necrobia rufipes* is recorded eating eggs and young larvae, and there are parasitic mites (*Pyemotes* spp.) and eugregarine amoebae internally.

Control. Pitfall traps are successful for use in population monitoring.

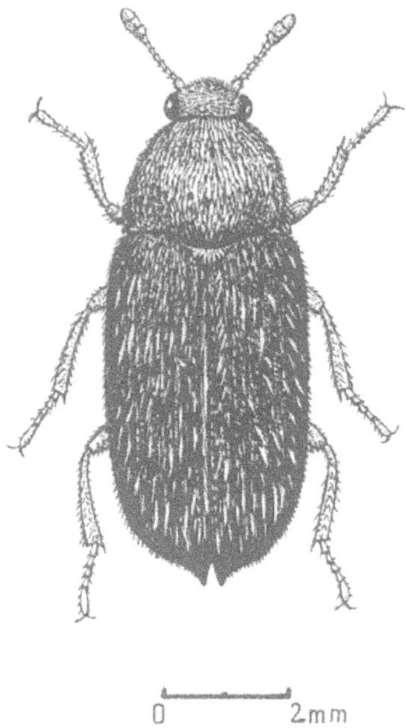

Figure 14.36. Dermestes maculatus (Hide beetle) (Coleoptera; Dermestidae), adult.

A few other species of *Dermestes* are regularly encountered in produce stores, including:

Dermester ater Degeer (= *D. cadaverinus* Fab.)

Usually found on copra; a known predator of fly puparia; less often found on animal materials. Adults have golden setae dorsally and laterally, and the larvae have straight urogomphi.

Dermestes carnivorous Fab.

Not recorded from Africa, but basically similar to *D. maculatus.*

Dermestes frischii Kugelann

Adults are very similar to *D. maculatus*, but are lacking the acute spines on the inner apex of each elytron. They are found more often on drying or dried marine fish, and they are quite salt tolerant; larvae can develop on fishmeal with a sodium chloride content of 25%. The life cycle can be completed in 34 days, at 30°C and 75% RH; presence of salt will retard development.

Dermestes haemorrhoidalis Kuster
Dermestes peruvianus L. de Castelnau

These are thought to be Neotropical in origin, and have now spread to North America and Europe. Both infest animal products and are found in the UK; some development can occur on wheatgerm. *D. haemorrhoidalis* has a preferred temperature range of 25-30°C, but *D. peruvianus* prefers a lower temperature (20-25°C).

Trogoderma granarium Everts (Khapra beetle) (Figure 14.37)

Pest status. A very serious international pest of stored grain with phytosanitary regulations enforced to curb further redistribution.

Produce. This is the only truly phytophagous dermestid beetle; the larvae feed on oilseeds, cereals, cocoa, and, to some extent, pulses and cereal products; adults rarely eat.

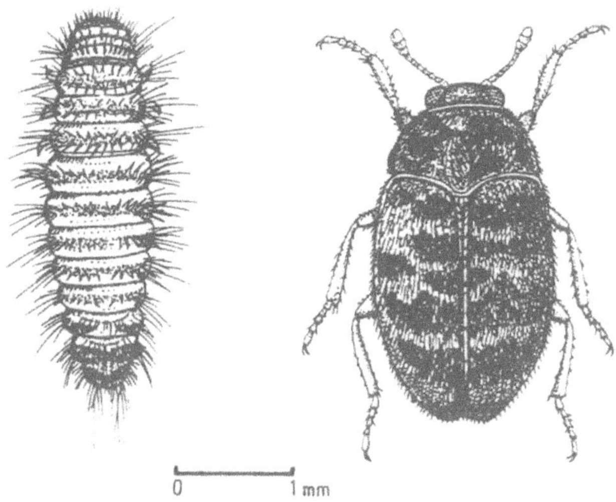

Figure 14.37. Trogoderma granarium (Khapra beetle) (Coleoptera; Dermestidae), larva and adult beetle.

Damage. The feeding larvae hollow out grains and seeds, and cause a rapid loss of produce; larval exuviae contaminate the produce.

Life history. The number of eggs laid is small – on average, 35 per female over a period of 3-12 days (at 40°C and 25% RH), and the female then dies.

Larval development does not take place at less than 21°C, but will proceed at very low humities; at 35°C and 75% RH, it is completed in 18 days; males moult 4 times and females 5, though this is variable. Some larvae are able to enter facultative diapause when conditions are adverse – without food they can survive for about 9 months, but with food they can live for 6 years. Fully grown larvae measure about 5 mm, and have no urogomphi.

Figure 14.37a. Larvae of Khapra beetle feeding on oilseed cake (courtesy of ICI).

The pupa stays inside the last larval exuvium in the produce, and takes 3-5 days to develop (at 40°C and 25°C respectively).

The virgin female remains inside the exuvium for a day, then emerges and secretes a sex pheromone. After copulation, oviposition starts immediately at 40°C, but at 25°C there is a pre-oviposition period of 2-3 days. The female dies shortly after oviposition is completed, and the male a few days later, so neither adult lives for more than about 2 weeks. The beetles are small (2-3 mm), females larger, oval in shape, mottled dark brown and black; they have a median ocellus, and antennae with a fairly distinct club of 3-5 segments. Adults seldom, if ever, drink or eat; they have wings, but are not recorded to fly.

Under optimum conditions (warm and moist, 35°C and 75% RH), the life cycle can be completed in 18 days. The known geographical distribution shows most occurrance in warm dry areas, and it is concluded that in moist locations there is too much competition from other fast-breeding species.

Distribution. Of Indian origin, this species is recorded mostly in hot, dry locations – predictably in regions with a dry season of at least 4 months when conditions are hotter than 20°C (mean) and drier than 50% RH. Now it is pantropical, but reportedly seldom established in parts of S.E. Asia, South Africa, Australia, and most of South America.

Natural enemies. The predacious bug *Amphibolus* and many parasites are recorded, including mites, Hymenoptera (Chalcidoidea), and Protozoa.

Control. Legislative control is practiced internationally through inspection and quarantine measures, and in the USA and parts of Africa, former populations have been eradicated by fumigation, and the pest is being kept at bay. Fortunately, the adults are short-lived and do not fly, and the number of eggs laid is small, so the explosive reproductive and dispersal potential is less than might be expected, but the capacity for population increase is still very considerable.

Chemical control is usually repeated fumigation with methyl bromide, for in most cases population eradication is the aim, not just population depletion. Larvae that are diapausing are far less susceptible to insecticides than usual, and thus are difficult to kill. Population monitoring can be achieved using commercially available female sex pheromone traps.

Trogoderma inclusum Leconte (= *T. versicolor* (Creutzer))

Found in Europe and in the USA, but generally cosmopolitan. It eats vegetable matter and a wide range of agricultural products, animal feeds, processed foods, and the bodies of dead insects. It is a pest of malt in the UK. The larvae can diapause at temperatures between 20 and 25°C; populations are generally quite small. Adults normally do not feed, so all the damage is done by the larvae. Optimum conditions for larval development are 30-35°C and 50-73% RH.

The pupa remains inside the last larval skin; pupation takes 3 days at 40°C, and 13 days at 20°C. Adults are short-lived; at 73% RH, males live for 10 days at 40°C, and 38 days at 20°C, on average, and mated females for 8-38 days.

Trogoderma glabrum (Herbst.)

A widespread pest in the USA on a wide range of foodstuffs of both plant and animal origin, but preference is shown for animal feeds (especially mixed feeds such as cat and dog food). They can thrive on whole grain cereals. As with other *Trogoderma*, the adults do not feed, and the damage is done exclusively by the larvae. Survival and development rates vary greatly according to ambient conditions and the food available, but optimum conditions seem to be about 32°C and 70% RH, when development takes about 30 days.

Trogoderma variabile (Warehouse beetle)

A widespread species most abundant in Canada and the USA, on a wide range of food materials; most prolific on animal feeds, processed grains, whole kernels of wheat and barley, pollen, and a range of grocery commodities (flour, noodles, oatmeal etc.). Dried pulses were a poor diet. An important penetrator species.

When reared at 30°C and 60-70% RH on wheatfeed, the development stages took 6 days (egg), 26 days (larval) and 5 days (pupal) respectively. Some larvae enter diapause and may remain for two years. It is reported infesting an unheated warehouse in Canada where it survived the winter period.

Trogoderma spp.

Three other species occasionally occur in food stores and cargoes; the larvae feed on cereals and insect corpses, and some can diapause, but they are usually only minor pests.

Anthrenus spp. (Carpet and museum beetles) (Figure 14.38)

These are small (1.5-4 mm) beetles, oval, with a rounded convex body, often with brightly coloured spear-shaped body scales; they have a median ocellus, and antennal grooves at the front of the thorax. Larvae are small, somewhat ovoid, with spear-shaped setae, and they lack urogomphi. The larvae eat keratin (wool, furs, carpets, etc.) and in food stores feed on the chitin of insect corpses; adults are found on flowers eating pollen and nectar.

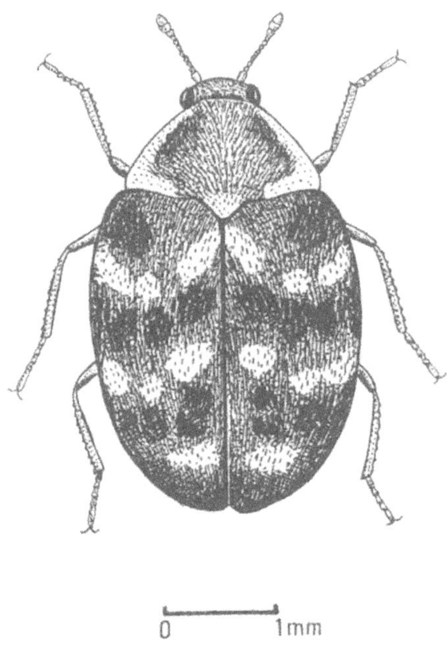

Figure 14.38. Anthrenus sp. (Carpet beetle) (Coleoptera; Dermestidae), adult beetle.

Four species are recorded in food stores:

Anthrenus coloratus Reitter, from southern Europe, North Africa, and India;
Anthrenus flaviceps (Leconte), widespread;
Anthrenus museorum (Linn.) (Museum beetle), Holarctic Region;
Anthrenus verbasci (Linn.) (Varied carpet beetle), worldwide.

Other species are known, but they feed on insect remains in spiders' webs.

Attagenus spp. (Fur and Carpet beetles) (Figure 14.39)

Larger beetles, some 3-6 mm long, with a rounded ovoid body, dark brown to black; they have a median ocellus, and a 3-segmented antennal club. Larvae are bristly, with a banded appearance and a characteristic terminal tuft of long golden bristles.

In the wild, these beetles are usually found in birds' nests and animal lairs. The larvae feed on keratin and chitin (wool, fur, feathers, and insect remains), but have been recorded to eat dried meat, dried egg yolk, flour, bran and cereal products. When found in stored grain, they are usually feeding on insect remains.

The three main species of economic importance are:

Attagenus fasciatus (Thunberg) (= *A. gloriosae)* (Fab.), pantropical;
Attagenus pellio (Linn.) (Fur/Carpet beetle), cosmopolitan (temperate);
Attagenus unicolor (Brahm) (Black carpet beetle) (= *A. megatoma* (Fab.)) (= *A. piceus* (Oliver)), a cosmopolitan species.

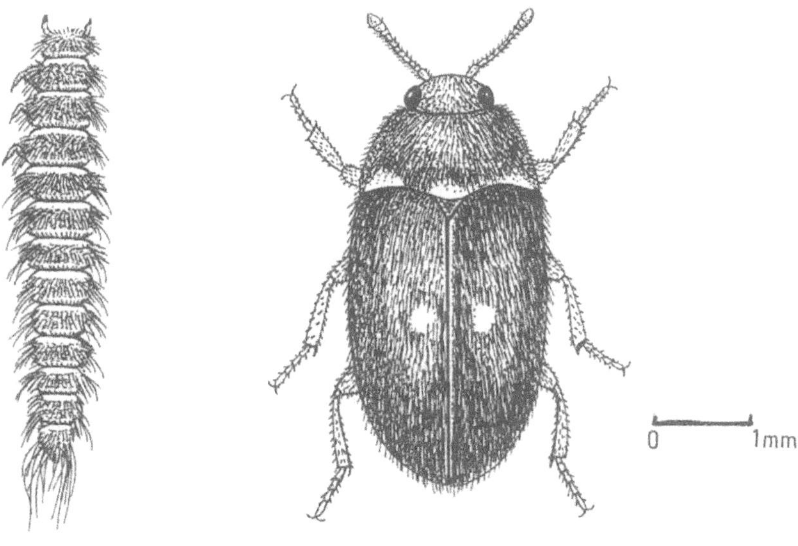

Figure 14.39. Attagenus pellio (Carpet beetle) (Coleoptera; Dermestidae), larva and adult beetle.

Two other species of *Attagenus* are found in North Africa, and North and Central America, but are not very important.

Phradonoma spp.

Small ovoid beetles, 1.5-3 mm long, similar to *Trogoderma*; several species are recorded from the Mediterranean Region and India as minor pests on groundnuts, coffee beans, cotton seed cake, and bones.

Thorictodes heydeni Reitter

A small brown beetle, 1.5 mm long, with tiny eyes, and is wingless. Larvae are similar to those of *Dermestes*, and 2-3 mm long, with urogomphi curved posteriorly. It is pantropical, but not recorded from South America; usually found on cereals and stored pulses.

14.10.6 Family Anobiidae (1,000 species)

A group that is mainly tropical in distribution; they are small oval beetles, with serrate antennae, and the head strongly deflexed under the prothorax; most are wood borers. The larvae are scarabaeiform with three pairs of thoracic legs, and minute spinules on some of the folds on the thorax and abdomen (as distinct from the Ptinidae).

Several species may bore in the structural timbers of produce stores and warehouses, and two species are important pests of stored products.

Lasioderma serricorne (Fab.) (Tobacco beetle) (Figure 14.40)

Pest status. A sporadically serious pest in many situations in the warmer parts of the world, especially damaging to high value commodities. A penetrator species, probably the most ubiquitous stored produce pest.

Produce. A wide range of produce is damaged by this pest, including cocoa beans, cereals, tobacco leaf, cigarettes, oilseeds, pulses, cereal products, spices, dried fruits, and some animal products. Other commodities may be damaged by the larvae boring prior to pupation, and by emerging adults. A consignment of brassières was badly damaged during shipment as cargo from Hong Kong to Holland in 1974.

Damage. Direct damage by feeding larvae is the eating of the food material, and there is contamination by frass. Indirect damage is caused when the fully fed larvae leave the food material in order to pupate in a cell fixed to some solid substrate, and later the adults emerge from the cocoons and may bite their way out through cardboard boxes and wrappings as they depart.

Life history. About 100 eggs are laid per female; deposited loosely in the food, they hatch after 6-10 days.

The larvae are very active when young, but become sluggish as they age; there are 4-6 instars, and the later ones are white and scarabaeiform. The final stage larva makes a pupal cell or cocoon out of food fragments, and attached preferably to a firm substrate, sometimes boring a short distance in wrapping material etc. to pupate. Larval development on a good diet takes 17-30 days.
Pupation takes 3-10 days, and is followed by a pre-emergence maturation period also of 3-10 days.

Adults are small (2-3 mm), brown, rounded to elongate beetles, with smooth elytra, and the head strongly deflexed under the pronotum when alarmed; antennae are 11-segmented, and segments 4-10 are serrate. Adults drink but do not feed, although they will tunnel through materials to leave the cocoon – they are often recorded making extensive holes in cigarette packets and other cardboard cartons. Adults live for 2-6 weeks, and they are active fliers, especially in the evenings.

The life cycle can be completed in 26 days at 37°C, but takes 120 days at 20°C and 70% RH. At 17°C it is recorded that development ceases, adults are killed after 6 days at 4°C.

Distribution. A pantropical species , found in heated stores in temperate countries, very widespread; one of the most abundant and widely distributed stored foodstuffs pests.

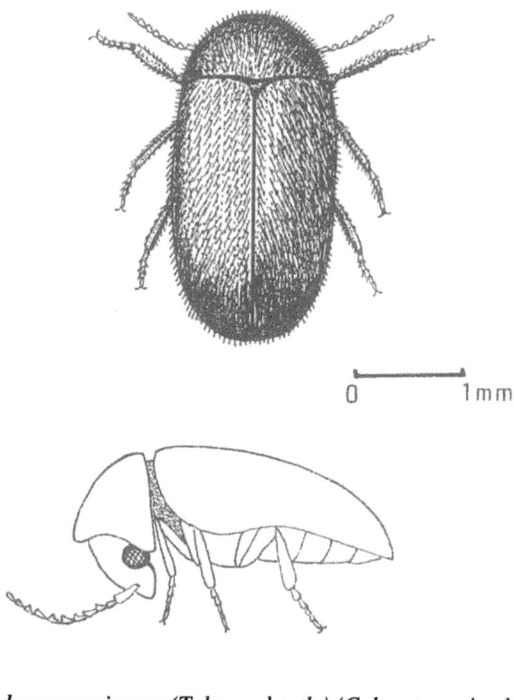

Figure 14.40. Lasioderma serricorne (Tobacco beetle) (Coleoptera; Anobiidae), adult beetle dorsal view, and profile showing deflexed head.

Natural enemies. Several beetles are recorded as predators, including *Tenebriodes, Thaneroclerus* (Cleridae), and some Carabidae. Predatory mites eat the eggs, and parasitic Hymenoptera include the Pteromalidae *Anisopteromalus calandrae, Lariophagus* and *Choetospila*, the Eurytomid *Bruchophagus*, and the Bethylidae *Israelius* and *Cephalonomia.*

Control. As a tropical pest, it can be killed in high value commodities by cooling or refrigeration; in 1963, infestations of expensive animal feed supplements in Hong Kong were successfully killed by brief storage in large commercial cold stores at -20°C.

Light traps have been used with some success in tobacco warehouses, but are more often used for population monitoring, as also are sticky traps. The usual pesticides are generally effective against this insect.

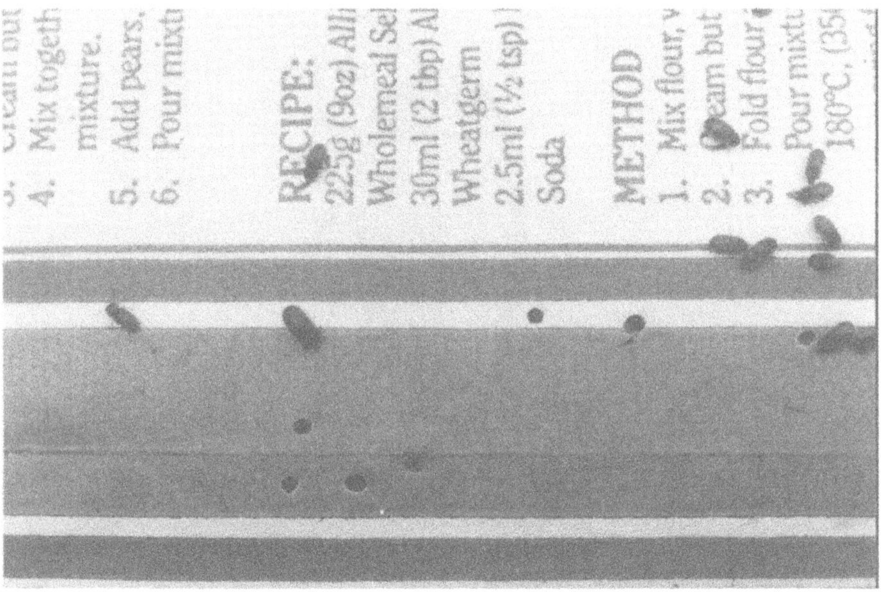

Figure 14.41. Flour bag with Tobacco Beetle emergence holes.

Stegobium paniceum (Linn.) (Biscuit or Drugstore beetle) (Figure 14.42)

Pest status. Sporadically serious on a wide range of stored foodstuffs, seeds and spices, and processed goods; generally important in cooler situations than *Lasioderma.* A penetrator species.

Produce. Polyphagous on many different foodstuffs and grains; often recorded on spices and herbs, biscuits, chocolates and other confectionery (hence the American name of Drugstore beetle), tobacco, leather, books and textiles.

Damage. This is mostly direct produce loss through eating, but high value commodities are ruined commercially by only slight damage and contamination.

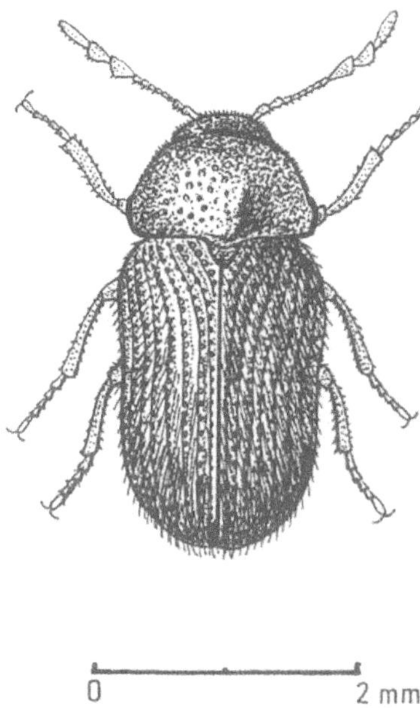

Figure 14.42. Stegobium paniceum (Biscuit or Drugstore beetle) (Coleoptera; Anobiidae), adult beetle and infested coriander seeds; Calcutta, India.

Life history. Females lay a maximum of 75 eggs, and live for 13-65 days. Adults are recognised by the loosely segmented antennal club, and longitudinal striae on the elytra. Optimum development conditions are reported to be 30°C and 60-90% RH, when the life cycle is completed in about 40 days; but development proceeds at temperatures between 15 and 34°C, and at humidities above 35%.

Distribution. Worldwide in warmer regions and in heated premises in temperate countries (such as the UK), but less abundant in the tropics than *Lasioderma*; a more temperate species.

Natural enemies. It is suggested that they are much the same as those recorded for *Lasioderma.*

Anobium punctatum (Degeer) (Common furniture beetle) (Figure 14.43)

A small brown beetle, 2.5-5 mm long, body more or less cylindrical and head right under the prothorax. The larvae tunnel in wood, either dead tree branches or trunks, or in structural timbers in domestic situations; they attack hardwoods and softwoods used as rafters and flooring, and some buildings can be seriously weakened after heavy infestations. They do not appear to attack dried foodstuffs inside stores, as distinct from some other timber borers that will attack dried cassava and other materials.

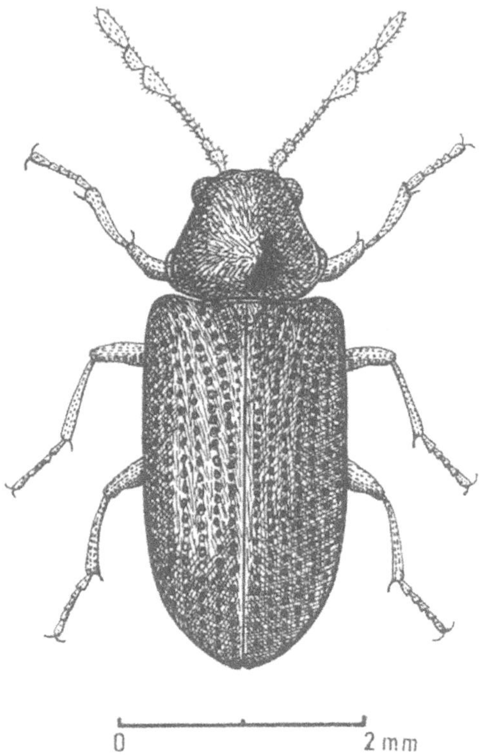

Figure 14.43. Anobium punctatum (Common furniture beetle) (Coleoptera; Anobiidae), adult beetle.

Xestobium rufovillosum (Degeer) (Death-watch beetle)

A larger beetle, up to 7 mm long, the female larger. In the UK and Europe, it infests hardwoods (mostly oak) in the presence of fungal infection. In domestic premises it is only recorded in timbers, and so is occasionally encountered in barns and food stores of timber construction.

Catorama spp.

Several species occur in Central America in both tobacco and stored grains, but the damage is apparently slight.

14.10.7 Family Ptinidae (Spider beetles) (700 species)

These beetles have a rounded body, small rounded thorax, long legs and long antennae (11 segments and no club), and they resemble spiders, hence the common name. The group is basically temperate, but some species occur in the tropics. Most are scavengers, and in the wild are usually found in old birds' nests and the like. Infestations are usually small and sporadic, and often associated with food debris, but occasionally a serious attack on stored foodstuffs can be observed in stores with a mixed content.

A total of 24 species have been recorded as minor pests of stored products; a key to eight species is given in Hinton and Corbet (1955). The literature is apparently rife with misidentifications.

Ptinus tectus Boieldieu (Australian spider beetle) (Figure 14.44)

Pest status. A regular but usually minor pest in many parts of the world, especially on mixed produce; more important in Europe than in North America.

Produce. Recorded on wheat in Australia, flours and meals, spices, and the debris of a wide range of plant and animal materials; it prefers a food with a high content of group B vitamins. Also found in birds' nests. At pupation, there may be physical damage to cardboard containers, sacking or even wood.

Damage. The feeding larvae bite the produce with their mandibles and eat the food, and the adults do some damage as they feed. Fully-fed larvae leave the produce to search for a pupation site, and will bite their way through cardboard containers, sacking and even wood. This species is thought to be probably the most damaging of the spider beetles known.

Life history. Each female lays about 100 eggs (other species lay fewer), many singly; they are small and sticky (0.5 mm long) and are soon covered with debris; they are laid over a period of 3-4 weeks; eggs hatch after 5-7 days at 20-25°C.

Larvae are white, fleshy grubs with a curved body and small legs, and body surface is covered with small fine golden setae. They pass through three instars in some 40 days or more. Fully-fed, last instar larvae leave the produce to find a site for pupation, and at this time they often bore through cardboard cartons, sacking and other wrapping materials, and they may make a pupation chamber in wood.

Pupation takes place inside a flimsy but tough silken cocoon, inside the pupal chamber. Development usually takes 20-30 days, and the young adult usually remains inside the cocoon for 1-3 weeks.

Adults are sub-globular in shape, with a rounded prothorax, 3-4 mm long, reddish-brown in colour, and the elytra having rows of pits. They eat and drink readily, and live for some months; other species live for up to 9 months. In habits they are nocturnal, and they may be very active, especially at dusk. Both sexes fly well.

Under optimum conditions of 23-27°C and 70-80% RH, the life cycle can be completed in 60-70 days; breeding is often continuous, and three generations per year (or more) can be completed in North Western Europe.

Distribution. A cosmopolitan species now, but it is not common in the tropics; it invaded Europe around 1900 from either New Zealand or Tasmania.

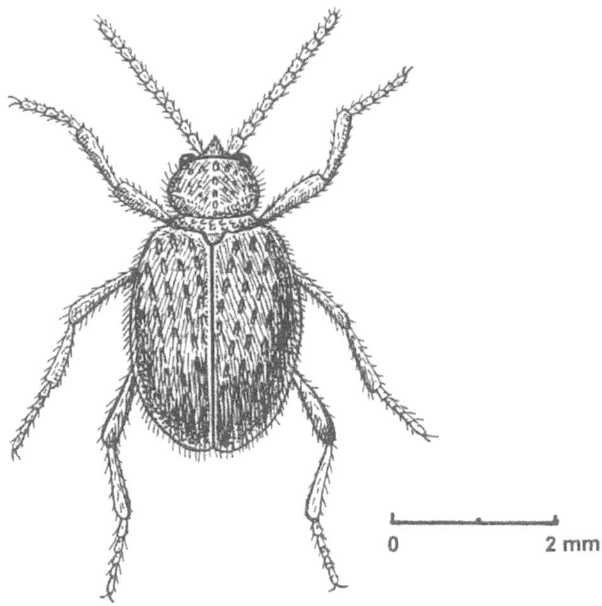

Figure 14.44. Ptinus sp. (Spider beetle) (Coleoptera; Ptinidae), adult beetle, Hong Kong.

The other species of Ptinidae that are most regularly encountered in food and produce stores are as follows:

Gibbium psylliodes (de Czenpinski). Pantropical; dark brown and shiny; 1.7-3.2 mm long; now regarded as essentially a European species.

G. aequinoctiale is mostly found in Asia and the New World, and is the more widespread of the two species.

Mezium americanum Laport. Cosmopolitan; head and prothorax golden, but abdomen shiny black.

Niptus hololeucus (Faldermann) (Golden spider beetle). A temperate species from West Asia and now widespread in Europe; body 3-4.5 mm long, and covered with golden-yellow setae. Several species are known. They prefer vegetable food and do not survive on fishmeal.

Pseudoeurostus hilleri (Reitter). Scattered records from Canada, Asia, Near East, but more common in Europe; brown in colour, 2-3 mm long.

Ptinus fur (Linn.) (White-marked spider beetle). Prothorax has two patches of white setae posteriorly. body 2-4 mm long, and white scales on the elytra; the last larval stage may be a resting facultative diapause. Several other species of *Ptinus* may be encountered, but are difficult to identify.

Trigonogenius spp. Several species (2-4 mm long) are widely recorded and may be quite common in East Africa.

(See Gorham, 1991; pp.177-180)

14.10.8 Family Bostychidae (Wood borers) (430 species)

A family of distinctive beetles, brown or black in colour, with a cylindrical body, deflexed head under a large rounded and ridged prothorax, and they are unusual in that the adults tunnel in wood (branches and tree trunks); three small species are found attacking stored products. The group is essentially tropical, but also to be found in the subtropics. The wood boring species may be found in food stores after having emerged from wooded timbers, packing cases, boxes and dunnage, and the like, and occasionally they are recorded attacking grain in storage and dried roots; and the casual boring of high value commodities can be of economic importance.

Prostephanus truncatus (Horn) (Larger grain borer) (Figure 14.45)

Pest status. A very serious pest of recent origin, after having been accidentally introduced into East Africa; now a serious threat to smallholder peasant farmers in Africa, especially since it can attack relatively dry grain in storage. A minor pest in native Central America.

Produce. A pest of stored maize and dried cassava tubers in on-farm stores mostly. Softer grains are especially vulnerable. It can damage a wide range of grains and seeds, but can only breed on maize and cassava.

Damage. Adults bore the grains and cassava and create a lot of dust. Females lay their eggs in tiny chambers cut at right angles to the main tunnel; some larvae develop inside the grains, and some in the dust. Thus both adults and some larvae feed on the produce and cause damage. Sometimes structural timbers may be bored. Grains are eventually hollowed out, and cassava roots bored to such an extent that they are reduced to dust internally (as a hollow shell); cassava losses of 70% after only 4 months' on-farm storage have been reported. The damage is done by the insects biting and chewing with their mandibles.

Life history. Eggs are laid in short side tunnels inside the grains in store; each female lays 50-200 or more eggs over a 12-14 day period; hatching occurs after 3-7 days at 27°C, or 4 days at 32°C.

Larvae are white and scarabaeiform with distinctly larger thoracic segments; they feed mostly on the dust (flour), and they moult 3-5 times (usually 3). Fully-grown larvae measure 4 mm, and development takes some 27 days at 32°C and 80% RH, when fed on maize grain. At 32°C, larvae developed when the grain mc was as low as 10.5%, and the time required was increased by 6 days; field studies in Tanzania showed maize with mc as low as 9% being heavily infested.

Pupation occurs inside a pupal case in the frass and powder, or inside the grains, and requires 5 days at 32°C. Adults are small dark cylindrical beetles, 3-4 mm long; the large hooded prothorax is densely and coarsely tuberculate; antennae of 10 segments have a large loose club of three segments. Their characteristic feature is that the elytral posterior declivity is steep, flat, and limited by a distinct carina posteriorly and the surface is tuberculate. Adults fly and may start field infestation of maize prior to harvest; they usually live for 40-60 days.

The life cycle on maize grains can be completed in 32 days at 27°C and 70% RH (25 days on cobs at 32°C), whereas on cassava it is recorded taking 43 days (Hodges, 1986).

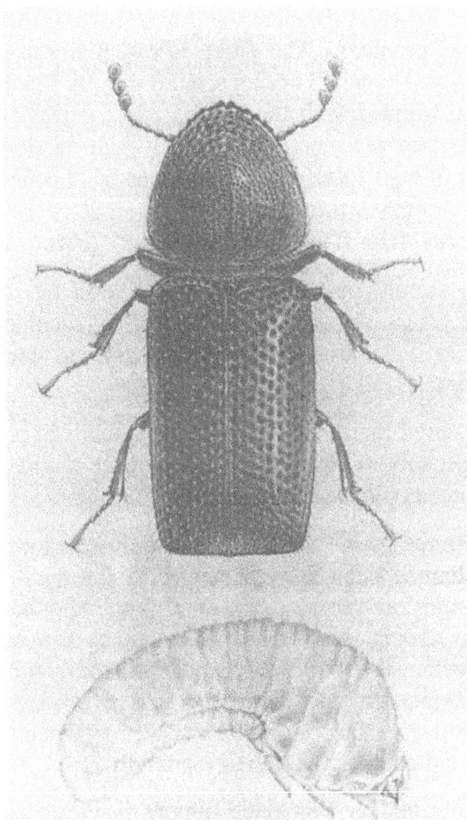

Figure 14.45. Prostephanus truncatus (Larger grain borer) (Coleoptera; Bostrychidae); dorsal view of adult beetle showing abrupt posterior termination (courtesy of ICI).

Distribution. Native to Central America, it has spread to South America, southern USA, and in the 1970s was accidentally introduced into Tanzania, and has since spread to Kenya, and in 1984 was found in Togo. The spreading distribution is a serious threat to smallholder farming in tropical Africa.

Natural enemies. A number of predacious and parasitic insects are found in infestations, but their precise relationships are not yet known – the topic is the subject of extensive research in the hope that this pest might be controlled biologically.

Control. The shelling of maize prior to storage does reduce the extent of the damage, as also does the growing of the old flinty varieties rather than the new high-yielding softer grained varieties. Male-produced aggregation pheromone is being used for population monitoring. ICI recommend 'Actellic Super', a mixture of pirimiphos-methyl and permethrin, for chemical control. Other chemicals currently being used are permethrin, deltamethrin and fenvalerate.

Several natural enemies are being released in Africa in an attempt to control populations – with some success. See Dick (1988)

Rhizopertha dominica (Fab.) (Lesser grain borer) (Figure 14.46)

Pest status. A serious pest of stored grains and other foodstuffs worldwide; a primary pest of stored grain; occasionally recorded infesting ripe cereals in the field. An important penetrative species.

Produce. Cereal grains in store, cereal products, flours, and dried cassava; to some extent pulses may also be eaten.

Damage. Eating of the grains and food material is the main damage; on intact grains, the adults show a preference for the germinal region; such selective damage can be quite serious economically.

Life history. Each female lays 200-500 eggs, either dropping them loosely into the produce or else laying them in crevices on the rough surface of seeds. More eggs are laid at higher temperatures, and oviposition may continue for up to 4 months. Hatching occurs after a few days.

The larva is white and parallel-sided, with a small head and quite prominent legs; the first instar has a distinctive median posterior spine. There are 3-5 larval instars, and development takes about 17 days (34°C and 70% RH) on wheat. Development is more rapid on cereal grains than on flours. Newly hatched larvae may feed on flour dust created by the adult beetles, but usually bore into the whole grains, which are eventually hollowed out. Larvae can develop on grain with a low mc – at 34°C they can develop on grain with a mc as low as 9%, although mortality is high. It is clear that there is interaction between mc and temperature in controlling the rate of larval development.

Pupation usually takes place inside the damaged grain, and takes 3 days at 34°C and 70% RH.

Adults are small (2-3 mm) dark cylindrical beetles, with the head hidden under the large hooded, tuberculate prothorax. Antennae are 10-segmented, with a large, loose, 3-segmented club. The abdominal sternites are used for sexual distinction. The elytra have rows of punctures with short setae; apically the elytra are rounded, and there are no other ornamentations, nor a terminal carina. Adults are long-lived, feed extensively, and fly quite well. They are seldom obvious in infestations as they are usually inside the infested grains, together with the larvae.

The life cycle is completed most rapidly when feeding on grains at a high temperature (3-4 weeks at about 34°C and 70% RH). Development is most rapid on broken kernels, then whole grains; flours are not preferred.

Distribution. Basically a tropical species, but cosmopolitan throughout the warmer parts of the world, and occurring in heated stores in temperate regions. During World War 1, infested wheat from Ausralia distributed this pest throughout the USA and many other countries.

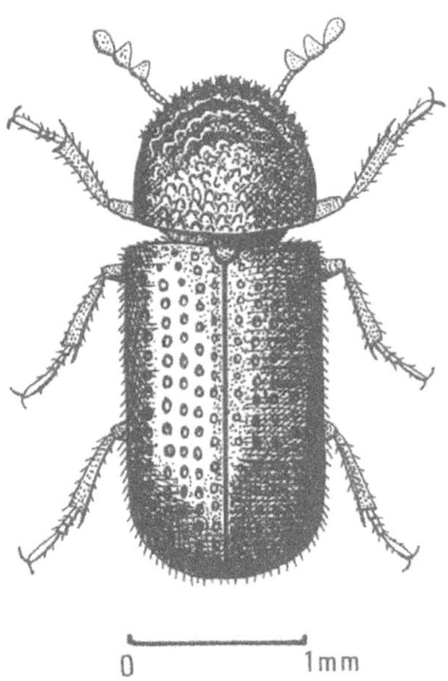

Figure 14.46. Rhizopertha dominica (Lesser grain borer) (Coleoptera; Bostrychidae), adult beetle, dorsal view.

Natural enemies. The usual stored product beetles parasites (Pteromalidae) are recorded, namely *Lariophagus, Chaetospila* and *Anisopteromalus.*

Control. Population monitoring for this pest usually relies on sack sieving and spear sampling, as the adults tend to be sedentary and will not move to traps. There is now a male-produced aggregation pheromone that might prove to be effective.

Dinoderus spp. (Small bamboo borers etc.)

Five species of these tiny beetles are recorded boring in dried cassava, as well as in bamboos and wooden structures. *D. minutus* (Fab.) seems to be the most important species, and is reported to breed in cassava, some maize varieties, and some soybeans. The genus is pantropical, but some species are confined to Asia. These are important penetrator species that damage packaging materials extensively.

Heterobostrychus spp. (Black borers)

These are large borers, measuring 7-13 mm, usually in wood, but in produce stores are recorded from structural timbers, cassava, potatoes, coffee beans, oilseeds and pulses, in S.E. Asia and parts of Africa.

Several other tropical and subtropical Bostrychidae are wood borers and may be encountered in food and produce stores in structural timbers, and occasionally the adults are recorded boring into stored products. These include: *Apate* spp. (Black borers), Africa, Israel, and tropical South America; *Bostrychopsis parallela* (Black bamboo borer), S.E. Asia; *Bostrychopolites* spp.; *Sinoxylon* spp.; *Stephanopachys* spp.; *Xylon* spp.; *Xyloperthella* spp.

14.10.9 Family Lyctidae (Powder-post beetles) (70 species)

Adults and larvae bore timbers and wood; some species are temperate and others subtropical; several species have been recorded associated with stored products but only one is at all common.

Lyctus brunneus (Stephens) (Powder-post beetle) (Figure 14.47)

This is thought to be of North American origin, but it is now cosmopolitan, and infestations in food stores are thought to have been due to their presence in plywood boxes. The beetles are elongate and variable in size, from 2-7 mm long. The larva is white and scarabaeiform, with a distinctive large spiracle on the eighth abdominal segment. Dried cassava and other root crops have been damaged in stores. Cypermethrin has been developed for timber protection.

14.10.10 Family Lophocateridae

An obscure small family, but *Lophocateres pusillus* (Klug) (Siamese grain beetle) is regularly encountered in stores in the tropics on a wide range of stored foodstuffs, recorded as being rice, other cereals, cassava, pulses, and groundnuts. But it is said to be only a minor pest associated with the primary pests. The adult is a small, flattened beetle, 2.5-3 mm long, like a tiny Cadelle (Figure 14.48).

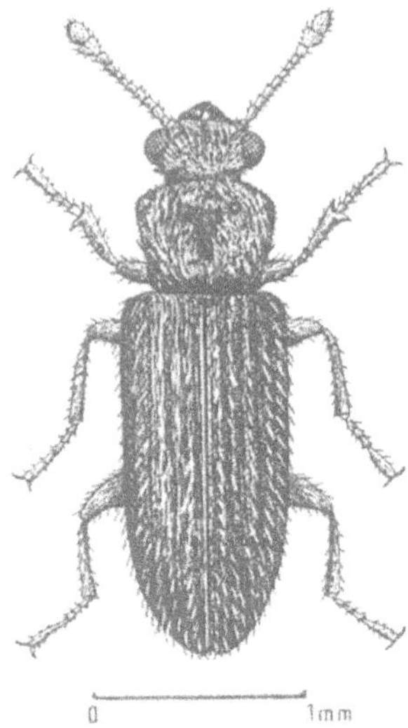

Figure 14.47. Lyctus sp. (probably brunneus) (Powder-post beetle) (Coleoptera; Lyctidae), adult beetle; Hong Kong.

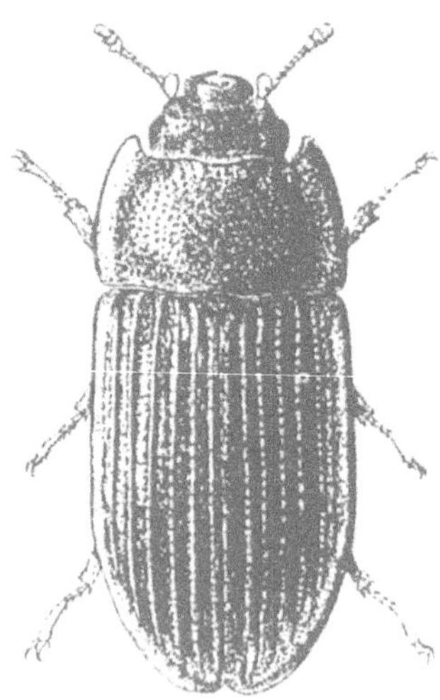

Figure 14.48. Lophocateres pusillus (Klug) (Siamese grain beetle).

14.10.11 Family Trogossitidae (600 species)

A tropical group of great diversity in both appearance and habits, but only one species is of any importance in the present context.

Tenebroides mauritanicus (Linn.) (Cadelle) (Figure 14.49)

One of the largest stored products beetles (5-11 mm long); dark brown or black in colour, and flattened in shape. Generally a minor pest on a wide range of produce, including both cereal grains and flours, and also oilseeds. Larvae often selectively eat the germ of grains. The larva is campodeiform and omnivorous (as is the adult), and will eat other insects in the stores. The last larval instar may bore into soft wood to construct a pupation site (a small chamber). Biting damage may be done to other soft containers and materials; it can be an important penetrator species; both adults and larvae feed, the adults readily eat other insects.

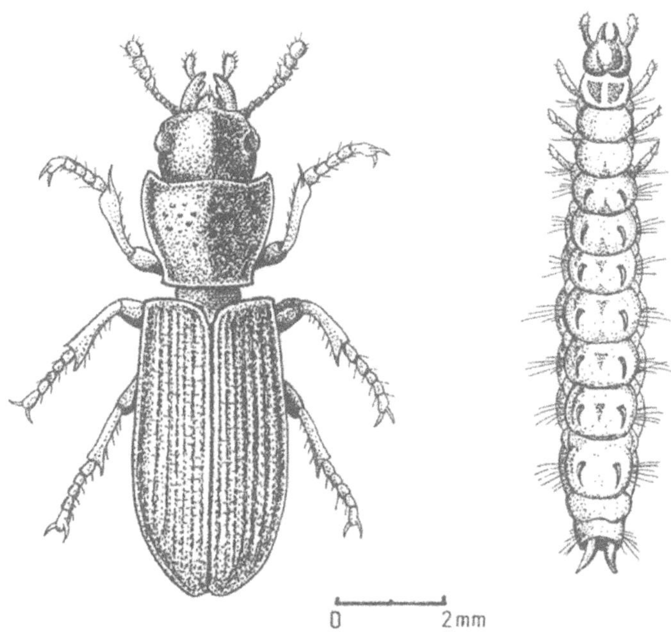

Figure 14.49. Tenebroides mauritanicus (Cadelle) (Coleoptera; Trogossitidae), adult beetle and larva.

Larval development is slow, taking 70-80 days in Washington in the summer, but it can take up to 300 days! Eggs are laid in batches (10-60 usually), either loosely or in crevices. Pupation takes place in a cell which may be hollowed out of softwood. Usually two generations per year in temperate regions, but may be three in the tropics.

14.10.12 Family Cleridae (Checkered beetles) (3,400 species)

Brightly coloured beetles, sparsely hairy, with a cylindrical prothorax narrowed posteriorly into a neck; found mostly in the tropics. They are basically predators, both adults and larvae, on larvae of wood-boring insects. Several species are recorded from stores, but only *Necrobia* survives on stored foodstuffs; in the wild, they eat animal corpses and the fly maggots they contain – a proteinaceous diet is clearly required.

Necrobia rufipes (Degeer) (Copra beetle; Red-legged ham beetle) (Figure 14.50)

A shiny green to blue beetle, 4-5 mm long, with basal segments of antennae, and legs red. A common species in the tropics and warmer parts of the temperate regions. Usually found feeding on copra, stored bacon and hams, and cheeses. Other insects in the produce are eaten, and it is also recorded on oil palm kernels, oilseeds, cocoa beans, spices, bones, dried fish, and some meat products. A mixed diet appears to be required. This species prefers warm (30-34°C) and dry conditions. Larval development takes about 17 days, and they leave the meat to pupate in dark dry crevices. Pupation takes 13 days.

Adults are very successful when reared on larvae of *Piophila casei* – they are longer-lived (280 : 130 days), and females lay many more eggs (900 : 140).

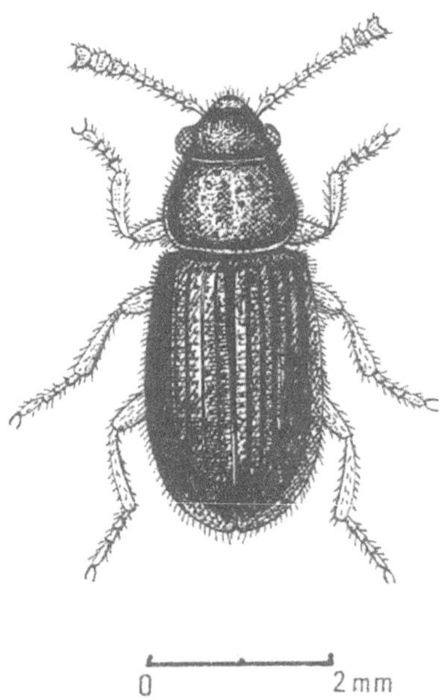

Figure 14.50. Necrobia rufipes (Copra beetle) (Coleoptera; Cleridae), adult beetle.

Control of infestations on drying fish in Africa was achieved using tetrachlorvinphos.

Figure 14.50a. Copra infested by Necrobia rufipes adults (courtesy of ICI).

Necrobia ruficollis (Fab.) (Red-necked bacon beetle)

This beetle has prothorax and basal quarter of the elytra red, and the rest of the body shiny blue or green; it is 4-6 mm long. It is also tropical, but most abundant in South America and Africa. Its diet includes fewer insects, and it is usually only found on animal produce.

Both species cause great annoyance to people handling infested copra or other products, and their presence alone on high value commodities such as hams or processed meats can lead to produce rejection and serious economic losses. The larvae bore into the meat, mostly in the fatty parts, and the adults are surface feeders.

Necrobia violacea (Linn.)

Metallic blue/green with dark legs, 4-5 mm long, now cosmopolitan and found on dead animals and dead fish mostly, and on dried meats.

Several other species are regularly recorded from stores, and occasionally they damage hams and skins (*Korynetes coeruleus* (Degeer)), or they prey on Anobiidae in stored foodstuffs (*Thaneroclerus buqueti* (Lefevre), and others).

14.10.12 Family Nitidulidae (Sap beetles) (2,200 species)

A large and diverse family, of variable form and habits, but most feed on plant sap or fermenting plant and animal material. A few species are pests of field crops, and a few are important on stored products; 22 species (mostly of *Carpophilus*) are recorded on foodstuffs. Most are found on dried fruit or on mouldy produce.

Carpophilus hemipterus (Linn.) (Dried fruit beetle) (Figure 14.51)

Pest status. A worldwide pest of importance on stored dried fruits.

Produce. Dried fruits mostly, but also on mouldy cereals (maize etc.); preferred fruits appear to be raisins, currants and figs.

Damage. Direct eating of the produce is the main damage, done by both adults and larvae. But this beetle flourishes when humidity is high, and often the produce is mouldy. They often act as vectors for fungi and bacteria that cause fruit spoilage. Sometimes the beetles are processed along with the fruit, and in canned fruit the contamination is economically very serious. The presence of this species is often an indicator of damp mouldy conditions in the store, especially if they persist long after harvest. With heavy infestations, contamination of the produce is a major hazard.

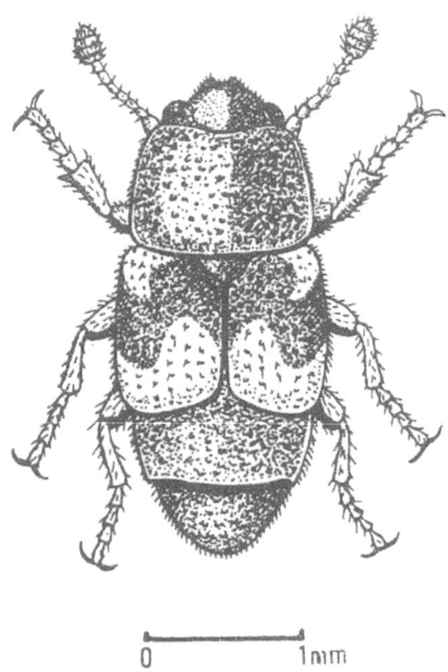

Figure 14.51. Carpophilus hemipterus (Dried fruit beetle) (Coleoptera; Nitidulidae), adult beetle.

Life history. Each female lays on average 1,000 eggs, which hatch in 2-3 days.

There are 3 larval instars which take 6-14 days to develop. The larvae feed on either the stored produce or the fungal mycelium; they appear to have difficulty in penetrating undamaged fruit. Larvae are campodeiform, with short legs and two pairs of small projections at the tip of the abdomen.

Pupation takes place in the produce or on the surface of bags, and requires 5-11 days.

Adults are flat and oval, about 2-4 mm long; dark brown with yellow patches on the elytra; the last two abdominal segments are uncovered by the elytra. They are active insects and fly well (flights of 3 km have been recorded), and they live for 3-12 months or more.

The life cycle can be completed in 12 days, at 32°C and 75-80% RH, but at 18°C it takes 42 days; there may be many generations per year.

Distribution. A cosmopolitan species, but is not common in the cooler temperate regions of the world.

Carpophilus dimidiatus (Fab.) (Corn sap beetle)

A cosmopolitan pest of a wide range of stored foods, especially dried fruits and cereals, but it rarely attacks intact grains. In the wild it feeds on decaying fruit and vegetables, and on sap exuding from injured plants. The most favourable conditions for development seem to be 22-25°C and 70% RH.

The other species of *Carpophilus* recorded from stored products include:

Carpophilus freemani Dobson. Pantropical; on maize, rice, Brazil nuts, fresh vegetables, and bones.

Carpophilus fumatus Boheman. Africa; usually a field pest of fruits, cotton and some cereals.

Carpophilus ligneus Murray. Cosmopolitan; on dried fruits, cocoa beans, oilcake and cereals.

Carpophilus maculatus (Murray). S.E. Asia, Australasia, Pacific, and West Africa; on dried fruits, cereals etc.

Carpophilus obsoletus (Erichson); Pantropical; on dried fruits, cereals etc.

Carpophilus pilosellus Mots. Pantropical; on a wide range of produce.

Carpophilus spp. (8) are recorded in different parts of the world.

Urophorus humeralis (Fab.) (Pineapple sap beetle)

This pantropical species has 3 abdominal segments visible beyond the truncate elytra. It is a field pest of pineapple, and can contaminate tinned pineapples, but is a minor pest in stores generally, on damaged maize, dried fruits and dates.

Brachypeplus spp.

Several species are recorded from East and West Africa, on maize cobs, cocoa and castor beans.

Several other species of Nitidulidae are recorded occasionally on stored cereals and spices.

14.10.13 Family Cucujidae (Flat bark beetles) (500 species)

A difficult family taxonomically, formerly including the Silvaniidae; they are small (1.5-2.5 mm) flattened beetles with long antennae; most live under loose tree bark where they feed on dead insect remains in spiders' webs, or else are predacious; some feed on plant debris, and a few are stored products pests (*Cryptolestes* species).

Cryptolestes ferrugineus (Stephens) (Rust-red grain beetle) (Figure 14.52)

Pest status. A common and widespread pest in the warmer parts of the world in stored grains. It is a secondary pest, for small larvae cannot penetrate intact grains but can attack even very slightly damaged ones. Often pest populations consist of more than one species of *Cryptolestes*.

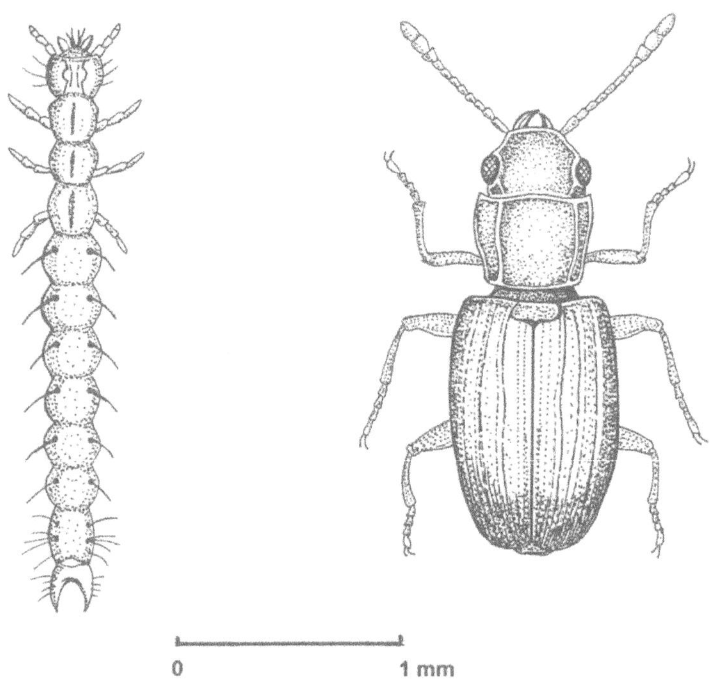

Figure 14.52. Cryptolestes ferrugineus (Rust-red grain beetle) (Coleoptera; Cucujidae), larva and adult beetle.

Produce. Stored grains of all types, and flours; often found in flour mills; also on dried fruits, nuts, oilcake, cassava and other produce.

Damage. Essentially secondary pests, but the larvae can penetrate grains if they are even only very slightly damaged. In grains they show a preference for the germ region, thus causing a loss of quality and reducing germination.

Life history. Each female lays up to 200 eggs, into the stored produce.
The larvae are campodeiform and have distinctive 'tail-horns'.

Adults are small (2.5 mm), flat and elongate, pale reddish-brown, with long filiform antennae (longer in the male); head and prothorax are somewhat disporportionally large and conspicuous; adults are winged but seldom recorded to fly.

The life cycle can be completed in 17-26 days, at 38°C and 75% RH, but takes 70-100 days at 21°C; optimum conditions appear to be about 33°C and 70% RH, when the life cycle takes 23 days. Apparently, the species can successfully overwinter in temperate climates.

Distribution. A cosmopolitan species found throughout the warmer regions of the world, and also in some parts of more temperate areas in the UK, Europe and North America.

Control. Adults cannot climb clean glass, so pitfall traps can be used for population monitoring; and they can be caught in refuge traps.

The other species of *Cryptolestes* recorded from stored products include:

Cryptolestes capensis (Waltl). North Africa and Europe; will tolerate rather drier conditions than the other species; mostly found in flour mills.
Cryptolestes klapperichii (Horn). Malaysia; common on stored cassava chips.
Cryptolestes pusilloides (Steel & Howe). Australia, East and South Africa, South America; it prefers moist conditions, and feeds on a wide range of foods including cereal grains and wheat products.
Cryptolestes pusillus (Schonberr). Humid tropics; in grain stores and flour mills, also feeds on copra, nutmegs, coffee berries, bulbs and cotton seed.
Cryptolestes turicus (Grouvelle). A temperate species in Europe and the UK, this species predominates in flour mills, but in the USA it is recorded from intact grain; also recorded on nutmegs, copra, cocoa and chillies, and in flour mills.
Cryptolestes ugandae (Steel & Howe). East and West Africa; on maize, sorghum, millet, cowpeas, cassava and groundnuts, especially at high humidities.

14.10.14 Family Silvanidae (Flat grain beetles) (400 species)

A small group with diverse habits; some are phytophagous, and some are predacious and often found under loose tree bark; two species are common on stored foodstuffs, and two more are of insignificance as pests. The beetles are small (2-4 mm), flattened, elongate and parallel-sided.

Oryzaephilus mercator (Fauvel) (Merchant grain beetle)
Oryzaephilus surinamensis (Linn.) (Saw-toothed grain beetle) (Figure 14.53)

Pest status. These two closely related species are regular pests of stored foodstuffs in all parts of the world, but most abundant in warmer regions. Their small size enables them to hide easily, and light infestations may be overlooked; they are basically secondary pests.

Produce. *O. mercator* shows a preference for oilseeds, and *O. surinamensis* for cereals and cereal products; also found on copra, nuts, spices, and dried fruits.

Damage. Eating the produce is the main form of direct damage, but the larvae bore into damaged grains to feed selectively on the germ, and they attack the germ region of intact grains. Packing materials may be damaged.

Life history. Each female lays 300-400 eggs, usually singly, at a rate of 6-10 per day; hatching requires 4-12 days.

Larvae are white, elongate, flattened; fully grown, they measure 4-5 mm; there are 2-4 larval instars according to conditions. They are mobile, and will bore into damaged cereal grains or attack the germ region of whole intact grains; they penetrate packing materials quite easily.

Pupation takes place in the produce, and requires 5-15 days, usually.

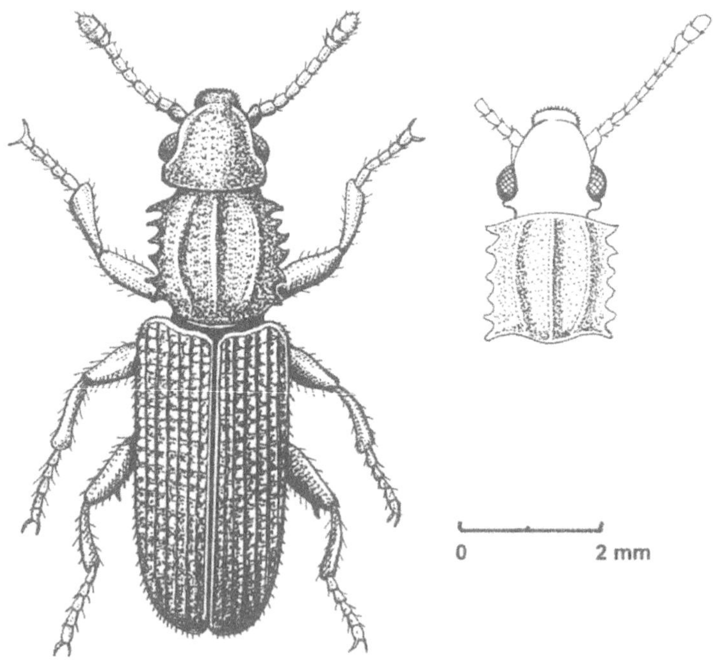

Figure 14.53. Oryzaephilus surinamensis (Saw-toothed grain beetle) (Coleoptera; Silvanidae), adult beetle; also head silhouette of O. mercator.

Adults are small slender brown beetles, 2.5-3 mm long, with short clubbed antennae; the prothorax has six tooth-like lateral projections. The two species are separated by the length of the temple, this being much shorter in *O. mercator*. Adults feed and are quite long-lived; they are very active and wander widely, and they have wings but seldom fly. *O. surinamensis* is recorded to survive periods of low temperature (down to 0°C) and to be able to overwinter in unheated stores in temperate countries, but *O. mercator* is more tropical.

The life cycle can be completed in 20 days at 37°C, or 80 days at 18°C; optimum conditions are recorded for *O. surinamensis* as being 30-35°C and 70-90% RH, and for *O. mercator* as 30-33°C and 70% RH; but *O. surinamensis* is apparently more tolerant of extremes of conditions of temperature and humidity, and is even said to survive exposure to sub-zero conditions for several days.

Distribution. *O. mercator* is widespread throughout the tropics and subtropics, while *O. surinamensis* is quite cosmopolitan, including temperate regions of the world.

Natural enemies. They are preyed upon by *Xylocoris* bugs and *Pyemotes* mites, parasitized by *Cephalonomia* wasps, and attacked by various protozoa and viruses.

Control. They are able climbers and escape from pitfall traps, and generally avoid refuge and bag traps.

Their mobility and small size enable them to hide successfully, and they may be difficult to reach by pesticide application.

Ahasverus advena (Waltl) (Foreign grain beetle) (Figure 14.54)

Pest status. This is basically a scavenger, feeding on animal and plant detritus, including dead insects and damaged foodstuffs, and fungal mycelium. A cosmopolitan species, most abundant when the produce is mouldy.

Produce. A wide range of damaged produce is fed upon, including cereal grains, flours, cereal products, cocoa, groundnuts, copra, palm kernels, dried fruit, herbs, spices and roots. It can be used as an indicator of damp storage conditions.

Damage. A secondary pest at most, and more often just a scavenger, or else feeding on the fungal mycelium of mouldy produce.

Life history. This species is not well studied, but the life cycle can be completed in 22-30 days at 30°C and 70% RH, and apparently it does not breed at less than 65% RH.

Pupation takes place in a chamber attached to the substrate, and takes 4-7 days. Adults are small (2-3 mm) but with a wider body, and the square prothorax bears only single apical teeth; they are active and strong fliers, and long-lived (160-200 days).

Several other species of Silvanidae are occasionally recorded from cocoa beans, groundnuts, copra, nuts, rice and other produce, in parts of Africa and South America.

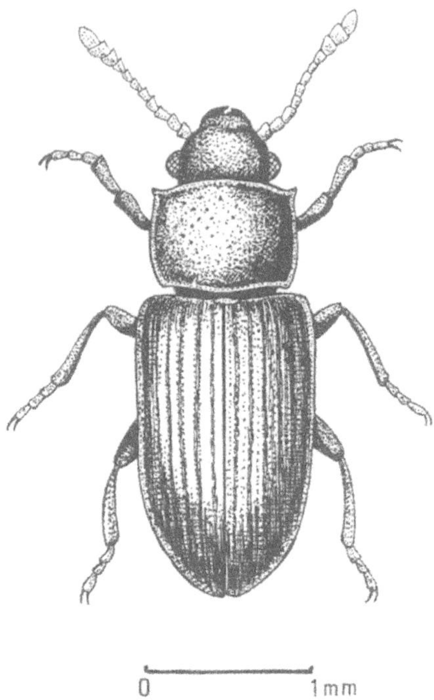

Figure 14.54. Ahasverus advena (Foreign grain beetle) (Coleoptera; Silvanidae), adult beetle.

Cathartus quadricollis (Square-necked grain beetle)

A cosmopolitan pest found in wheat, barley, rice, cacao, tobacco, oil palm kernels, ripe and dried fruits. In the USA, it attacks maize in the field. Not an important species, but occasionally encountered in food stores. Development can be completed in 20 days at 28°C and 80-85% RH; viability is negligible below 20°C or above 30°C, and at any humidity less than 65%.

14.10.15 Family Cryptophagidae (Silken fungus beetles) (800 species)

A small group of fungus-feeding beetles; most are small (1.5-4 mm) and pubescent; many live in birds' nests and animal lairs; others live in flowers, under bark and in fungi; body coloration varies considerably, and usually changes after death.

Two genera are recorded in stored produce; their presence serves as an indicator of damp and unhygienic storage conditions.

Cryptophagus spp.

Small beetles with a downy pubescence, and the prothorax has a flattened apical projection and a lateral (marginal) tooth at mid-level. At least seven species are recorded, on a wide range of produce including mouldy cereals, nuts and fruits.

Larvae are characteristically campodeiform. Some species are more temperate in distribution, and several occur in the UK.

Henoticus spp.

These beetles have a series of lateral teeth on the prothorax, and no flattened apex; at least one species is subtropical, and another is temperate.

14.10.16 Family Languriidae (400 species)

A small group of phytophagous beetles, mostly found in Asia and North America; the adult beetles are to be seen on flowers and leaves, usually. A few species are recorded in stored produce throughout the world, but are of little direct importance. The two species mostly encountered are *Cryptophilus niger* (Heer) and *Pharaxontha kirschii* Reitter.

14.10.17 Family Lathridiidae (Plaster beetles; Minute brown scavenger beetles) (600 species)

A small but widespread group of fungus beetles, tiny in size (1-3 mm) and brown or black in colour. The name 'plaster beetle' comes from their often being found on damp mouldy walls in old buildings. The number of species recorded from produce stores totals about 35 in seven genera, but they are difficult to identify. The species most regularly recorded are probably:

Adistemia watsoni (Wollaston).
Aridius spp.
Cartodere constricta (Gyllenhal).
Corticaria spp.
Dienerella spp.
Lathridius 'pseudominutus' group.
Thes bergrothi (Reitter).

A key to these species is given in Mound (1989).

14.10.18 Family Mycetophagidae (Hairy fungus beetles) (200 species)

A small group, resembling small Dermestid beetles, densely pubescent, 1.5-5 mm long, brown or black, and often with spots on the elytra. They are fungus-feeders, and three genera are found in food stores where their presence indicates damp conditions and mouldy produce; these are listed below:

Litargus balteatus (Leconte). Cosmopolitan; on a wide range of nuts, cereals, pulses, cocoa etc.

Mycetophagus quadriguttatus Mueller. A temperate species, recorded on damp grain or grain residues.

Typhaea stercorea (Linn.) (Hairy fungus beetle) (Figure 14.55). Cosmopolitan, but most abundant in the tropics, on a wide range of mouldy foodstuffs and cereals.

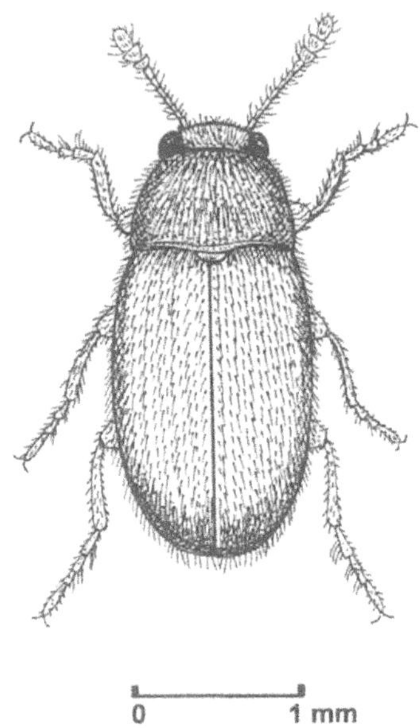

Figure 14.55. Typhaea stercorea (Hairy fungus beetle) (Coleoptera; Mycetophagidae), adult beetle.

14.10.9 Family Tenebrionidae (15,000 species)

A large and widespread group, found throughout tropical and temperate regions (as different species); there is some variation in body shape amongst adults, but the larvae are remarkably similar – they are called false wireworms because they resemble wireworms (Elateridae). Most feed on decaying plant material, but a few on living plants, and some are predacious. They are small to moderate in size (3-12 mm), and many are rather elongate and parallel-sided; antennae have 11 segments (usually) and often a distinct club. The eyes are partly divided horizontally by a backward projection of the side of the head (the genal canthus); and the tarsal formula is 5:5:4.

More than 100 species are recorded from stored foodstuffs worldwide but only a few are of importance. Most species are not suited to the dry conditions that characterise good produce stores.

Alphitobius diaperinus (Lesser mealworm)

A cosmopolitan insect commonly associated with damp or mouldy grain and cereal products. Abundant in (chicken) broiler houses, especially in the UK, where both adults and larvae can be seen eating carcasses of dead birds. Under good conditions, the life cycle can be completed in 40 days at 38°C to 90 days at 16°C, but adults can live for a year or more. Adult females tend to lay 1-4 eggs per day for the whole of their life. Adults are 5-7 mm in length, and resemble *Tribolium* in appearance. A major pest in parts of India.

Two other species of *Alphitobius* are known from stored foodstuffs.

Blaps spp. (Churchyard beetles)

Larger beetles, 15 mm long; with a more rounded body form; cosmopolitan; recorded on cereals; generally rather rare but very conspicuous.

Blapstinus spp.

A few species recorded in the West Indies and Central America on cereals and also on dried fruit and cereal products.

Cynaeus angustus (Larger black flour beetle)

A fairly common pest in farm-stored grain in the North Central states of the USA. Eggs are laid singly, up to a total of about 400 per female. Maize is the preferred host, and undamaged kernels can be penetrated. Development takes about 55 days at 30°C and 14% mc; fungus is an important element in the diet.

Gnathocerus cornutus (Fab.) (Broad-horned flour beetle) (Figure 14.56)

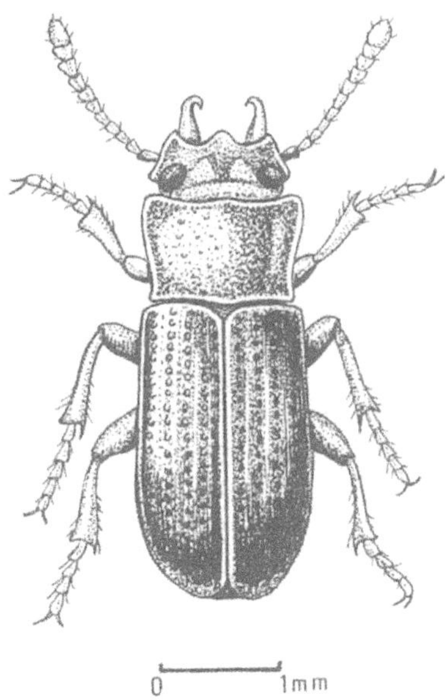

Figure 14.56. Gnathocerus cornutus (Broad-horned flour beetle) (Coleoptera; Tenebrionidae), adult male beetle.

A minor secondary pest of cereals, widespread around the world in both tropical and temperate regions; also found in flour mills; recorded on a wide range of produce. The male beetle has a pair of broad mandibular horns; body length is 3.5-4.5 mm; colour red-brown; females resemble *Tribolium*; both adults and larvae eat other insects, especially the pyralid moths (*Cadra* etc.). Development is successful between 15° and 32°C; larval stages occupy about 39 days, with pupation a further 10. Adults live for 8 months under optimum conditions, and the females lay up to 360 eggs.

Gnathocerus maxillosus (Fab.) (Narrow-horned flour beetle)

Mentioned separately because this species is restricted to the tropics; the male horns are slender. Often recorded from maize, and occasionally found on ripe crops in the field. Also found on wheat, other cereal grains, seeds of various types, pulses and nuts.

Larval instars number 7-9, and development takes 30 days (27°C). Temperature range is 17.5-35°C, and successful development can occur at 7%-100% RH according to temperature. At 30°C, the development period is 35-40 days, at 20°C it is 115-132 days.

Gonocephalum spp. (Dusty brown beetles)

A large genus of flattened brown beetles, moderate in size at 12 mm, found in the tropics and subtropics; some are pests of field crops, some are more predacious, and some recorded from stores on cereals and oil palm kernels.

Latheticus oryzae Waterhouse (Long-headed flour beetle) (Figure 14.57)

Pest status. A serious pest of stored cereals in S.E. Asia, and other parts of the tropics; but it is essentially a secondary pest.

Produce. Stored cereals and cereal products, but less damaging on maize; also found on oilseeds and rice bran.

Damage. A secondary pest on intact cereal grains, but serious damage is done to cereal products and flours. Indirectly, as with other Tenebrionidae, the infestation is accompanied by a persistent unpleasant odour, due to the secretion of benzoquinones from a pair of defensive glands on the abdomen.

Life history. The number of eggs laid seems to be small – observed females laid only 5.6 on average over a three day period; eggs hatched in 3-4 days, under optimum conditions of 35°C and 85% RH. Larval development through 7 instars took 15 days, followed by a pupal period of 3-4 days. The lower limits of temperature and humidity for development appear to be 25°C and 30% respectively.

The adult is small (2.5-3 mm long); the terminal antennal segment is smaller than the others, and the head is longer in proportion to the body than in *Tribolium*; the body is yellow-brown.

Distribution. Generally distributed throughout the tropics.

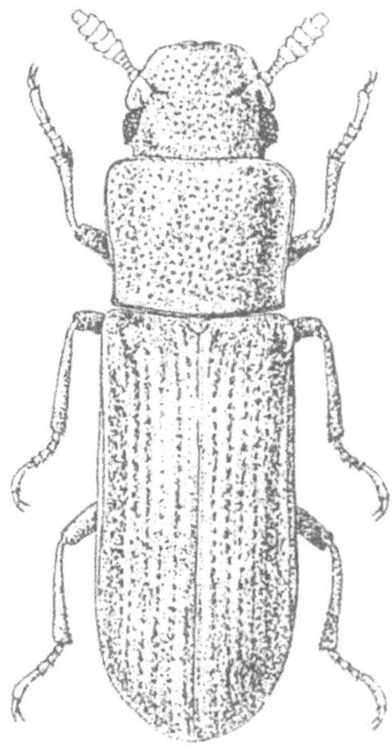

Figure 14.57. Latheticus oryzae, adult; length 3 mm.

Palorus spp. (Flour beetles) (6 species)

These are minor pests, but are regularly encountered in the tropics on a wide range of produce, including cereal grains, cereal products (meals, flours etc.); some species appear to be of Oriental origin, and others African. Most are found in association with *Sitophilus* infestations, and they appear to feed on the frass produced by the weevils.

Two species are more abundant than the others. They are:

Palorus ratzeburgii (Small-eyed flour beetle) is cosmopolitan and can be found in the UK in maltings in grain, flour, meal and barley; it thrives at temperatures between 20 and 35°C, when development can be completed in 55 to 18 days; pupation takes 4 days.

Palorus subdepressus (Depressed flour beetle) nearly cosmopolitan, but absent from the northern temperate region, and recorded from most cereals, flours and nuts, as well as pepper and ginger. Common in granaries, warehouses, feed and flour mills.

Tenebrio molitor (Linn.) (Yellow mealworm beetle) (Figure 14.58)

Pest status. A very conspicuous insect, but with a slow rate of reproduction, found only in temperate countries, so not often of much importance as a pest, especially since populations are typically quite small.

Produce. Grain debris, and cereal products, and material of animal origin, including dead insects. May be found in chicken houses.

Damage. Direct eating of farinaceous materials, and indirectly – they impart a strong odour to the infested foodstuffs. But in food stores they seldom occur in large numbers, so the damage done is very limited.

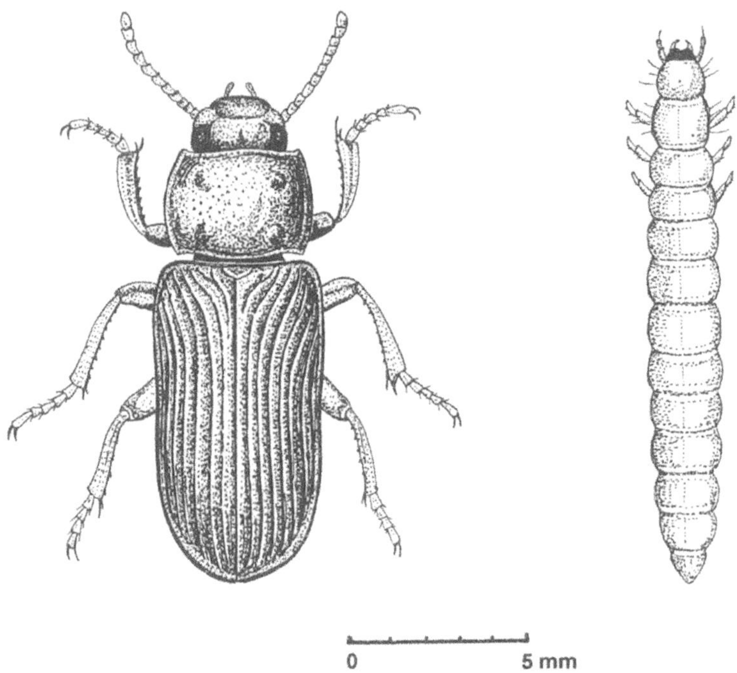

Figure 14.58. Tenebrio molitor (Yellow mealworm beetle) (Coleoptera; Tenebrionidae), adult beetle and larva (mealworm); Hong Kong.

Life history. Each female may lay up to 500 eggs, singly or in small groups; the eggs are sticky and soon become covered with debris. Hatching requires 10-12 days at 18-20°C.

The larvae are slow to develop; they pass through 9-20 instars, taking 4-18 months; fully grown, they measure up to 28 mm. The body is yellow-brown in colour, smooth and cylingdrical; it is referred to as a mealworm – this species is often reared commercially as food for cage birds and reptiles in captivity.

The pupa lies in the foodstuff, and requires some 20 days for development. At 25°C, larval development times can be reduced to 6-8 months, and the pupal stage to 9 days.

Adults are large (12-16 mm), dark reddish-black, flattened beetles, somewhat caraboid in appearance, but with a smaller head and short thick antennae; they live for 2-3 months.

The life cycle usually varies from 280 to 630 days.

Distribution. A cosmopolitan temperate pest occurring in small numbers throughout Europe, northern Asia and North America. Larvae are reared as animal food in Europe and North America.

Tenebrio obscurus (Fab.) (Dark mealworm)

A darker species but basically very similar to *T. molitor*. Both species overwinter as larvae, but this can develop a little faster – the larval stage can be appreciably shorter.

Tribolium castaneum (Herbst) (Rust-red flour beetle) (Figure 14.59)

Pest status. A major international pest of stored cereals and various foodstuffs, causing considerable financial losses; it has the highest rate of population increase recorded for any stored products pest, and adults are long-lived.

Produce. A wide range of commodities is fed upon by both the larvae and adults, including cereals, cereal products, nuts, spices, coffee, cocoa, chocolate, dried fruits and sometimes pulses. Dried milk and hides may also be damaged. Flours and other milled products are preferred foods.

Damage. Larvae and adults are secondary pests of cereal grains, and both show a preference for the germinal part of the grains; they penetrate deep into the stored produce. Cannibalism and predation are practiced by this species; eggs and pupae are cannibalized by adults, and both adults and larvae prey on all stages of Pyralidae in stored products and also on *Oryzaephilus*. Seed-borne fungi may also be important in their diet.

Life history. Each female can lay, on average, 150-600 eggs (up to 900 recorded), according to temperature (at 25°C and 32°C respectively); she lays 2-11 eggs per day for several months, or a year; hatching requires 2-3 days under optimum conditions (35°C and 75% RH).

Larvae are typically tenebrionid in appearance, have two terminal curved urogomphi; they usually pass through 7-8 instars (can be 5-11), and can pupate after 13 days.

Pupal development can be completed in 4-5 days.

Adults are small, flat, elongate, red-brown beetles, 3-4 mm long; they are winged and fly well. The beetles live for 6 months or more (up to 3 years), feeding continuously and mating frequently.

The life cycle can be completed in as little as 19-20 days, under optimum conditions (35°C and 75%+ RH); diet and climate are the main regulating factors;

predation of moth eggs, for example, increases the rate of development and reduces mortality. On wheatfeed, it can develop at temperatures from 20°C to 40°C, and RH 10-90%. This is a primary colonizer, and is often the first species to appear in a harvested crop in store, where infestation is caused by flying adults, although they are weak fliers.

Distribution. Worldwide in the warmer regions and regularly invading temperate countries where it survives for a while. It is the most commonly recorded invader in imported grain and foodstuffs in the British Isles. It is thought to have originated in India.

Natural enemies. Cannibalism is an important population controlling factor; male beetles show a preference for pupae, and the females for eggs. Many enemies are recorded; eggs are preyed upon by mites (*Blattisocius*), and larvae and adults are attacked by *Xylocoris* bugs. Mites (*Pyemotes* and *Acarophenax*) parasitize flour beetles, as do the wasps *Rhabdepyris* and *Cephalonomia.* Several parasitic amoebae are regularly recorded.

Control. Population monitoring can use pitfall traps, as well as bag and refuge traps; and they can be baited with the male-produced aggregation pheromone.

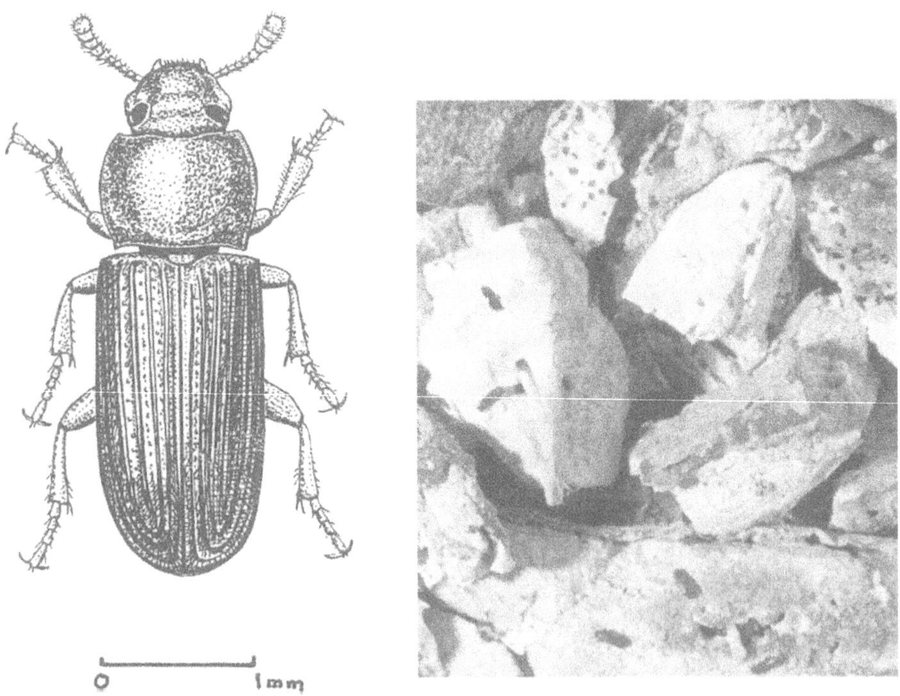

Figure 14.59. (a) Tribolium sp. (Red flour beetle) (Coleoptera; Tenebrionidae), adult beetle and (b) infested dried cassava tubers (courtesy ICI).

Tribolium confusum J. du Val (Confused flour beetle)

This is very similar to the previous species, but distinguished by features of the eyes and antennae. The species differs in that it thrives under slightly cooler conditions (2.5°C lower minima and maxima); it is a less successful species than *T. castaneum*, and there is often competition between the two. The centre of origin is thought to be Ethiopia; this species has spread further north, and is thus less abundant in the hot tropics. It may lay more eggs, and adults may live longer, than *T. castaneum*. Adults have wings but are not recorded flying.

Tribolium destructor Uyttenboogaart (Dark flour beetle)

A larger species, 5-6 mm long, black or very dark brown in colour, but otherwise it resembles the other two species. Found mostly in Europe and cooler (upland) regions in the tropics (e.g. the Highlands of Ethiopia, Kenya, Afghanistan).

Tribolium spp. (Flour beetles)

Several other species (at least six) are recorded from stored foodstuffs in different parts of the world, both in the tropics and in Northern Europe. Clearly in *Tribolium* infestations of foodstuffs, there may often be several closely related species involved, and they may be very difficult to distinguish.

14.10.20 Family Anthicidae (Ant beetles) (1,700 species)

A cosmopolitan group of phytophagous beetles, usually found in decaying vegetation; the adults have a narrow 'neck', and bear a resemblance to ants. Several species of *Anthicus* have been recorded from a wide range of stored products, but they are regarded as being unimportant.

14.10.21 Family Cerambycidae (Longhorn beetles) (20,000 species)

This enormous cosmopolitan family includes the main wood-boring species that attack trees, both living and dying. A few species are occasionally found in food stores and buildings when they have emerged from timbers used in the construction of the buildings or from packing cases. The larvae that tunnel in the wood generally take a year to develop, but the larger species and the species in cooler temperate regions may take up to 2-3 years, so infested timbers used in building construction can yield adult beetles months or even years later. A few species are polyphagous, but many Cerambycidae are quite host specific; many different species have been occasionally found in stores and buildings, depending upon the type of wood being used, but three species have probably been found more often than others.

Chlorophorus annularis (Fab.) (Bamboo longhorn) (Figures 14.60, 14.61)

One of the smaller species, at 14 mm (excluding antennae), feeding on sugarcane and the larger species of bamboo; S.E. Asia to Japan, and the USA.

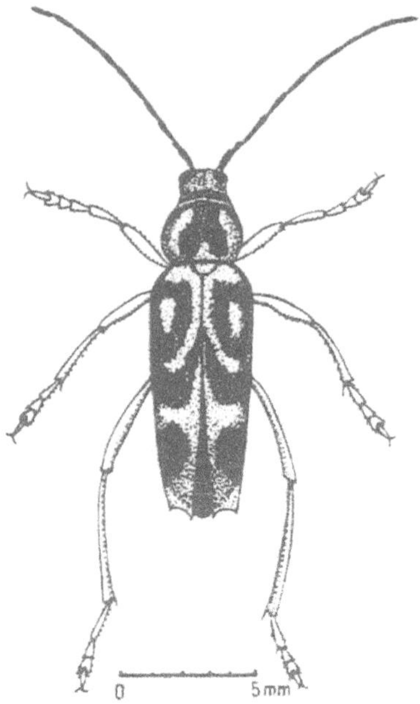

Figure 14.60. Chlorophorus annularis (Bamboo longhorn beetle) (Coleoptera; Cerambycidae), adult beetle; Hong Kong.

Figure 14.61. Bamboo pole bored by longhorn beetle larvae, which eat out the internode; from ceiling fixtures in a hotel restaurant, Hong Kong.

Hylotrupes bajulus (Linn.) (House longhorn beetle)

Attacks a wide range of conifers; Europe, Mediterranean Region, and South Africa.

Monochamus spp. (Pine Sawyers etc.) (Figure 14.62)

Several species bore in *Pinus* in Europe, Asia, and North America.

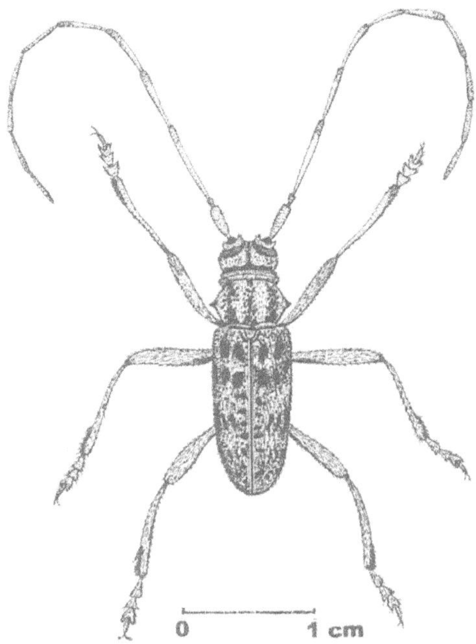

Figure 14.62. Monochamus sp. (Pine longhorn beetle; Pine sawyer) (Coleoptera; Cerambycidae), larvae found in domestic pine timbers; Hong Kong.

14.10.22 Family Bruchidae (Seed beetles) (1,300 species)

An interesting worldwide group, most abundant in the tropics, whose larvae develop inside seeds; most hosts belong to the Leguminosae, but other families are used – Southgate (1979) recorded 24 other families as hosts. Most of the species attack the growing crop, but they get carried into stores in the ripe pods and seeds, and some species can continue their development on the dry seeds in store, whereas others cannot and the infestation dies out in the store.

Adults are characterised as small stout beetles, with short elytra that do not completely cover the abdomen; the pronotum tapers anteriorly into a neck; male antennae are often serrate; eyes are typically deeply emarginated; hind femora are thickened and sometimes toothed. Adults are short-lived and take little or no food, but will drink water and nectar.

The species which are pests of stored products attack only pulses (the edible seeds of leguminous plants); some species are quite polyphagous, but others are found only on one host. Most species avoid the chemical toxins that develop in pod walls and seed testa by feeding only on the cotyledons.

Different species appear to have evolved on different continents, but now the main pest species are widely distributed throughout the world.

Acanthoscelides obtectus (Say) (Bean bruchid) (Figure 14.63)

Pest status. A serious stored products pest, adapted for life and reproduction in the dry conditions of produce stores, although many infestations may start in the field on the ripening seeds; it is multivoltine in produce stores on pulses.

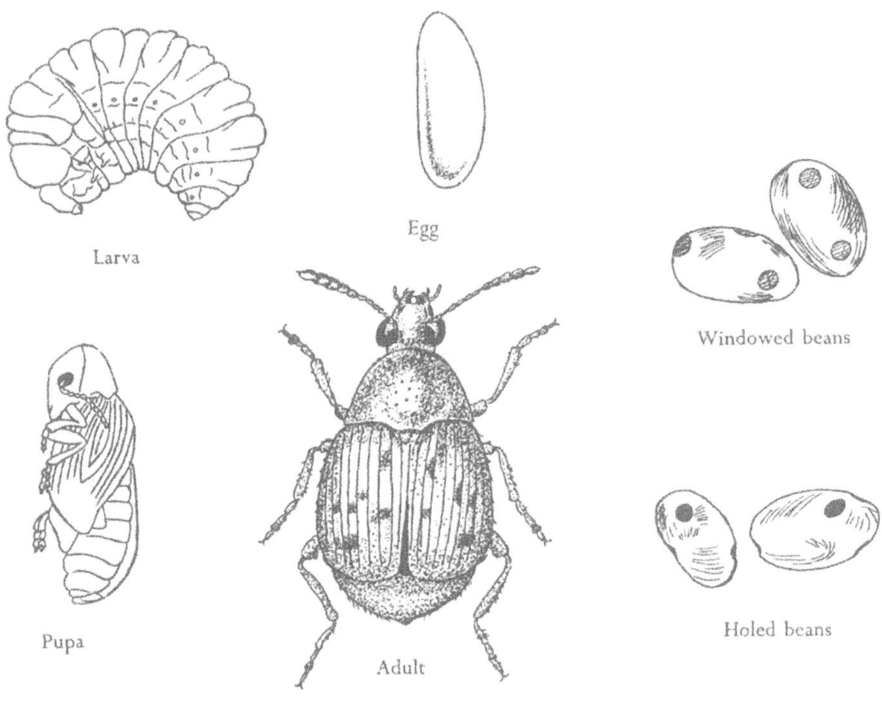

Figure 14.63. Anthoscelides obtectus (Bean bruchid) (Coleoptera; Bruchidae), drawing showing immature stages, from specimens in Nairobi, Kenya.

Produce. Most serious on *Phaseolus* beans, but it is recorded damaging many other different pulses in storage.

Damage. Direct eating of the cotyledons – there may be several larvae per seed; the rapid rate of development results in a high potential for population growth, and accumulated damage can be very extensive (Figure 14.64).

Life history. Eggs are laid either loosely in the produce or on pods in the field, or in cracks in the bean testa; each female lays on average 40-60 eggs (more than 200 are recorded); hatching takes 3-9 days. Many infestations start in the field, and the larvae feed on the ripening seeds; they also feed on fungi in the grain.

Larval development through four instars takes 12-150 days, according to conditions. Optimum conditions are about 30°C and 70% RH, but development will proceed slowly at temperatures as low as 18°C. The larvae are white, curved, thick-bodied and legless, and are found inside the pulse seeds.

Pupation takes place within a small cell inside the bored seed, behind a thin'window' composed almost entirely of testa (for easy emergence of the adult); pupation usually takes 8-25 days.

Adults are small, 2-3 mm long, stout, brownish-black with pale patches on the elytra; legs and abdomen are partly reddish-brown. The eyes are distinctly emarginate, and the hind femora have a large ventral spine, and two or three smaller ones distally. On unthreshed beans, the emerging adults will bite their way out through the pod wall (Figure 14.66). Adults fly quite well, and field infestations are started by adults flying from stores; but adults do not feed usually, and are short-lived.

Figure 14.64. Phaeolus beans attacked by Bean bruchid in storage; Harer, Ethiopia.

The life cycle can be completed in only 23 days, and so this species has a great potential for rapid population growth. Typically, there are often one or two field generations, followed by continuous breeding on the dried seeds in storage – six generations per year is usual in the Mediterranean Region. Because of its temperature tolerance, it can be found in the cooler highland areas in the tropics, and also in some temperate regions.

Distribution. Native to the New World (Central and South America), it has now spread throughout warmer regions of Europe and Africa, but its precise status in Asia is not clear, and it is not yet recorded from Australia.

A total of about 300 species of *Acanthoscelides* is known in the New World, and a few are found in food stores.

Natural enemies. Several parasitic Hymenoptera (Pteromalidae) are regularly recorded from most larvae of Bruchidae in stores, including *Anisopteromalus* and *Dinarmus.*

Control. Field infestation can be minimized by the use of clean (or fumigated) seed, or crop hygiene (especially destruction of crop residues) or, as a last resort, by insecticide application.

Flight traps can be used for monitoring adult bruchids, but pitfall traps are unsuitable as the adults can climb smooth surfaces.

Because of the size of pulse seeds (and hence the air spaces between), fumigation tends to be very successful.

Bruchidius spp.

Several species are regularly found on field crops in several parts of Africa and the Mediterranean Region, and they may be carried into produce stores; but they do not breed in the stores so the infestations die out. The adults resemble *Callosobruchus*; the host plants are mostly forage legumes (clovers and vetches).

Bruchus spp. (Pulse beetles)

Another group to be found on pulse crops in the field in temperate regions, and they are sometimes carried into produce stores; but they do not attack dried seeds, and store infestations peter out, although some adults may hibernate inside the seeds; but this is more common in the crop remnants left in the field.

The adults are 3-4 mm long, and their colour pattern is often somewhat variable; they are usually totally host-specific. The main species concerned are as follows:

Bruchus brachialis (Vetch bruchid). On vetches in North America.

Bruchus chinensis (Linn.) (Chinese pulse beetle). On pulses in China.

Bruchus ervi Froelich (Mediterranean pulse beetle). Mostly on lentils in the Mediterranean Region.

Bruchus lentis Froelich (Lentil beetle). On lentils etc. in the Mediterranean Region.

Bruchus pisorum (Linn.) (Pea beetle). (Figure 14.65) On pea crops; now cosmopolitan.

Bruchus rufimanus Boheman (Bean beetle). On field beans (*Vicia faba*) in Europe and Asia.

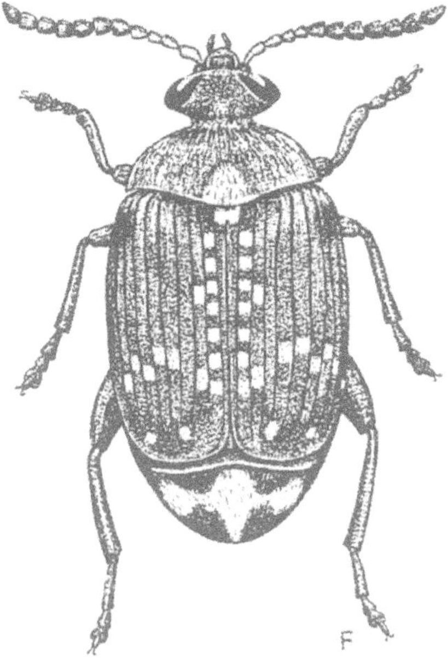

Figure 14.65. Bruchus pisorum (Pea beetle) (Coleoptera; Bruchidae), adult beetle from England.

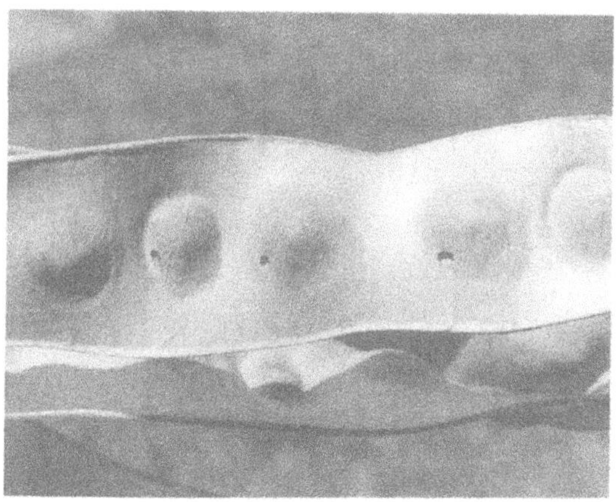

Figure 14.66. Pod of Acacia, with Bruchid emergence holes; Libya.

Callosobruchus spp. (Cowpea bruchids)

C. chinensis (Linn.) (Oriental cowpea bruchid) (Figure 14.67)
C. maculatus (Fab.) (Spotted cowpea bruchid) (Figure 14.67)

Pest status. These are serious pests of pulse crops, initially in the field and later in produce stores, throughout the warmer parts of the world.

Produce. Chickpea, lentil, cowpea, and *Vigna* spp. are the main hosts; neither species choose *Phaseolus*, and occasionally other pulses are attacked.

Damage. Larvae bore within the cotyledons and eventually hollow out the seed within the testa; typically, 1-3 larvae bore per seed. Infestations start in the field, and eggs are laid on the surface of maturing pods; later eggs are laid on the seed surface. Dried pods which are closed are resistant to attack.

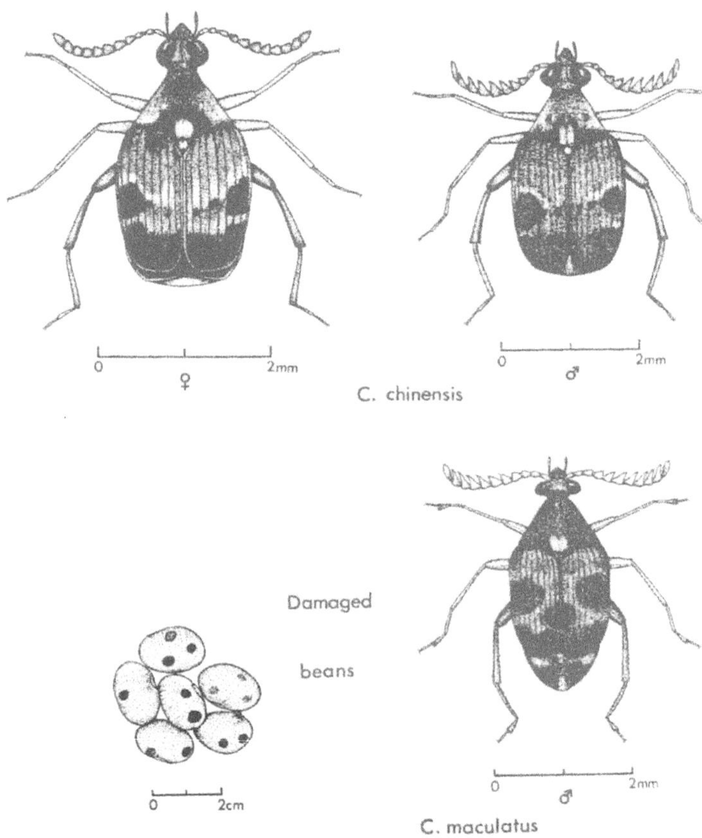

Figure 14.67. Callosobruchus chinensis (female and male) and C. maculatus (male) (Cowpea bruchids) (Coleoptera; Bruchidae) and beans with 'windows' and emergence holes.

Life history: Eggs are laid stuck on to the developing pod in the field, or on to the surface of seeds in dehisced pods, or on to seeds in store. Up to 100 eggs are laid per female, glued firmly on to the seed surface; incubation takes 5-6 days.

The hatching larva bites through the base of the egg, directly through the testa and into the cotyledons. The larva is scarabaeiform, and the 5 instars develop in about 20 days, the whole time being spent within the one seed. Optimum conditions for development are about 32°C and 90% RH.

Pupation takes place inside the seed in a chamber covered by a thin window of testa material, and requires about 7 days.

Adults are small brownish beetles, 2-3 mm long; *C. chinensis* are rather square in body shape, but *C. maculatus* are more elongate. Antennae are pectinate in the male and slightly serrate in the female; the hind femora have a pair of parallel ridges on the ventral edge, each with an apical spine (tooth). The markings on the elytra vary somewhat, but the dark patches can be quite conspicuous. The eyes are characteristically emarginated. As with the other species of Bruchidae, the elytra do not quite cover the tip of the abdomen. Adults fly quite well (usually up to one kilometer), but they do not feed on stored products, and thus are short-lived (up to 12 days, usually).

The life cycle can be completed in 21-23 days under optimum conditions (32°C and 90% RH), but at 25°C and 70% RH it takes 36 days; 6-7 generations per year are usual.

Distribution: *C. chinensis* is of Asian origin, where it is still the dominant species, and *C. maculatus* is thought to be African, but both are now widely distributed throughout the warmer parts of the world, but *C. chinensis* is not established in Australia.

Other species of *Callosobruchus* recorded as pests include:

Callosobruchus analis (Fab.) In parts of Asia; on *Vigna* spp. very similar to *C.maculatus*.

Callosobruchus phaseoli (Gyllenhal) Africa, parts of Asia and South America; on *Vigna* and *Dolichos labab*.

Callosobruchus rhodesianus (Pic.) An African species to be found on cowpea mostly; optimum conditions are 30°C and 70% RH when development takes about 25 days.

Callosobruchus sibinnotatus (Pic.) West Africa; on *Vigna subterranea*.

Callosobruchus theobromae (Linn.) India; on field crops of pigeon pea.

Caryedon serratus (Oliver) (Groundnut beetle) (Figure 14.68)

Pest status. A pest of groundnuts, especially when stored in their shells, in the warmer parts of the world, and also damaging to the pods of various tree legumes.

Produce. Groundnut is the main host, but also found in tamarind (*Tamarindus indica*) pods, and the pods of several other tree legumes (*Acacia, Cassia, Bauhinia* etc.).

Damage. The seeds inside the shell or pod are eaten by the developing larvae.

Life history. Eggs are laid stuck on to the outside of the pod, on groundnuts after harvest whilst drying in the sun. The hatching larva bores directly through the shell, and feeds upon the seed inside the pod.

The full-grown larvae usually leave the pod and pupate in a thin cocoon on the outer surface.

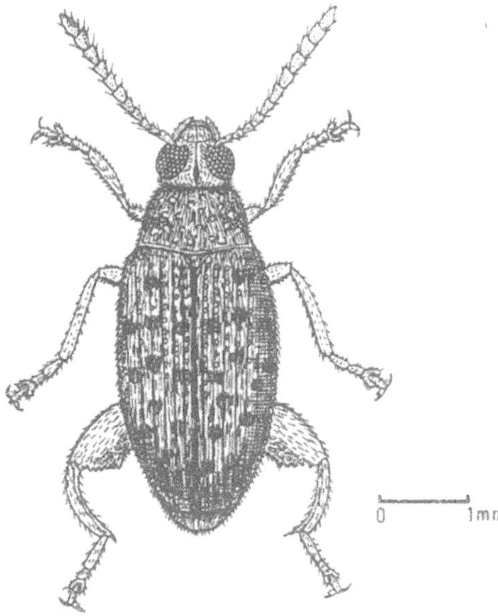

Figure 14.68. Caryedon serratus (Groundnut beetle) (Coleoptera; Bruchidae), adult.

Figure 14.69. Groundnuts bored by Caryedon serratus (courtesy ICI).

The adult is large (4-7 mm long), reddish-brown with small dark spots on the elytra; this species belongs to the group with normal eyes (i.e. non-emarginate); the hind femora have a comb of spines (1 large, plus 8-12 small ones distally), the tibiae are strongly curved.

The life cycle is completed in 40-41 days, under the optimum conditions of 30-33°C and 70-90% RH.

Distribution. Thought to be Asiatic in origin, but now widely distributed throughout the warmer parts of the world; however, it is only recorded as a serious pest of stored groundnuts in West Africa. Other species occur on other legumes in India.

Specularius spp.

Several species occur in parts of Africa, on cowpea, pigeon pea, and various wild legumes; it is thought that they are univoltine as stored products pests, but some have apparently been bred on dried pulses in laboratories.

Zabrotes subfasciatus Boheman (Mexican bean beetle)

Pest status. A major pest of beans in certain parts of the tropics.

Produce. Phaseolus beans are the usual hosts, but sometimes recorded on cowpeas, Bambarra groundnut (*Vigna subterranea*), and other legumes.

Damage. The pods are bored and the seeds eaten by the developing larvae.

Life history. The eggs are laid stuck on to the pods or on the testas of beans, and the larvae feed on the cotyledons. The optimum conditions for development are recorded as being about 32°C and 70% RH; the temperature limits are 20°C and 38°C.

The adult beetles are oval in shape, small (2-2.5 mm), with long antennae. The hind femur is without spines, but there are two moveable spurs (calcaria) at the apex of the hind tibiae.

The life cycle takes 24-25 days under optimum conditions.

Distribution. A New World species, and most important in Central and South America, but now widely distributed throughout the tropics, especially in Africa, India and the Mediterranean Region.

14.10.23 Family Anthribidae (Fungus weevils) (2,400 species)

A tropical group, most abundant in the Indo-Malayan Region, and the adults look like bruchids with a small snout and long clubbed (but not elbowed) antennae; most species are associated with dead wood and fungi, but one genus is of importance agriculturally and in food stores.

Araecerus fasciculatus Degeer (Coffee bean or Nutmeg weevil) (Figure 14.70)

Pest status. Quite a serious pest locally in many parts of the tropical and subtropical world, particularly on high value commodities.

Produce. Nutmegs, coffee beans, cocoa beans, cassava, maize, groundnuts, Brazil nuts, spices, dried roots, some processed foodstuffs, and various seeds. In Hong Kong, field infestations of seeds of ruderal nasturtiums were very heavy, and the adults would fly to kitchen windows seeking food.

Damage. Direct eating is the main damage; seeds are destroyed, and on dried cassava tubers destruction can be severe. On high value produce such as coffee and cocoa beans, nutmegs etc., contamination is often more important than the actual eating damage. Coffee cherries may be attacked in the field.

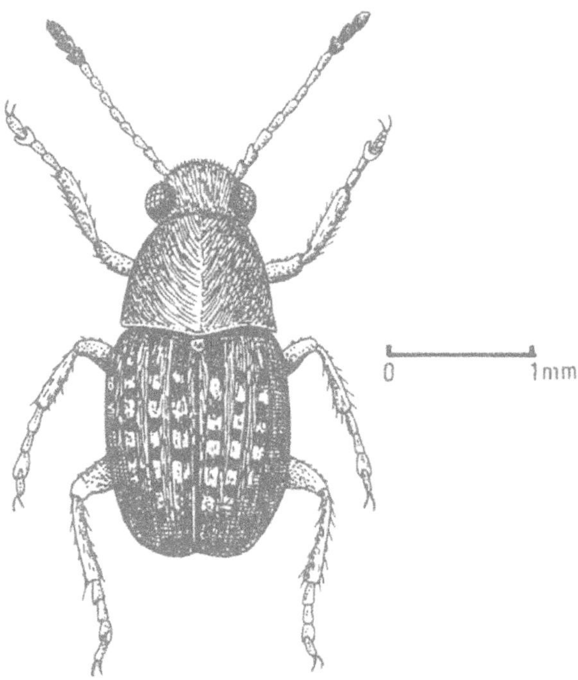

Figure 14.70. Araecerus fasciculatus (Coffee bean weevil) (Coleoptera; Anthribidae), adult beetle.

Life history. About 50 eggs are laid (on coffee beans) per female, singly on the coffee cherries or seeds.

The white legless larvae burrow within the seed, each larva usually spending its entire life within the same seed; pupation takes place there. In coffee cherries, the larvae feed initially on the pulp and then attack the seed. Optimum conditions for development are thought to be about 28°C and 80% RH (or more) – low humidities (less than 60% RH) have a serious adverse effect on development.

Figure 14.71. Coffee beans (dried) attacked by Araecerus during storage (courtesy ICI).

The adult is a stout little beetle, 3-5 mm long, grey-brown with small pale marks on the elytra, with quite long, clubbed (but not elbowed) antennae. They live for up to 4 months and fly well; in Hong Kong, they could be seen flying from the hillsides into flats and houses during the day, and often large numbers accumulated on kitchen windows.

The life cycle is completed in 30-70 days according to conditions, but in cooler regions (e.g. Hong Kong) there is quiescence over the winter period. In the UK, populations in unheated warehouses usually die out during the winter.

Distribution. Found throughout the world in the tropics and subtropical regions, and in a few temperate locations; often quite a localized distribution.

Other species of *Araecerus* known to occur in stored produce include:

Araecerus crassicornis (Fab.). In legume pods in Indonesia.

Araecerus levipennis Jordan. (Koa Haole seed beetle). Found in Hawaii.

Araecerus spp. Several species are recorded in India.

14.10.24 Family Apionidae (1,000 species)

One of the smaller groups regarded as being weevils, but now separated from the true weevils (Curculionidae). The family is worldwide, and characterized by having clubbed antennae that are non-geniculate. A few genera are important as agricultural pests of field crops, and some will be carried into produce stores, where development will be completed. The two genera of concern are *Apion* and *Cylas*.

Apion spp. (Seed weevils) (Figure 14.72)

A very large genus with many species, found completely worldwide; they are small (1.5-3 mm) black, long-snouted weevils with a globular body. The adults are often to be found in flowers, and they oviposit into the developing ovaries so that the larvae develop inside the seeds – one per seed. The host plants are mostly legumes, and ripe pods or threshed seeds are taken into storage, sometimes with larvae or pupae inside; development is completed in store, and the tiny black adult weevils emerge after biting a hole in the seed testa (and maybe also pod wall); but there is no further development in the store on the dry seeds.

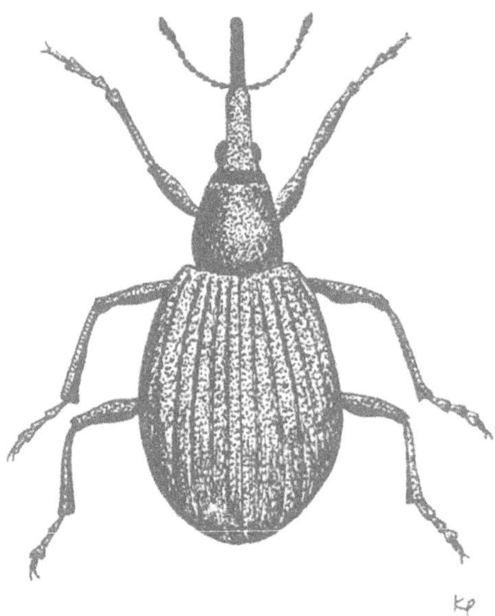

Figure 14.72. Apion sp. (Pulse seed weevil) (Coleoptera; Apionidae), adult weevil, body length 2 mm; Cambridge, UK.

Cylas spp. (Sweet potato weevils) (Figures 14.73, 14.74)

Specific to *Ipomoea*, there are several species of *Cylas* that bore in the tubers of sweet potato or in the stems of the climbers. The larvae bore in the stems and tubers, and they pupate in the gallery produced; adults are also to be found in the tunnel gallery. When infested sweet potato tubers are harvested, they are dried and taken into stores, where weevil development continues. Most tubers are sold fresh in the tropics where they are grown, and usually not stored long, if at all, but there is an ever-growing export trade to Europe and other more northern regions which does involve storage. Heavily infested tubers are easily recognised, usually by being wizened, and there is often rotting of the tissues, so such tubers are rejected and not stored; but slight infestations are not so obvious.

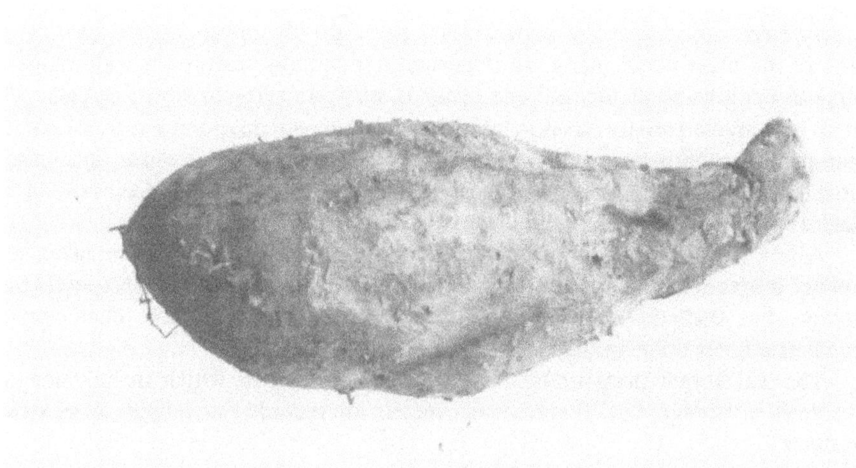

Figure 14.73. Sweet potato tuber showing effect of weevil infestation (right side); South China.

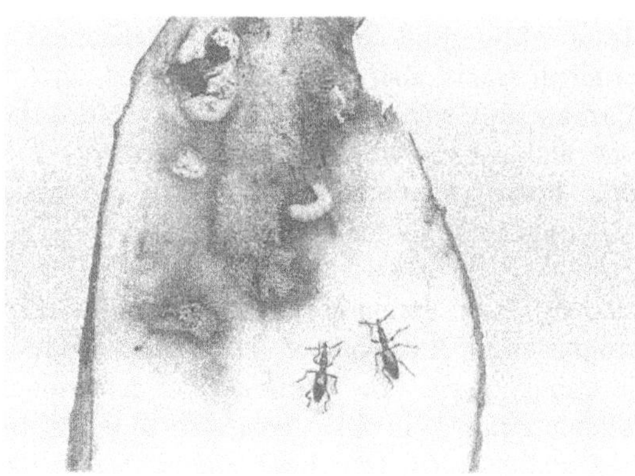

Figure 14.74. Cyclas formicarius (Sweet potato weevil) (Coleoptera; Apionidae), tuber cut to show infestation, one larva and two adult males; South China.

In parts of Asia and Africa, tuber infestation is very widespread, and so a large proportion of the tubers lifted and stored have weevils inside, but they are specific to *Ipomoea*, and there is no cross-infestation in the stores. During storage, large populations can develop if conditions are right, as they breed in harvested tubers.

Cylas puncticollis only occurs in Africa, but *C. formicarius* is extensively pantropical (Africa, South and East Asia, Australasia, Southern USA, Central and South America.

14.10.25 Family Curculionidae (Weevils proper) (60,000 species)

A very large group, quite worldwide, and most are phytophagous and feed on all parts of the plant body; many are important agricultural pests, on a wide range of crops and cultivated plants, and one genus is very damaging to stored cereals. The group is characterized by having clubbed, geniculate antennae and a rostrum (snout) bearing the mouthparts distally. Many species have a very elongate and narrow rostrum (shorter in the male), and these long-nosed weevils bite a deep hole in the plant tissues to lay eggs singly in these excavations.

A few species develop in nuts, fruits or roots, and are likely to be taken into produce stores where the final development takes place, and the adult weevils may emerge, but there is no breeding in the store. Nuts will have characteristic emergence holes bored in the shells.

The real storage pests are the three species of *Sitophilus* which are so damaging to cereals in storage, but 30 species of weevils are recorded worldwide from stored products.

Sitophilus oryzae (Linn.) (Rice weevil) (Figure 14.75)

Pest status. One of the most destructive primary pests of stored grains worldwide in warmer regions; very important and very damaging.

Produce. Capable of infesting all cereal grains, but recorded as favouring wheat and rice, and other small grains; also on flours, pasta and other cereal (farinaceous) products. In the Eastern Highlands of Ethiopia, however, this species is recorded to be the dominant one, and it is very common on maize there – more so than on local wheat and sorghum. Food preferences are somewhat variable and complicated in this genus; some strains have been bred on pulses.

Damage. Direct damage is the eating of the cereal grain; usually one larva hollows out one small grain during its development, but in maize (Ethiopia), several larvae can develop inside a single grain. The importance of contamination of grain varies with locale and circumstances. In S.E. Asia, Africa and China, the presence of weevils is expected and tolerated to a large extent, but in the temperate region supermarkets, contamination by adults (and pupae inside grains) can cause serious reduction in value. Contamination of produce can be serious due to the accumulation of uric acid, when grain can be rendered unpalatable. Adults fly into the ripening crops in fields that are close to the stores, and many infestations start in the field (Figure 14.76).

Life history. The female weevil bites a tiny hole in the grain surface and lays an egg within, sealing the hole with a special waxy secretion; 150-300 eggs can be laid per female, and incubation takes about 6 days at 25°C. Eggs are laid at temperatures between 15 and 35°C, and at all moisture contents above 10%. The females continue to lay eggs for most of their long lives.

The white legless larva feeds inside the grain, excavating a tunnel; it passes through 4 instars in about 25 days (at 25°C and 70% RH). At low temperatures, development is slow, taking 100 days at 18°C. Fully grown larvae measure about 4 mm. Pupation takes place inside the grain, and the young adult is quite conspicuous as a dark patch under the testa. The adult eats its way out of the grain, leaving a characteristic circular emergence hole.

The adult is an elongate dark brown little weevil, 2.4-4.5 mm long, with four reddish-brown patches on the elytra. It is an active insect, diurnal usually, and flies readily; crops in the field are at risk for up to about one kilometer distance from grain stores. Adults feed and are long-lived (4-12 months), and the females lay eggs for most of their adult life, although 50% may be laid within the first 4-5 weeks.

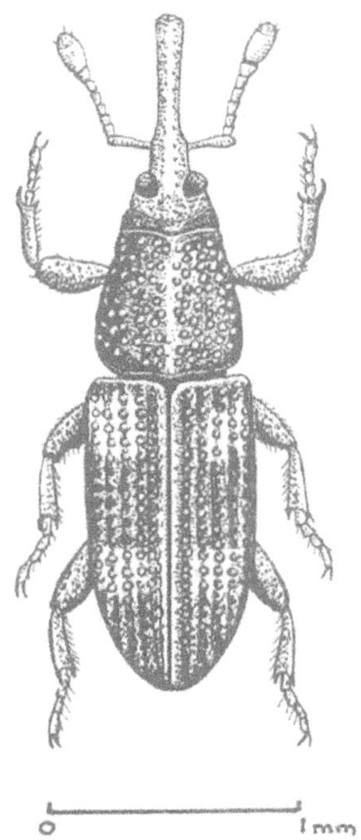

Figure 14.75. Sitophilus oryzae (rice weevil) (Coleoptera; Curculionidae), adult weevil; from rice in Hong Kong.

Figure 14.76. Sitophilus oryzae, maize cob one month after harvest; showing emergence holes of first (field) generation weevils; Alemaya, Ethiopia.

The life cycle is completed in 35 days under optimum conditions, but under cooler or drier conditions, development is protracted. Different varieties of maize also influence the rate of development; in Africa, the old flinty varieties are generally less favoured as food, whereas the newer softer grained high-yielding varieties are usually preferred.

Distribution. Completely tropicopolitan and generally abundant everywhere; and it regularly occurs in cooler temperate regions in imported produce.

This pest has only relatively recently been separated from *S. zeamais*, and the only reliable diagnostic character is recorded to be the male genitalia (and, to a lesser extent, the female) (TDRI, 1984). The literature concerning these two species is very confusing, as it is seldom known which species is being referred to; both species are broadly sympatric.

Natural enemies. The three species of *Sitophilus* are regularly parasitized by parasitic Hymenoptera (Pteromalidae), including the three most common species *Anisopteromalus calandrae, Lariophagus distinguendus* and *Chaetospila elegans.*

Sitophilus zeamais Mots. (Maize weevil)

Now distinguished from *S. oryzae* on the basis of the shape of the male aedeagus and a sclerite in the female genitalia; it is thought that this species is slightly the larger in size, but there is much variation in size in most populations. It is also thought that this species prefers maize, and in Indonesia, surveys recorded this species mostly on milled rice, with *S. oryzae* mostly on rough paddy rice (TDRI, 1984). Dietary preferences have not really been established between these two closely related species, but some striking infestation differences are easily observed in some countries.

In the literature, the nomenclature for the *Sitophilus* complex has been very confusing, and many names have been used for these three species.

Sitophilus granarius (Linn.) (Grain weevil) (Figures 14.77, 14.78)

This species is reported to have the same environmental optima for development as the other two, but cannot compete with them in the tropics, and since it can tolerate lower temperatures (down to 11°C), it has become established throughout the world in temperate regions and in cooler upland areas in the tropics (e.g. in Addis Ababa in Ethiopia). The wings are vestigial and the beetles cannot fly, and therefore do not cause field infestations. A morphological character is that the punctures on the prothorax are oval in shape, whereas in the other two species they are circular; in some specimens there are no coloured patches in the elytra.

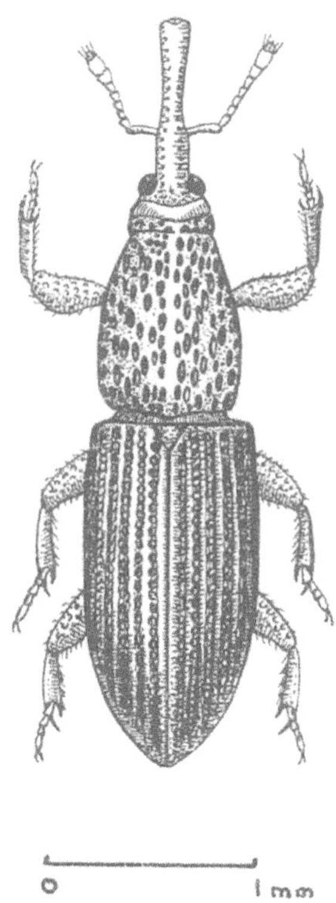

Figure 14.77. Sitophilus granarius (Grain weevil) (Coleoptera: Curculionidae), adult weevil.

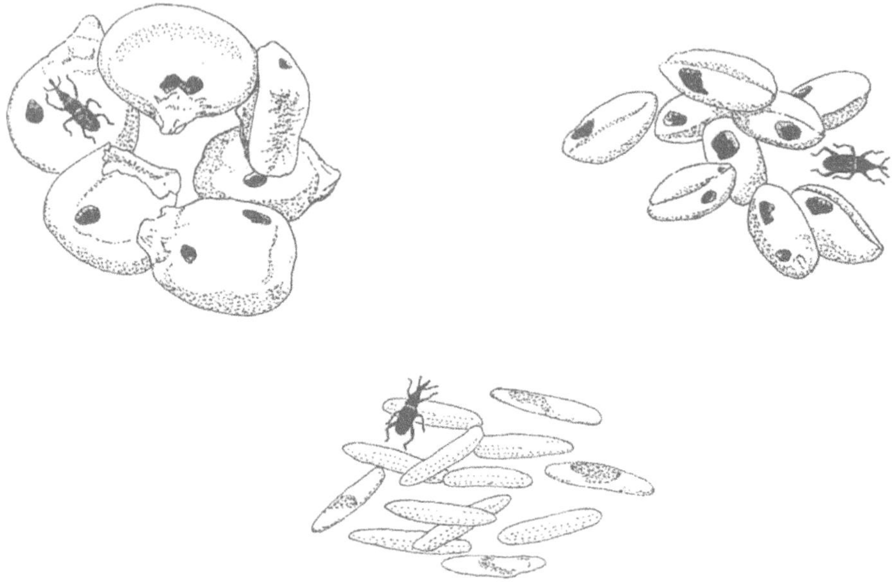

Figure 14.78. The three most common grains attacked by grain weevils (mostly Sitophilus zeamais); maize, wheat and rice.

Sitophilus linearis (Herbst)

A pest of tamarind pods in India; the adult resembles *S. oryzae*, but can be distinguished morphologically.

Other weevils that are sometimes found in produce stores, more or less regularly, are listed below:

Balanogastris kolae (Desbrochers) (Kola nut weevil)

A robust weevil which feeds on the nuts of *Kola acuminata* in parts of the tropics; larvae develop inside fallen nuts, or nuts on the tree, and are carried into produce stores where development is completed.

Catolethus spp.

These beetles are thought to inhabit rotting wood, but two species have been observed several times in large numbers in stored maize on farms in Central America.

Caulophilus oryzae (Gyllenhal) (Broad-nosed grain weevil) (Figure 14.79)

Recorded on stored maize in Central America, and on ginger in the West Indies; now also regularly recorded on maize in southern USA; field infestations of ripening grain are quite common.

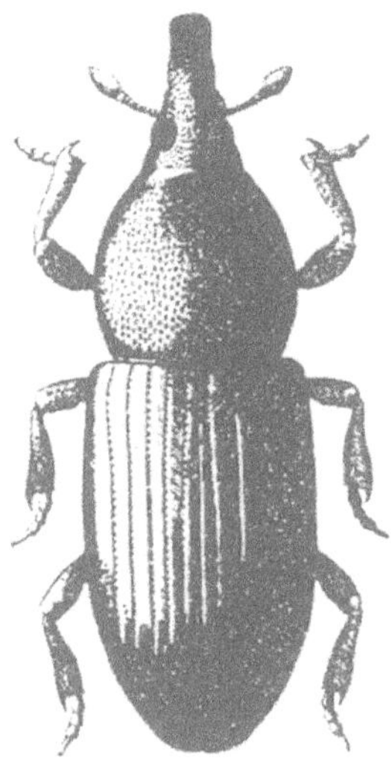

Figure 14.79. Caulophilus oryzae (Broad-nosed grain weevil)

It attacks soft or damaged grains, and is unable to breed in dry, hard, uninjured grain.

Ceutorhynchus pleurostigma (Marshall) (Turnip gall weevil) (Figure 14.80)

The larvae make globular galls in the root of turnips and in the stems of Cruciferae – lightly infested plants may be taken into vegetable stores, but heavily infested plants are pretty obvious. Pupation usually takes place in the soil, so the small black adult weevils (3 mm long) are not likely to be seen in the stores.

Curculio nucum Linn. and spp. (Hazelnut weevils) (Figure 14.81)

There is one common species in Europe, and two in North America – they develop inside hazenuts. Nuts in storage are often found with a distinct 2 mm hole in the shell and no kernel inside. The larva feeds on the kernel, and when fully grown it bites the emergence hole in the shell, and then leaves to pupate in the soil; typically, in Europe, the adult emerges the following May, so adults are not likely to be found in stores. *Curculio sayi* is the small chestnut weevil of the USA, and similar damage may be found on sweet chestnuts in storage. Nuts in storage are likely only to have the larval emergence hole, but no insect inside, and no kernel!

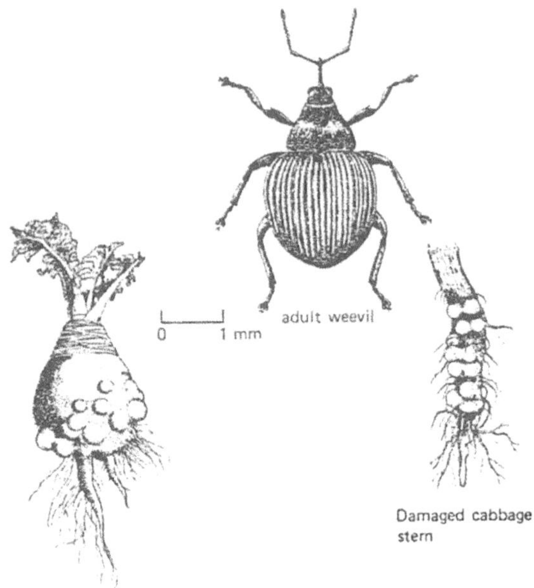

Figure 14.80. Turnip root attacked by larvae of Ceutorhynchus pleurostigma (Turnip gall weevil) (Coleoptera; Curculionidae); UK.

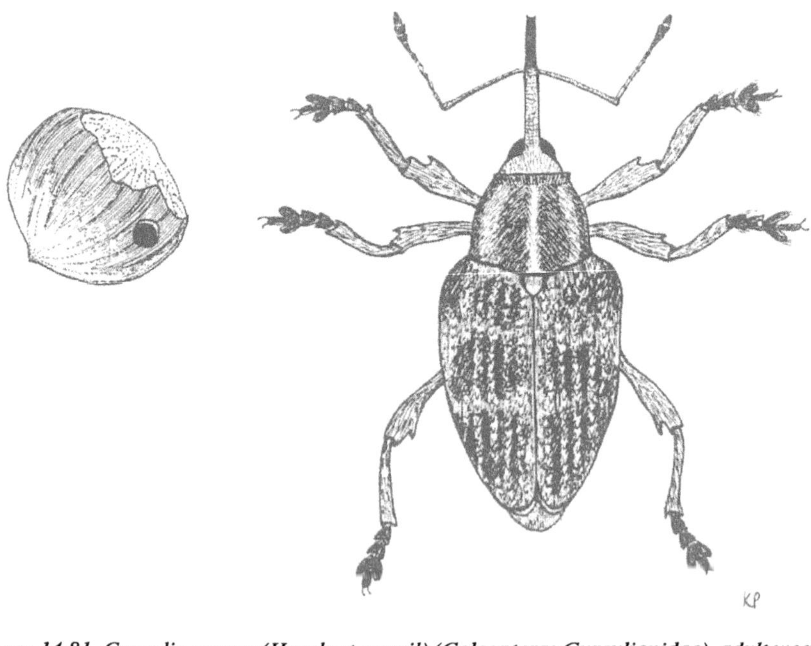

Figure 14.81 Curculio nucum (Hazelnut weevil) (Coleoptera; Curculionidae), adult weevil, and infested nut ((from France) with emergence hole.

Lixus spp. (Beet and Cabbage weevils) (Figure 14.82)

These weevils develop inside the stem and roots of beet and various Cruciferae, and so will sometimes be found in vegetable stores; typically, an infested stem is quite swollen.

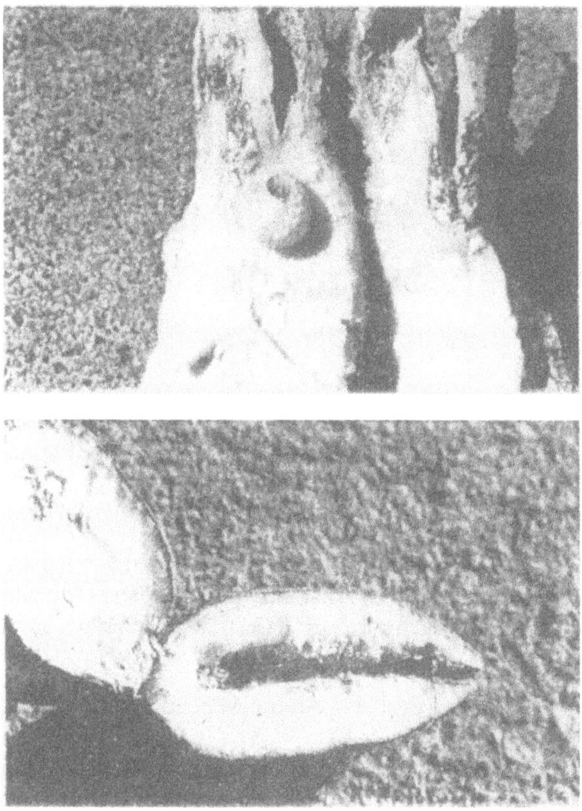

Figure 14.82. Lixus anguinus (Cabbage weevil) (Coleoptera; Curculionidae), larva and tunnel in root of wild radish (species uncertain); Alemaya, Ethiopia.

Sipalinus spp. (Pine weevils and borers) (Figure 14.83)

There are several species of weevils whose larvae bore in wood. Development is slow, and often takes many months, for the insects are quite large, measuring up to 20 mm. Wood used in packing cases or as structural timbers may yield the occasional adult weevil. These species show a preference for *Pinus*, and are found throughout Europe and Asia.

Other genera are found in other species of tree. In Europe, the small *Euophryum* weevils (2-5 mm) are found in a wide range of both soft and hard woods, provided that the wood is being attacked by one of the wet-rot fungi.

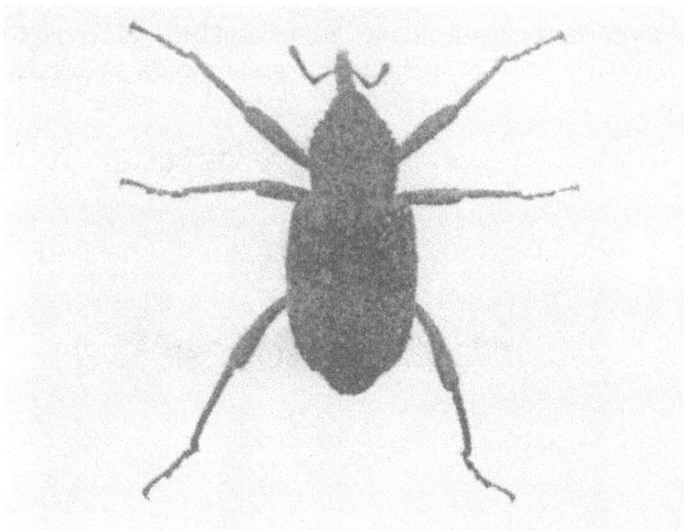

Figure 14.83. Sipalinus sp. (Pine weevil) (Coleoptera; Curculionidae), adult weevil (body length 14 mm); one of several species of weevil whose larvae tunnel in timber and may be found in buildings; Hong Kong.

Sternochetus mangiferae (Fab.) (Mango seed weevil)

In parts of the Old World tropics, mango fruits may be inhabited by weevil larvae which bore into the stone to eat the seed, and in a few cases the adult weevil may emerge from the fruit in storage or transit. Infestation symptoms are few.

14.10.26 Family Scolytidae (Bark and Ambrosia beetles)

A group of tiny (1-3 mm) dark beetles, cylindrical in body shape, with head deflexed for burrowing through wood and bark and other plant tissues. Most of the burrowing is done by the adults constructing breeding galleries. The group is mainly of concern to forestry, and causes devastating damage to some tree species. A couple of species are sometimes encountered in produce stores, and one can be serious in South America.

Coccotrypes dactyliperda (Date stone borer)

A primary pest of unripe fruits of the date palm in the Mediterranean region, and adults occasionally emerge from ripe fruits in storage or in packaged fruits.

Hypothenemus hampei (Ferrari) (Coffee berry borer) (Figure 14.84)

Essentially a pest of ripening coffee cherries on the bush – adults bore into the fruit and lay eggs inside the tunnel; up to 20 larvae develop inside one berry. Some larvae are carried in the drying beans into produce stores, and pupation takes place inside the larval gallery, so adult beetles (1.6-2.5 mm) may be found in the produce. Basically an East African pest; recently introduced into South America, this insect is causing serious damage in coffee plantations in Brazil. This species has also been recorded attacking *Phaseolus* and *Vigna* beans.

Hypothenemus liberiensis Hopkins is recorded from maize in Nigeria, and other species are occasionally recorded in produce stores in parts of the tropics.

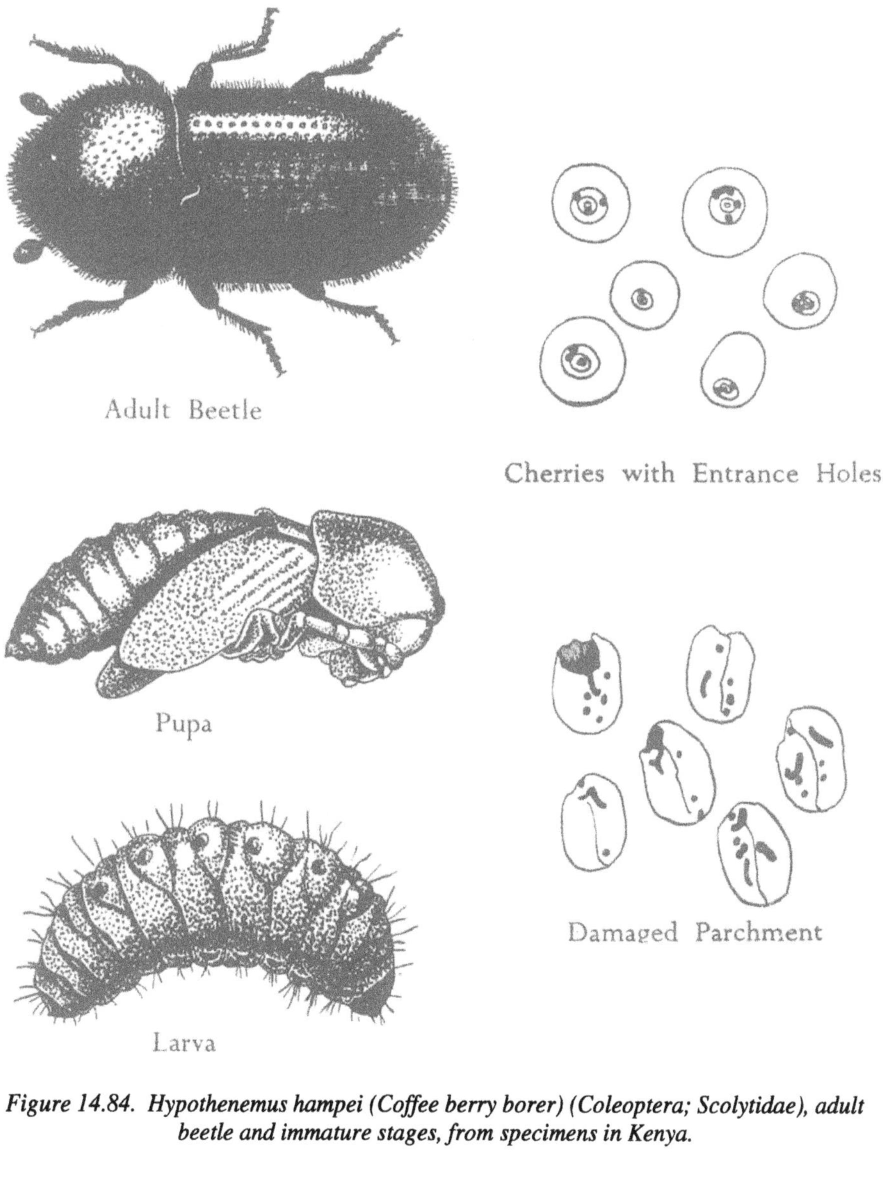

Figure 14.84. Hypothenemus hampei (Coffee berry borer) (Coleoptera; Scolytidae), adult beetle and immature stages, from specimens in Kenya.

Pagiocerus frontalis (Fab.)

A species widely recorded throughout South America and the West Indies, but reputed (TDRI, 1984) to be a serious pest only on certain soft-grained maize varieties in the Andes. It is said to have been bred on the dried seed of avocado.

14.11 ORDER DIPTERA (True flies) (112 families, 85,000 species)

A large group of insects, very small to moderate in size, and characterized by having only a single pair of wings, and the hindwings are modified into slim club-like balancing organs (halteres); mouthparts in the adult are modified for sucking fluids, and sometimes for piercing (usually forming a proboscis). Larvae are often worm-like, with head often reduced; some have biting mandibles; and they are terrestrial, aquatic or parasitic in habits. The group is of most importance as medical and veterinary pests, especially when transmitting bacteria, viruses, parasitic protozoa, or parasitic worms, and some are important agriculturally as pests of growing plants and fruits.

None are pests of stored grain, and as such the group is often omitted from lists of stored products insects. But a few species are predacious on other insects, and a number feed on proteinaceous materials, usually of animal origin, and saprozoic species will feed on the corpses of rats and mice in stores. Prepared foodstuffs are at risk from several species of flies carrying parasitic micro-organisms, and fruit flies and other phytophagous species infest fruits and some vegetables.

When produce stores are located together with dwellings, there may be a number of flies which feed on dung, decaying rubbish and the like outside the stores, and which fly into the stores for shelter or to hibernate. Several species are known as cluster flies, for they invade dwellings in the autumn for the purpose of hibernation over winter. But many of the synanthropic species have adapted to feed on the same foods as man (to some extent) and they will infest these foodstuffs in storage. Synanthropy is the co-existence of flies with humans, over a period of time. Different categories are defined by the degree of association and the reasons for the association – shelter, food etc. Some species live almost entirely inside human dwellings, whereas others can be independent for part of their time, and others are associated with livestock either in pastures or stables.

Some species deliberately seek shelter in buildings, such as some adult mosquitoes which spend the day resting in dark corners inside buildings, and at night they prowl for their prey; adult *Psychoda* (Moth flies) spend almost all their time resting in cool dark buildings.

The group is divided into three distinct suborders – the more primitive Nematocera with filiform antennae; the Brachycera with short stout antennae; and the Cyclorrhapha with short, 3-segmented antennae bearing a subterminal arista – the latter group is regarded as being the most advanced, and contains most of the synanthropic species.

14.11.1 Family Mycetophilidae (Fungus midges) (2,000 species)

A worldwide group of tiny to small flies whose larvae feed on rotting vegetation, and some are predacious. Species of *Mycetophila* live gregariously in mushrooms and other fungal fruiting bodies. In parts of the Far East, and in other regions, dried mushrooms of several species are stored in great quantities, and fresh mushrooms may be stored for a short while, so mushroom flies may well be encountered regularly.

14.11.2 Family Sciaridae (Dark-winged fungus gnats) (300 species)

A small group whose larvae feed in soil on organic material of all types, and several species (*Sciara* etc.) regularly infest mushrooms. *Pnyxia scabiei* is the Potato scab gnat of the USA and parts of Europe; the larvae attack damaged tubers in the field and in store; females are wingless, and may overwinter in the potato stores as well as the larvae on the tubers.

14.11.3 Family Cecidomyiidae (Gall midges) (500 species)

A worldwide group of minute flies – they have delicate bodies and long filiform antennae. The larvae are phytophagous mostly, and feed on all parts of the plant body, and most species induce a characteristic gall on the host plant. Plant leaves, fruits and seeds are the parts most commonly attacked, and, clearly, species that infest cereal grains, hop fruits, sunflower seeds, clover seeds, sesame capsules, pea pods and the like are to be expected in produce stores occasionally.

Contarinia sorghicola (Coq.) (Sorghum midge) (Figure 14.85)

A pantropical species that infests sorghum – the larval instars live individually or gregariously inside the sorghum grains, and usually development is completed while the crop is in the field, but pupae may aestivate or hibernate in the cocoon, and so the tiny (2 mm) orange-bodied midges may sometimes emerge in the store. But there is no breeding in the store, and the adults can do no damage except for a little produce contamination.

14.11.4 Family Scenopinidae (200 species)

Scenopinus fenestralis (Linn.) (Window fly)

A small dark fly with red legs, to be found on the windows of produce stores. The larva is long and thin, with a distinct head capsule, and is found in stored grain and debris where it preys on the larvae and pupae of beetles and moths. It is found regularly in granaries and produce stores in most of Europe, and is recorded to be common throughout North America, and distributed more or less worldwide through commerce.

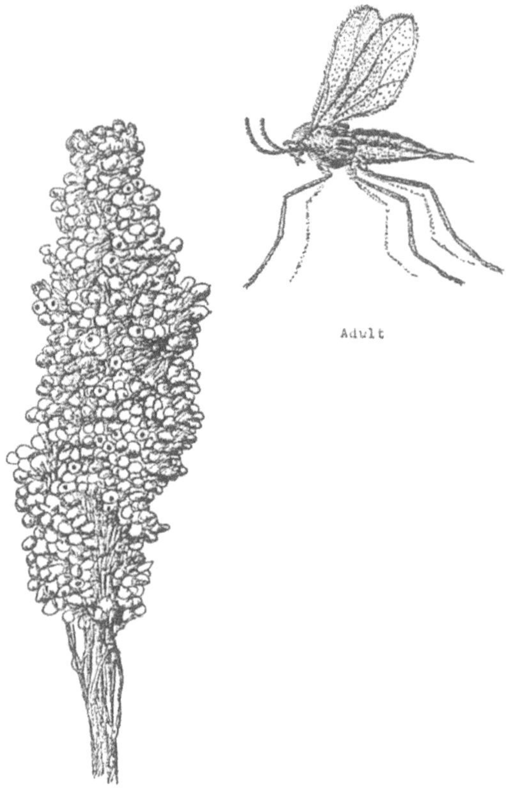

Figure 14.85. Contarinia sorghicola (Sorghum midge) (Diptera; Cecidomyiidae), adult female midge and infested panicle; specimens from Kenya.

14.11.5 Family Phoridae (Scuttle flies)

A number of species are found in domestic situations where larvae feed on dead animals and rotting vegetation; a few species (*Megascelia* spp. etc.) are to be found in mushrooms.

14.11.6 Family Syrphidae (Hover flies etc.)

A distinctive group of medium-sized flies, either brightly coloured or dark and bristly – some clearly mimic wasps, and others, bees. The adults can be seen feeding on flowers in the sunshine. Larvae are very diverse in habits; many are predacious or saprozoic, but a few are phytophagous. A few species are of importance as bulb flies, and the larvae (maggots) in bulbs are often carried into produce stores. The pupae of some predacious species are often collected from pea foliage and contaminate frozen peas.

Eumerus spp. (Lesser bulb flies) (Figure 14.86)

Found throughout Europe, Asia and North America, as at least six species. The larvae bore in the bulbs of onions, and various flowers belonging to the Liliaceae (Figure 14.86), and occasionally in carrot, potato and ginger rhizomes.

The adults are small shiny brownish flies, about 6 mm long. Larval attacks are often associated with fungal and bacterial rots. Sometimes the larvae are regarded as secondary pests; in reality, it is likely that they are primary pests, but are attracted by rotting tissues so that they will attack bulbs already rotting.

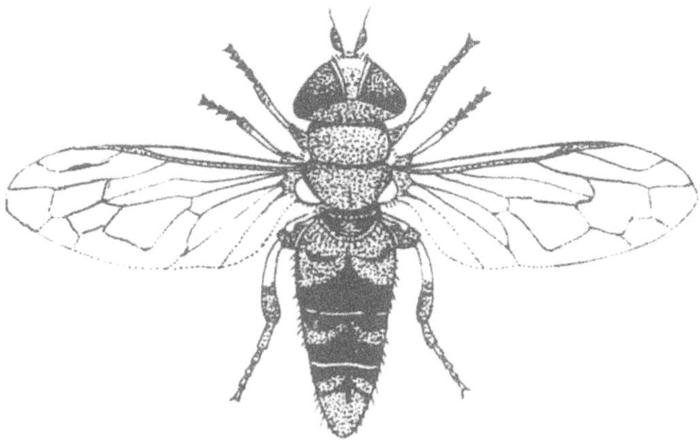

Figure 14.86. Eumerus sp. (Small bulb fly) (Diptera; Syrphidae), adult fly.

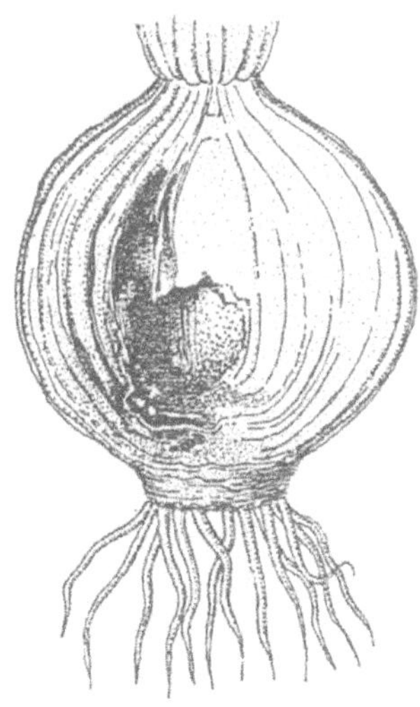

Figure 14.87. Bulb (Liliaceae) infested by tunneling maggots, and showing internal damage.

Merodon spp. (Large narcissus flies) (Figure 14.88)

These are large (13 mm) furry, fat-bodied flies with a distinctive banded body; two species are of particular importance in Europe and parts of North America. The larvae are usually single (one per bulb) inside the bulbs of *Narcissus* and other Amaryllidaceae, and they are definitely primary pests, but are usually only recorded in light infestations. Bulb fly larvae may be specifically controlled using methyl bromide on bulbs.

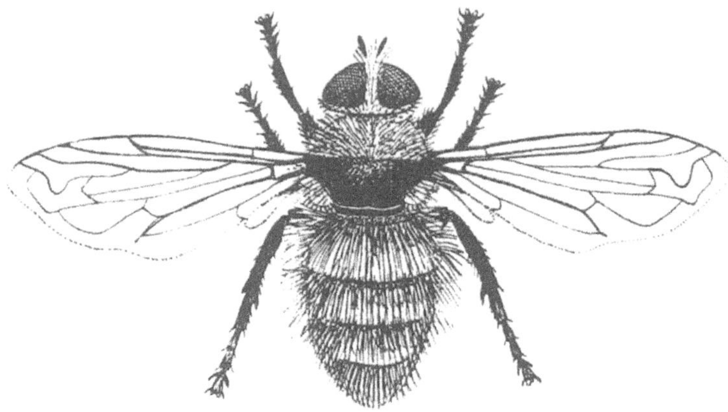

Figure 14.88 Merodon equestris (Large narcissus fly) (Diptera; Syrphidae), adult fly, body length 10 mm.

14.11.7 Family Lonchaeidae (300 species)

A small family whose larvae are mostly scavengers in rotting vegetation or dung, but a few are phytophagous and recorded from grasses or fruits. Two species are regularly recorded from fruits and may occasionally be found in fruit stores. *Lonchaea aristella* Beck. (Fig fly) infests figs in the Mediterranean region, and *Lonchaea laevis* (Bez.) (Cucurbit Lonchaeid) is recorded from fruits of some Cucurbitaceae in parts of Ethiopia, and tomato in Kenya.

14.11.8 Family Drosophilidae (Small fruit flies, Vinegar flies) (1,500 species)

These are small yellow flies with red eyes, and they are attracted to the products of fermentation. Their presence in a produce store is usually an indication of the presence of over-ripe fruits. But a few species have a different life style.

Drosophila spp. (Small fruit flies) (Figure 14.89)

This is the main genus in the family, with many species; adults are yellowish with a banded abdomen, 3-4 mm long, and with distinctive red eyes. The larvae are usually to be found in ripe or over-ripe or damaged fruits; it is thought they are probably feeding on the yeasts and other fungi present, rather than on the actual fruits. The pupae are characteristic in being elongate, brown, and with two protruding anterior spiracular horns. Contamination of fruits and fruit products can be a serious matter, by either the tiny white larvae, or the more conspicuous small adults.

They have a very high biotic potential as they can complete a generation in 8-10 days during the warmer months in Europe and North America. They are completely cosmopolitan and many species are known, some more tropical and others more temperate. About eight species of *Drosophila* have been recorded as pests, often because of their swarming habits near a suitable food source.

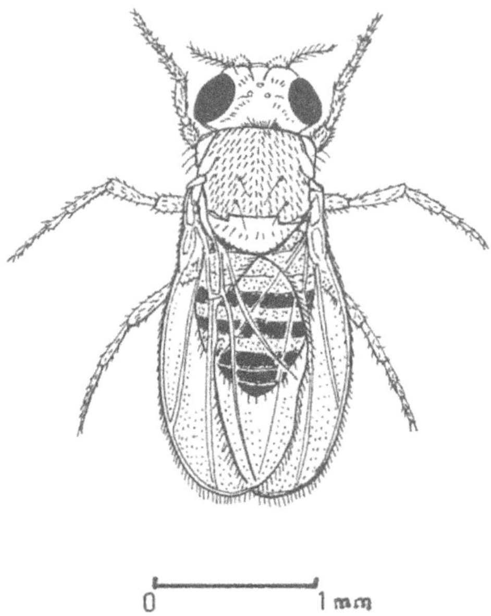

Figure 14.89. Drosophila sp. (probably melanogaster) (Vinegar or Small fruit fly) (Diptera; Drosophilidae), adult fly, Hong Kong.

14.11.9 Family Psilidae (100 species)

A small group, but of interest agriculturally since two species of *Psila* are pests of some importance.

Psila rosae (Fab.) (Carrot fly) (Figure 14.90)

This fly is common throughout Europe and much of North America, and in some localities it is a very serious pest – the larvae tunnel (usually superficially) in the root of carrot and parsnip, and more internally in the stem and leaf petioles of celery

(Figure 14.91). Since these roots are often kept in storage, this species can be found regularly in vegetable stores. *Psila nigricornis* Meigen, the Chrysanthemum stool miner, is also recorded from carrot roots in parts of Europe and in Canada.

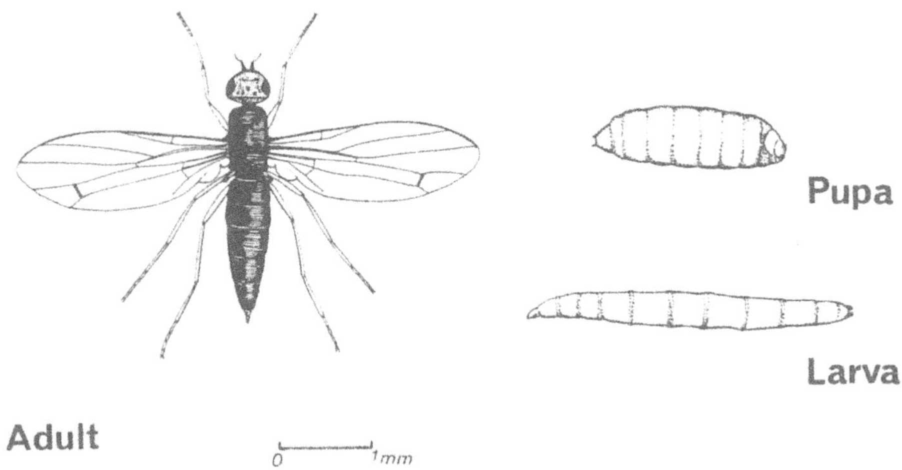

Figure 14.90. Psila rosae (Carrot fly) (Diptera; Psilidae), adult and immature stages, specimens from Feltwell, Norfolk, UK.

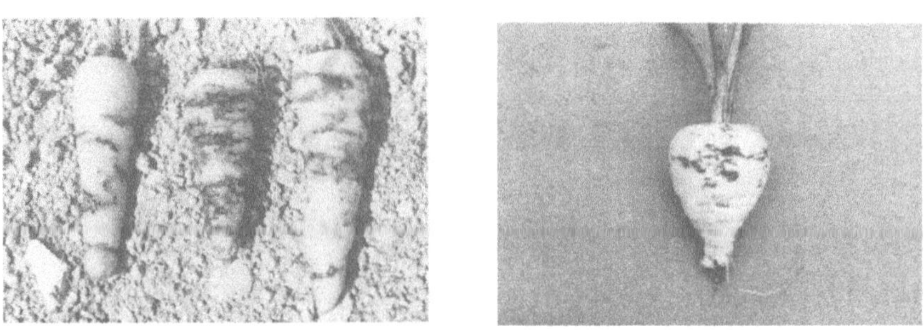

Figure 14.91. Carrot fly damage to carrots (a), and parsnip (b); Cambridge, UK.

14.11.10 Family Anthomyiidae (Root flies etc.)

A group of some importance agriculturally, but only rarely encountered in vegetable stores. The family is very closely related to the Muscidae, some species are synanthropic, and most larvae are basically saprophagous.

Delia antiqua (Meigen) (Onion fly) (Figure 14.92)

A serious field pest of onions, the larvae bore inside the bulb, although often salad onions are more heavily attacked than the ware crop (Figure 14.93). A Holarctic species that in some localities occurs in dense populations, causing considerable crop losses. Bulb onions taken into store for drying and subsequent storage may be infested, and these infestations are usually associated with fungal or bacterial rots, so damage to the stored crop could become very extensive.

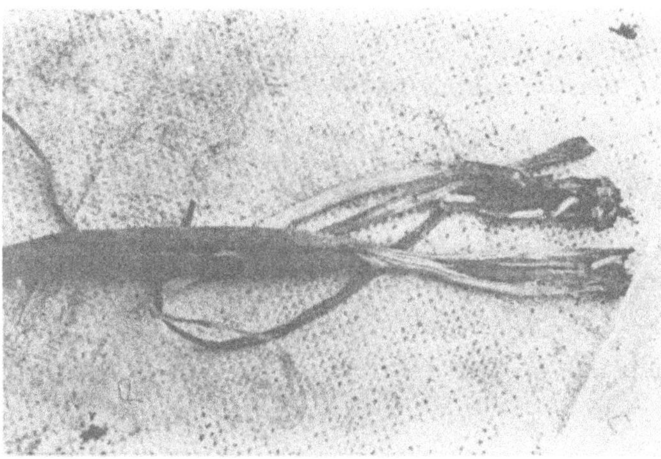

Figure 14.92. Salad onions attacked by maggots of Onion fly; Sandy, Bedfordshire, UK.

Delia radicum (Linn.) (Cabbage root fly) (Figure 14.94)

The larvae are large white maggots, and they bore in the roots of Cruciferae, and are very serious field pests in some regions. They are Holarctic in distribution, but as far as harvested vegetables in storage are concerned, only root crops such as swede and turnip are likely to be affected. Brussels sprout buttons and radishes are sometimes infested, although these are seldom stored for more than a short time, typically being sold fresh. Sprouts for freezing which are infested with maggots can be a very serious problem.

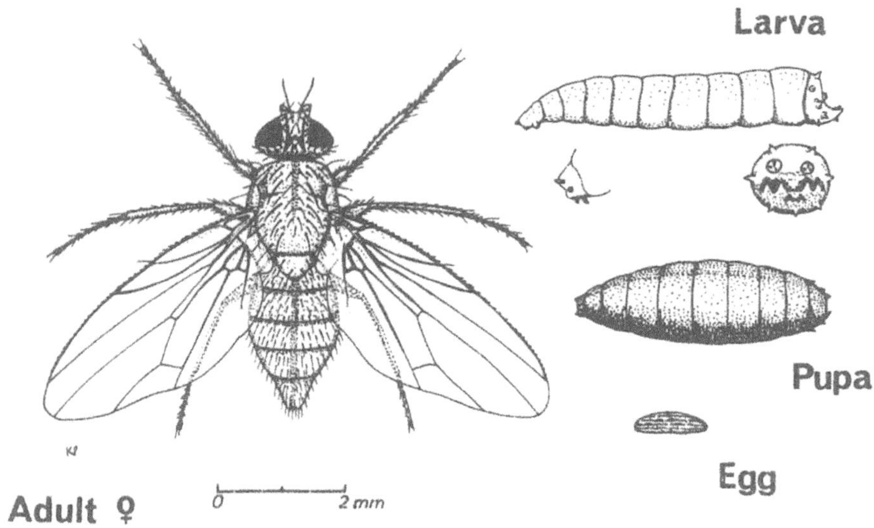

Figure 14.93. Delia antiqua (Onion fly) (Diptera; Anthomyiidae), adult female and immature stages; specimens from Sandy, Bedfordshire, UK.

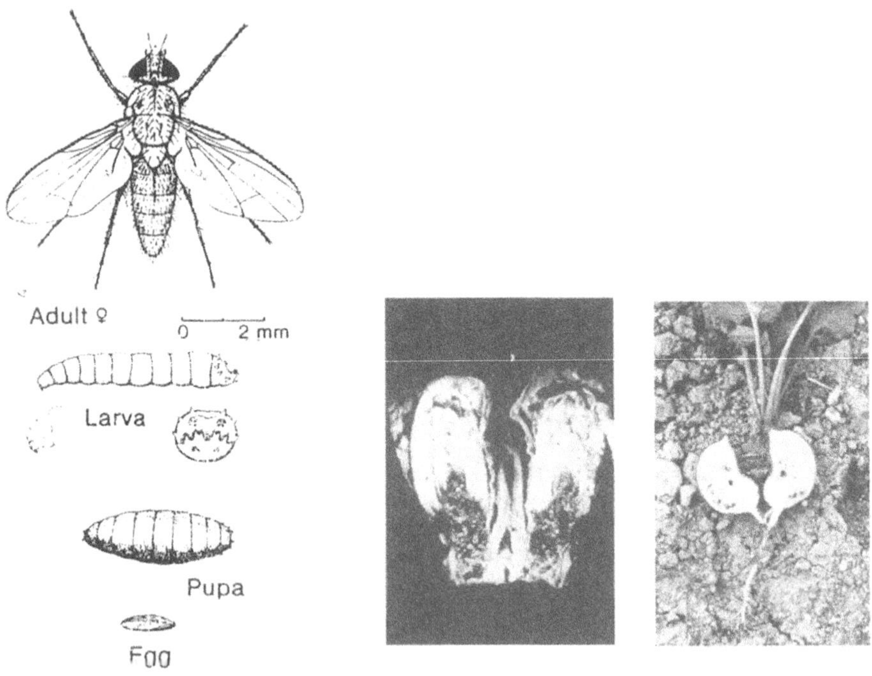

Figure 14.94. Delia radicum (Cabbage root fly) (Diptera; Anthomyiidae), adult female and immature stages; also showing infested Brussels sprout button, and infested radish; Cambridge, UK.

14.11.11 Family Muscidae (House fly etc.)

A large group of closely related species found completely worldwide. The superfamily Muscoidea (including the following three families) is thought to be an ancient group that originated from a compost-feeding ancestor, and has since evolved into groups which are phytophagous, blood suckers, and carrion feeders. Many species are synanthropic, and have accompanied man on his travels, and as he spread around the world.

Several of the synanthropic species are to be found on stored foods on domestic premises, but the house fly is typical of the group, and is by far the most important species.

Musca domestica (Linn.) (House fly) (Figure 14.95)

Pest status. A major domestic, and medical and veterinary, pest throughout the cooler parts of the world, both spoiling food, causing irritation, and acting as vector for many pathogenic organisms.

Produce. Adults feed (by enzyme regurgitation) on most of the proteinaceous foods eaten by man, as well as sugary foods, and also on faeces and decaying organic materials of all types. Larvae are essentially saprophagous, and will feed on a wide range of decaying organic materials, and will also feed on dried meats, dried fish, and some of the dried animal products used in Oriental medicine. The main infestation criterion is basically where the female fly chooses to lay her eggs, since there is little selection of foodstuffs by the larvae; a certain level of moisture is usually required in the food material. Dried stored foodstuffs would probably not be regarded as ideal food for the larvae, but infestations are of regular occurrence.

Damage. Direct eating of the food material, together with the effects of enzyme regurgitation and faecal bacteria causing food spoilage, are the main damage effects, but contamination can be a serious factor. The vomit spots produced by the feeding adults spread a vast range of pathogenic organisms (typhoid, dysentery, infantile paralysis etc.). Larvae accidentally swallowed in the food material sometimes survive in the human gut causing intestinal myiasis, with symptoms of pain, nausea, vomiting etc.

Life history. Each female lays 600-900 eggs, in batches of 100-150; each egg is elongate, about 1 mm long, and white; incubation takes only 3 days under warm conditions; The oviposition period varies from 4-12 days.

The larva is a typical cyclorrhaphous maggot; white, headless and legless, with a tapering anterior end bearing the mouth and mouth hook complex, and, externally, the small projecting anterior spiracles; the broad posterior bears the terminal spiracular plate. In the Muscoidea, most larvae are identified by the shape of the anterior spiracle, and the shape of the two posterior spiracles, together with the pointed tubercles arranged around the spiracular plate. The sculpturing on the egg chorion can also be characteristic. There are three larval instars, and the fully grown larva measures 10-12 mm, in 3-60 days according to temperature. The larvae feed by a scraping action of the mouth hook, and the fragments of food are sucked up by the pharynx.

Pupation occurs within an elongate to oval brown puparium (formed by the last larval exuvium), typically in the soil at a depth of 5-60 cm; the pupal period varies from 3-30 days, according to temperature.

The adult is a greyish fly, 7-8 mm long, with four longitudinal black stripes on the thorax, and males have two yellowish patches at the base of the abdomen. Adults are diurnal, very active, and long-lived, feeding on liquid foods using the sponge-like proboscis, or on solid food after regurgitation (vomiting) of enzymes on to the food. This is the way bacteria and other pathogenic organisms are transmitted by the flies.

The life cycle can be completed in 2 weeks under optimum conditions (30-35°C), and breeding is usually continuous. The rate of development is controlled by a combination of temperature, humidity, and food quality, but despite the high optimum temperature, this species is very temperature tolerant, and development will proceed over a wide range; in temperate regions, diapause is regularly practiced in overwintering hibernation.

Distribution. Cosmopolitan, but somewhat replaced in the hot tropics by *Musca autumnalis* (Face fly) and species of *Ophyra* which are called the Tropical house flies.

Natural enemies. Several species of Hymenoptera Parasitica are parasites of Muscoid larvae or pupae, including *Muscidifurax, Spalangia, Stilpnus* etc.

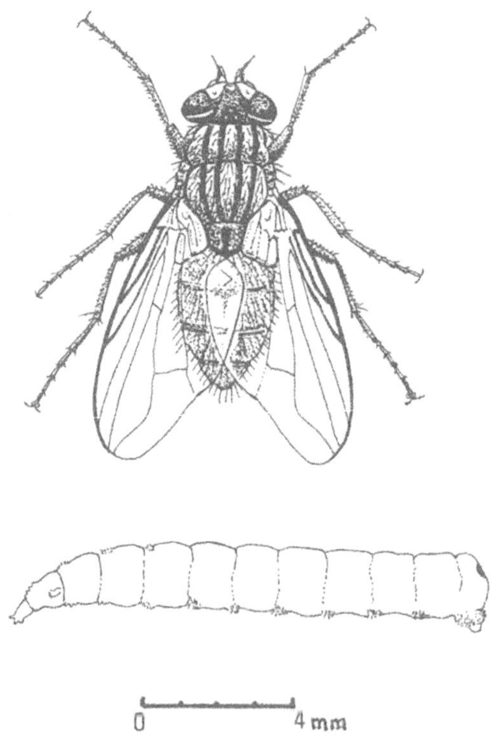

Figure 14.95. Musca domestica (House fly) (Diptera; Muscidae), adult fly and maggot.

Control. The synanthropic flies can best be controlled by a combination of hygiene, food protection, and judicious use of insecticides.

Control of breeding sites includes removal of rubbish, refuse heaps, dung heaps and the like.

Food protection includes the use of fly screens, and covers for vulnerable foodstuffs, and means of drying or else of reduction of atmospheric moisture levels.

Insecticides for use against fly larvae in rubbish heaps etc. include diazinon, dimethoate, malathion, or DDVP, or the more modern alternatives, pirimiphos-methyl and bioresmethrin. Fish for drying can be dipped in pirimiphos-methyl e.c. which gives good protection against fly infestation. In the past, adults (and larvae) were controlled very spectacularly using the organochlorine compounds, especially DDT and HCH, but resistance quickly developed and soon became very widespread. Other chemicals widely used are malathion, diazinon, dimethoate, fenthion, trichlorfon, and synergized pyrethroids. They can be used as either residual sprays or space sprays. Space sprays and aerosols are usually now formulated as a mixture of several chemicals, often including synergized pyrethroids that have a rapid knockdown effect. Resistance to insecticides must always be expected in local fly populations.

Other species of *Musca* to be found in human dwelling places include:

M. autumnalis (Face fly) – this looks like a large House fly; the female has a black abdomen and the male, orange (that of the House fly is yellowish). It is widespread in Africa, Asia and Europe; it invaded North America in 1952, and is rapidly spreading across the USA and Canada now. Mainly a pest of livestock, it breeds in fresh cow dung, but does invade domestic premises in large numbers, causing considerable nuisance.

M. vicinia (Egyptian house fly) – very closely related to the House fly, this is the dominant synanthrope in urban and rural areas of Egypt. Equine and human dung are the favoured sites for breeding, but domestic refuse can also be used.

Other species of Muscidae include:

Muscina spp. (False stable flies) – they look like large grey House flies, and enter houses in search of food. Adults carry pathogens on their body, and larvae can cause intestinal myiasis. First instar larvae are saprophagous, but later instars are predatory on larvae of other flies.

Ophyra spp.(6+) (Dump flies; Tropical house flies) – small shiny black flies found in all warm parts of the world in urban areas, especially around garbage dumps. The larvae are saprozoic on a wide range of organic materials. Adults enter houses and alight on foods, where they can pose a health hazard.

14.11.12 Family Fanniidae (more than 250 species)

This group is now separated from the Muscidae, but has a very similar biology, and a few species are recorded as infesting stored foodstuffs of various types. At least 12 species of *Fannia* are listed as being associated with man and human dwelling places, but these are mainly medical or hygiene pests.

Fannia canicularis (Linn.) (Lesser house fly) (Figure 14.96)

Pest status. Actually, now a more widespread domestic pest than the House fly, but less common in stored foodstuffs.

Produce. Decaying organic matter, such as over-ripe fruit, decaying vegetables, dung, fungi etc.; usually a semi-liquid medium is preferred for oviposition and larval development, so it is not often found in stored produce unless the material is badly decayed. As mentioned under *Sarcophaga*, in the Orient there is a great deal of production of proteinaceous sauces, and bean curd, soya sauce and oyster sauce in unsealed containers can be infested by fly larvae, which could cause intestinal myiasis when swallowed.

Damage. Usually only in produce already decayed, but can be found in liquids when contamination is the main offence, although intestinal myiasis can occur. The adults are seldom of direct importance as pests.

Life history. Eggs are laid in batches of up to 50 on suitable decaying or fermenting organic matter; the eggs have two small float-like appendages to keep them on the surface of liquids. They hatch in about 24 hours.

The larvae are characteristic in shape, being slightly flattened, and they have lateral processes bearing fine setae, used to propel the larvae through the semi-liquid medium (i.e. to swim). Larval development takes 7-30 days according to temperature.

Pupation takes place away from the food, but never deep in soil, inside the last larval exuvium, so the puparium has the characteristic larval form; pupation takes 1-4 weeks.

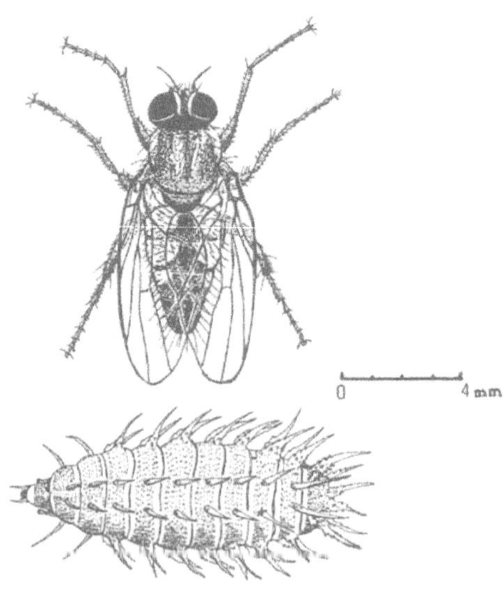

Figure 14.96. Fannia canicularis (Lesser house fly) (Diptera; Fanniidae), adult male fly and larva (maggot).

Adults are small grey flies, 6-7 mm long, with the thorax bearing three faint dark stripes; the female abdomen is ovoid and grey, but in the male it is narrow, tapering and blackish with pale yellow patches. In houses, the flies (usually male) are characteristically seen flying in erratic circles under pendant lamps. Since they settle less frequently on foodstuffs, they are usually less important as disease vectors in comparison with the other domestic flies.

This species prefers slightly lower temperatures than the house fly, and 24°C is recorded to be optimum; breeding is usually continuous, and the life cycle only takes 22-27 days.

14.11.13 Family Calliphoridae (Blowflies, Bluebottles, Greenbottles etc.)

A group of stout bristly flies, often metallic blue or green in colour; the antennal arista is markedly plumose, usually for its whole length; the dorsal surface of the abdomen is usually bare of bristles. The family is worldwide, but best represented in the Holarctic region. Within the group there has been a gradual evolution from saprophagous habits to animal parasitism, and several species are screw-worms.

Calliphora spp. (Blowflies, Bluebottles) (Figure 14.97)

Pest status. Several species are synanthropic and are serious hygiene and medical pests, and are of particular importance as food pests as they infest meat and fish products in domestic premises. Several genera are collectively known as blowflies – 'blowing' is the deposition of eggs on to exposed meat. In hanging game in Europe, and in countries where fish are air- and sun-dried, these flies can be very serious pests (Figure 14.98).

Produce. Exposed meats and dried fish are normally the produce attacked by these flies; meat products and prepared foodstuffs are at risk. In the wild, they infest animal corpses, and in produce stores they will breed on the bodies of dead rats and mice. In the Orient, it is traditional for shops to have dried fish, pressed ducks, dried squid, and several forms of dried meats (sausages etc.) on open display and at risk from blowflies.

Damage. Feeding on meats and fish causes putrefaction and spoilage; larvae can cause intestinal myiasis if ingested.

Life history. Eggs are laid on to the flesh; they are white, elongate and about 1.5 mm long; up to 600 eggs may be laid by one female during her lifetime. Hatching occurs after 18-46 hours at 18-20°C, but can occur after only 12 hours. The optimum temperature for development seems to be about 24°C.

Larvae are typical white maggots, and they feed on the flesh using their mouth hook; the three instars can be completed in 8-11 days; when fully grown, they measure 18 mm.

Pupation typically takes place in soil under the animal carcass, within an oval brown puparium, and requires 9-12 days.

Adults are stout bristly flies, 9-13 mm long, with a shiny blue coloration; they feed on liquids and also nectar from flowers, and are diurnal in habits. They are long-lived (about 35 days) and strong fliers; their flight makes a loud buzzing sound which is quite characteristic and may be disturbing in domestic premises. Flights of 4-28 miles have been recorded in experiments.

The life cycle can be completed in as little as 16 days when conditions are optimal; breeding is usually continuous.

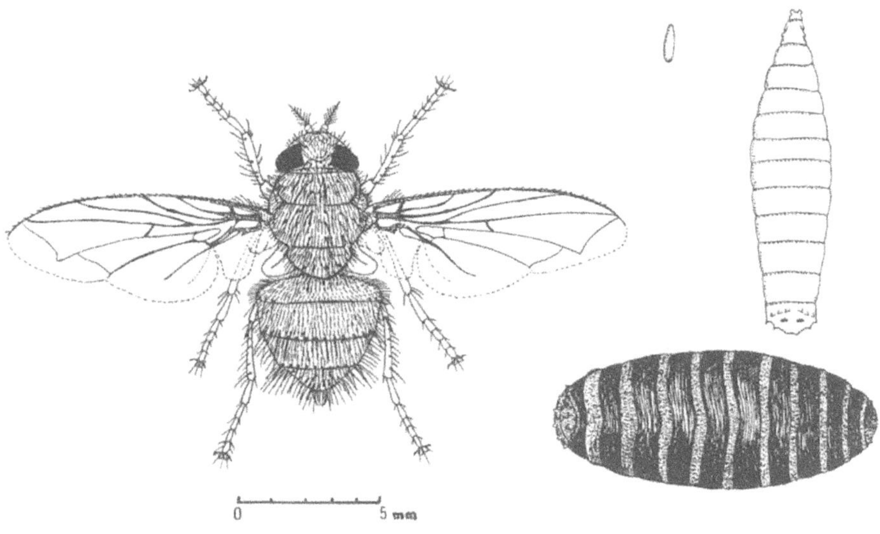

Figure 14.97. Calliphora sp. (Blowfly; Bluebottle) (Diptera; Calliphoridae), adult fly and immature stages.

Distribution. The genus is completely cosmopolitan, but more abundant in temperate regions, and to some extent replaced in the hot tropics by *Chrysomya* (Tropical latrine fly etc.). Most pest damage is attributable to only a few species (probably 5); not a large genus.

Lucilia spp. (Greenbottles) (Figure 14.99)

Pest status. These are also species whose larvae feed on flesh, and several are very important economically as they cause sheep-strike in Australia and other countries. They are not particularly synanthropic, and are most likely to be encountered on sun-dried fish and meats, or game left to hang.

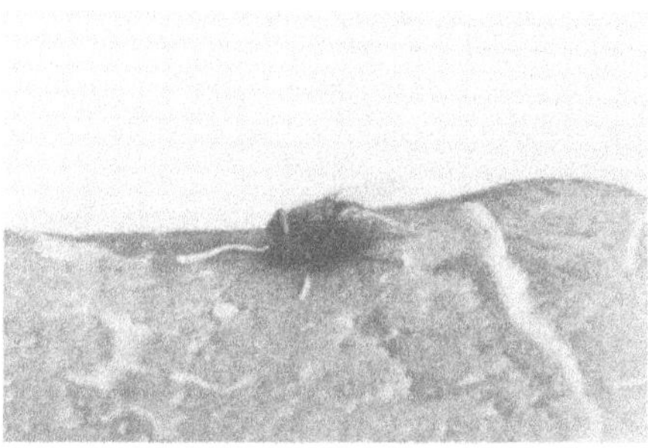

Figure 14.98 (a). Calliphora female laying eggs on a piece of meat (beef).

Figure 14.98 (b). Fly-blown meat; eggs of Calliphora on beef steak; Skegness, England.

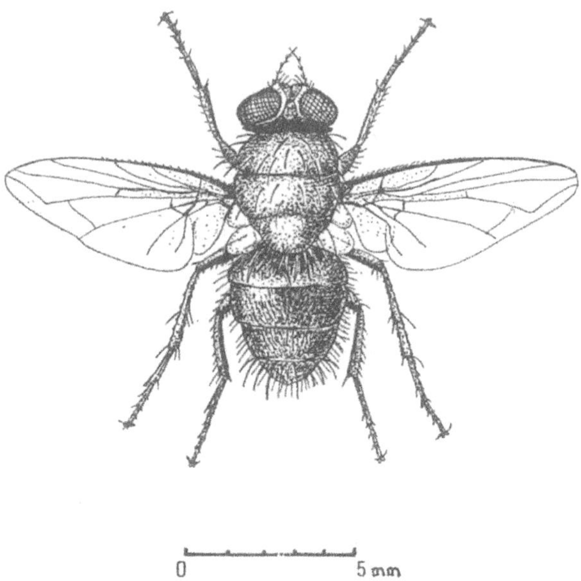

Figure 14.99. Lucilia sp. (Greenbottle) (Diptera; Calliphoridae), adult fly.

Produce. Only animal flesh is normally at risk – typically meats, bacon, dried fish and fishmeal, as far as stored products are concerned, or meat products.

Damage. Eating of the flesh induces putrefaction and spoilage, and the larvae, if ingested, can cause intestinal myiasis in man.

Life history. This is basically very similar to that of *Calliphora.* The adults differ in that they are mostly less bristly, and some are metallic green in colour.

Distribution. *Lucilia* contains a large number of species, and is quite cosmopolitan, but only a few are serious pests.

14.11.15 Family Sarcophagidae (Flesh flies)

This group is now separated as a distinct family, and the great majority of species belong to the single genus *Sarcophaga.*

Sarcophaga spp. (Flesh flies) (Figure 14.100)

Pest status. Several species are synanthropic, and can be serious domestic pests causing various types of human and animal myiasis.

Produce. Both meats and organic matter of a proteinaceous nature are susceptible. In Hong Kong, a serious case of human intestinal myiasis resulted from a poorly sealed bottle of soya sauce being infested. Decaying plant material can also be infested; the larvae are, basically, scavengers.

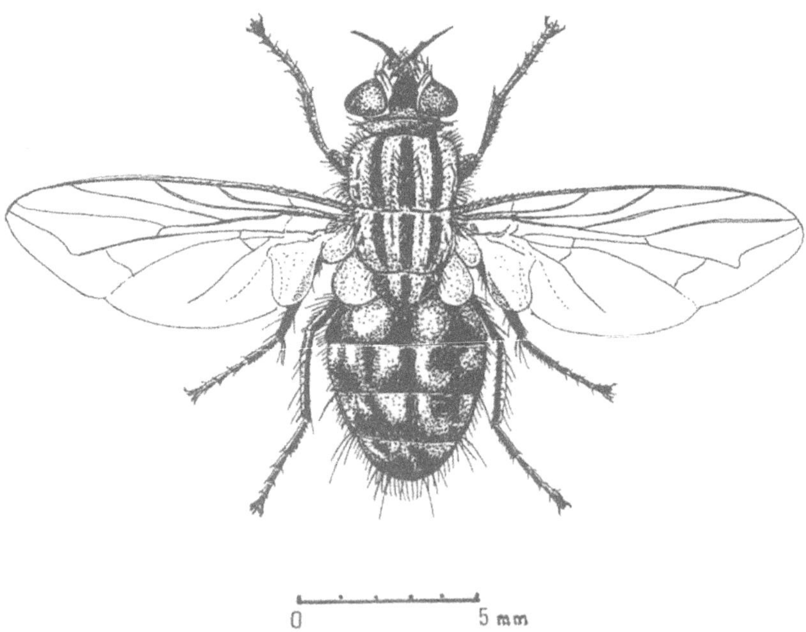

Figure 14.100. Sarcophaga sp. (Flesh fly) (Diptera; Sarcophagidae), adult fly.

Damage. Contamination of meats leads to spoilage and putrefaction, and larvae accidentally ingested can cause intestinal myiasis. Females lay first instar larvae, and infestation is instantaneous.

Life history. These flies are larviparous, and each female deposits 10-20 active first instar larvae on to a suitable substrate, the female being attracted by the smell. The substrate is usually faecal matter, both animal and human, but exposed meats are at risk.

The first instar larvae are active, and develop rapidly – the three instars can be completed in only 4 days.

Pupation takes 4-10 days, and the life cycle can be completed in 8-25 days under suitable conditions.

Adults are characteristic in appearance, grayish with a black-striped thorax and grey to black chequered abdomen with bristles at the posterior end; body length 10-14 mm; antennal arista plumose basally, but bare distally. There are many species, and they all look very similar. The adults are active in sunshine, and are to be seen on flowers feeding on nectar – they are quite long-lived.

Distribution. A cosmopolitan genus with many species, abundant throughout both the tropics and temperate regions. Nine species have been recorded as being a nuisance to man in all parts of the world, both warm and cool.

14.11.16 Family Agromyzidae (Leaf miners etc.) (1,800 species)

A large group of tiny flies whose larvae are mostly leaf miners in a very wide range of plants; they are of considerable importance agriculturally, but only occasionally encountered in produce stores. A number of *Liriomyza* and *Phytomyza* species mine leaves of vegetables and pupate at the end of the mine (Figure 14.101). Fresh Brassicas in vegetable stores are quite likely to have leaf miners *in situ,* and later, small black flies will emerge from the tunnels – they are 2-3 mm long (Figure 14.102). Dried cabbages could well contain viable pupae which would later yield adults. The larvae live in long winding tunnels (or, in some cases, in a blotch mine), and they pupate at the end of the tunnel with two small posterior spiracles protruding through the epidermis ventrally, and the small brown puparium is usually clearly visible.

Some species of *Melanagromyza* are Pulse pod flies, and larvae live inside the pods of beans, soybean and other pulses, and if intact pods are stored, these insects can be carried into the produce stores. Threshed pulses may show damage to the seeds caused by the feeding larvae.

14.11.17 Family Piophilidae (Skipper flies)

A small group of small metallic blue or black flies, with saprophagous habits. The larvae of some species are scavengers, others feed on a wide range of meats and cheese.

Figure 14.101. Brassica leaf showing larval tunnels and white pupae of Cabbage leaf miner (Phytomyza horticola); Hong Kong.

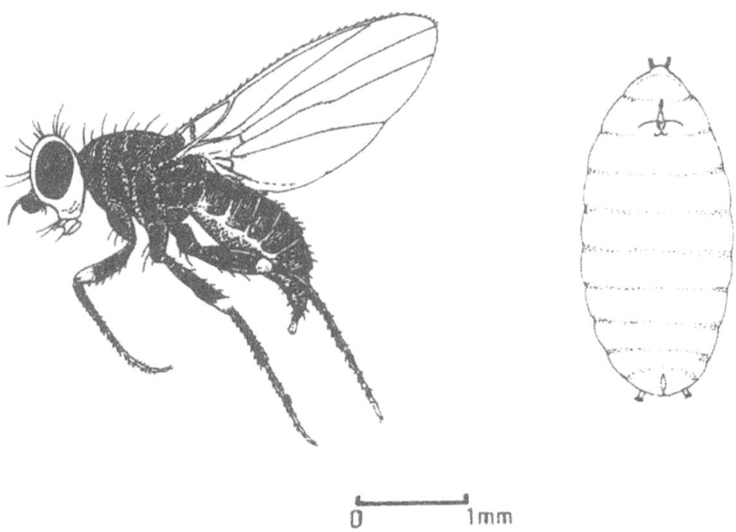

Figure 14.102. Phytomyza horticola (Pea and Cabbage leaf miner) (Diptera; Agromzidae), adult female fly and fully grown larva; from Chinese cabbage; Hong Kong.

Piophila casei (Linn.) (Cheese skipper) (Figure 14.103)

Pest status. Quite a serious pest of certain types of stored foods, but less important in recent years as more food stores become refrigerated.

Produce. Cheese, bacon, dried meats, salted beef and fish, dried fish, and other high protein foodstuffs, as well as dried bones. Animal carcasses and human cadavers have been infested.

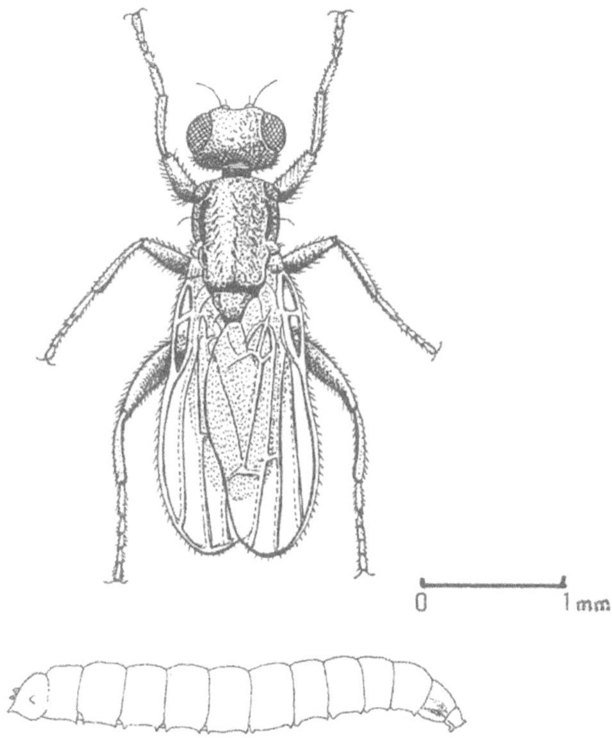

Figure 14.103. Piophila casei (Cheese skipper) (Diptera; Piophilidae), adult fly and larva (maggot).

Damage. Larval feeding causes the produce to spoil to some extent, but very likely the main damage is the possible intestinal myiasis which can be very painful and in extreme cases quite debilitating. This species of fly is probably the most serious in the world in causing human intestinal myiasis.

Life history. Eggs are laid on the produce by the female in response to the smell. The average number of eggs laid is about 150 per female.

The larvae tunnel deeply into the produce, and so many infestations are not readily obvious. They are called skippers because when fully grown they are able to jump (as much as 25 cm horizontally and 15 cm vertically). The larvae are very hardy, and have been passed through dog intestines experimentally remaining alive, and producing serious intestinal lesions.

Pupation takes place away from the food in a dark dry crevice.

The adult is a small blackish fly (about 4 mm long), with a yellow face and a rounded head; adults only appear in the warm season (summer), and they feed on juices from the larval food, but only live for 3-4 days.

The life cycle can be completed in as little as 12 days.

Distribution. A widely distributed species, but the precise limits are not at present known; in some regions, infestations may be quite local.

Control. The practice of hanging game animals and birds (partly to tenderize the flesh) and keeping cheeses until they are runny encourages this fly; generally, though, these habits are declining in most countries.

14.11.18 Family Tephritidae (Fruit flies) (1,500 species)

A large and worldwide group of considerable agricultural importance as the larvae develop inside fruits of all types, and some have other life styles. Some of the more important fruit pests are not yet cosmopolitan in their distribution, and there is both national and international legislation enforced to ensure that infestations are not imported into clean countries. The present tendency with fruit marketing in Europe and North America is to import various types of fruit from different regions at different times so as to ensure a continuous supply for retail purposes. Thus it becomes necessary to exercise greater care in the inspection of bulk fruit imports. The USA is particularly careful with fruit imports, for there is a local fruit production industry of great economic importance, and at regular intervals the exotic Medfly (*Ceratitis capitata*) gets accidentally introduced. Recent practices have included the routine fumigation of fruit cargoes with ethylene dibromide as a prophylactic measure. However, since it has been shown that ethylene dibromide can be carcinogenic, its use has been discontinued in the USA, and there is a search for a suitable alternative. The fruits most likely to be infested include *Citrus*, peach, melons and other Cucurbitaceae, and peppers (*Solanum* spp.).

The most dangerous species of fruit flies are the tropical and subtropical ones, for the temperate species are less damaging.

The Medfly can be taken as a typical example of the group, although several other species could be found in either local fruit stores in warmer countries, or in imported fruit stores in temperate regions.

Ceratitis capitata (Wied.) (Medfly) (Figure 14.104)

Pest status. A serious pest of subtropical fruits and other crops in many parts of the world, and potentially very serious in other countries where it is not as yet established.

Produce. A field pest of peach, citrus, plum, mango, guava, fig, and also cocoa and coffee (and many other fruits); its larvae can be carried into produce stores, but this pest is usually only univoltine in stores, and does not breed there.

Damage. Direct damage by eating the fruit usually results in the infested fruit being spoiled and unsaleable – larval tunneling is often associated with fungal and bacterial rots. In countries such as the USA, the potential damage, should the pest get established locally from imported fruits, is quite horrendous to contemplate; understandably, a single insect found would be sufficient for an entire shipment of fruit to be rejected.

Life history. Eggs are laid in batches under the skin of ripening fruits using the protrusible ovipositor; a total of 200-500 eggs are laid. Hatching occurs after 2-3 days.

The tiny white larvae bore through the fruit pulp and develop; there are three larval instars, and the development can be completed in 10-14 days. Typically, there are about 10 maggots per fruit, but larger numbers have been recorded. Infested fruits often fall prematurely.

Pupation usually takes place in the soil under the tree, in a brown oval puparium, and requires about 14 days.

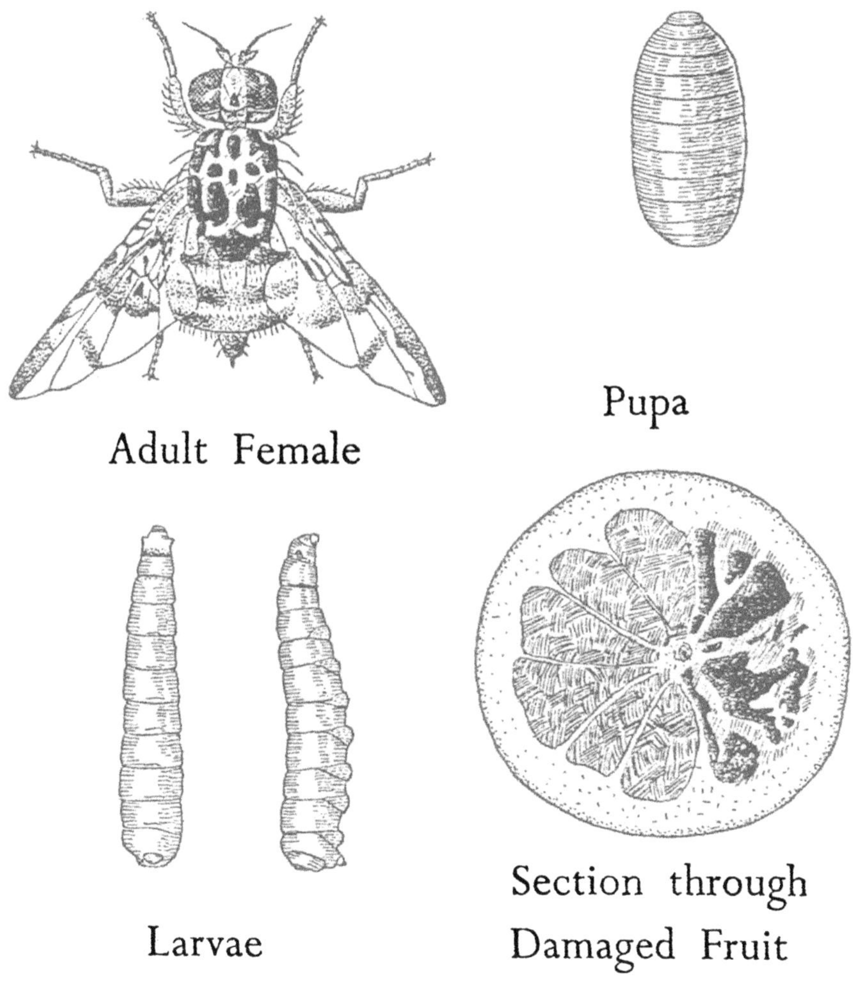

Figure 14.104. Ceratitis capitata (Medfly; Mediterranean fruit fly) (Diptera; Tephritidae), adult female and immature stages, and infested orange.

Adult flies are brightly patterned (black, yellow, white) with red to blue iridescent eyes, and mottled wings. Adults feed on sugary foods, and may live 5-6 months; they are 5-6 mm in body length.

The life cycle can be completed in 30-40 days, and there may be 8-10 generations per year in the field in warmer countries. In fruit stores, there is usually no subsequent breeding, and the fruit flies are univoltine. The great danger from some of these pests is that they may be accidentally imported into new countries in fruit cargoes.

Distribution. A subtropical species very widely distributed throughout the world, but absent from some regions, exterminated in others; there is extensive legislation to prevent its range increasing.

Control. Fruit cargoes for export are usually fumigated before leaving the country of origin, and are often fumigated again on importation for extra security. In Hawaii, fruit for export (papaya etc.) are either given hot-water treatment to kill fruit fly larvae, or irradiation using γ-rays. A newly-developed technique designed to replace the hot-water treatment is the high-temperature forced-air treatment which looks very promising.

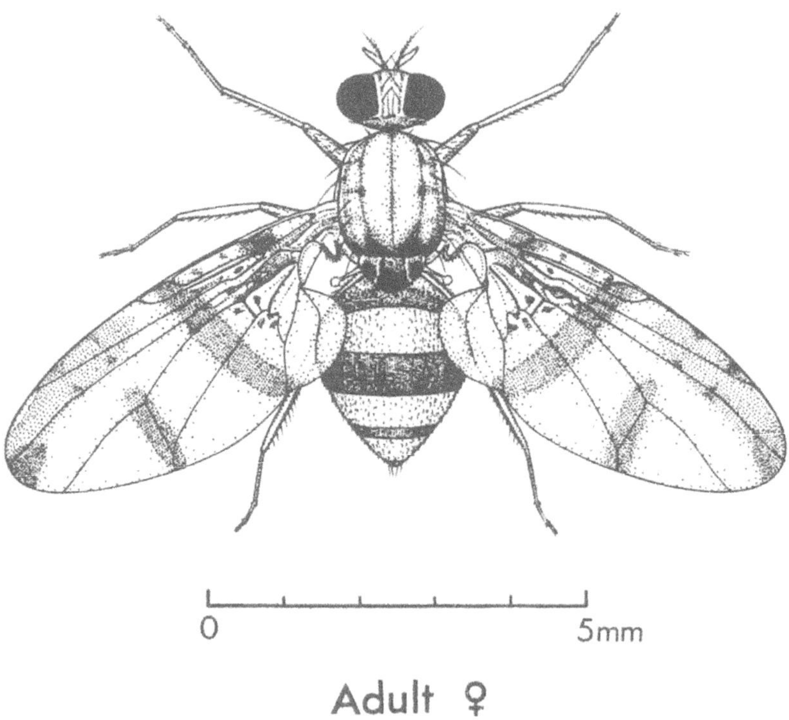

Figure 14.105. Ceratitis rosa (Natal fruit fly) (Diptera; Tephritidae), adult fly.

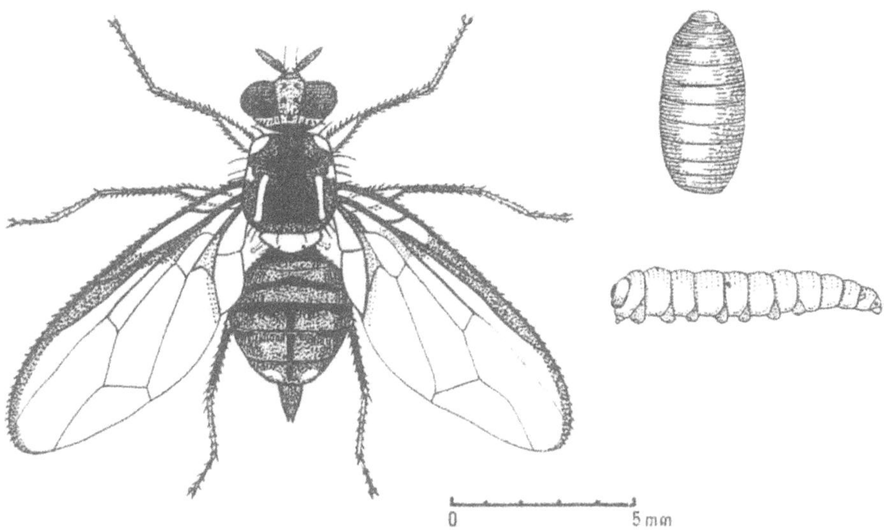

Figure 14.106. Dacus dorsalis (Oriental fruit fly) (Diptera; Tephritidae), adult female fly and immature stages.

Other fruit flies of international economic importance include:

Anastrepha spp. (3) Central and South America; on a wide range of fruits.
Ceratitis spp. (5) Subtropical Africa; on many different fruits (Figure 14.105).
Dacus spp. (10) Tropical Asia, Africa, Australasia, Hawaii; on many different fruits (Figure 14.106).
Pardalaspis spp. (2) (Solanum fruit flies). Africa; in fruits of Solanaceae.
Rhagoletis spp. (7) (Cherry/Apple fruit flies). Europe, Asia, USA; these are temperate species.

For further information on synanthropic flies, see Gorham (1991: pp. 195-205).

14.12 ORDER LEPIDOPTERA (Butterflies, Moths) (97 families, 120,000 species)

A very large group of insects, only a few are stored products pests, but those species are very important as they are abundant and widespread, and can be very damaging.

About 70 species of moths have been recorded infesting stored products, but the actual number of obligate stored products pests is low – less than a dozen. A number of other species feed on fruits and vegetables, and enter food stores/premises on or in these types of produce, but these are mostly field crop pests that are carried into produce stores where they may live for a while, and can be a nuisance economically.

The adults are characterized by having two pairs of large wings that overlap medially, and both wings and body are covered with tiny overlapping scales. The primitive biting and chewing mouthparts (mandibles) have gradually been lost during evolution and replaced by a long coiled (at rest) suctorial proboscis with which they suck nectar from flowers, and imbibe other juices.

The damaging stage is the larva; its eruciform body shape is characteristic of the Order, and it is known as a caterpillar. Larval mouthparts include mandibles used for biting and chewing, and in some groups the salivary glands have been modified to produce a continuous thread of silk, used mainly for making the pupal cocoon, but also as a larval life-line. Caterpillars have a well-developed head capsule, three pairs of thoracic legs ending in claws; the abdomen of ten segments typically bears short fleshy prolegs on segments 3-6, and a pair of terminal claspers. The prolegs bear rows of crotchets (tiny hooks) used for gripping. There are typically 5-6 larval instars. Each body segment has a conspicuous lateral spiracle, and the arrangement of the several bristles on each body segment (chaetotaxy) is usually specific, and can be used to identify the caterpillar (Hinton, 1943).

Eggs are globular or fusiform, upright, and distinctively sculptured, or flattened and ovoid; they are laid either singly, in small groups, or in large batches, and usually stuck firmly on to the substrate.

The pupal stage is usually concealed inside a silken cocoon often attached to the substrate or to packaging material, especially to the surface of sacks.

Most moths are nocturnal, and rely on the use of sex pheromones to bring the sexes together. Many of the stored products moths have had their sex pheromones synthesized, and they are available commercially for either population monitoring or for 'trapping-out' programmes.

Food preferences can be observed at times, but are often difficult to assess. A useful table has been compiled by Cox and Bell of the Slough Laboratory, MAFF, UK, as shown overleaf.

14.12.1 Family Tineidae (Clothes moths etc.) (2,400 species)

A large group of worldwide occurrence, but the literature may be confusing, for the family name was applied in the past to a greater assemblage than at present. Adults are characterized by having the head covered with rough scales; proboscis short or absent; labial palps porrect, and maxillary palps long and folded (adults do not feed at all); wings are typically slightly narrow with a long setal fringe. The larvae feed on dried animal material in the wild, and the clothes moths are important domestic pests. On stored products they are of less importance, but are regularly encountered, and a few species are phytophagous and can be damaging to stored grain and vegetable matter. The subfamily Tineinae contains several genera whose species all feed on animal materials and some of plant origin.

Table 14.1 Frequency of occurrence of moth species on stored commodities

Species	*High occurrence*	*Medium occurrence*	*Low occurrence*
Sitotroga cerealella	Wheat; corn	Rice; sorghum	Cereals; seeds; peanuts
Corcyra cephalonica	Rice; cocoa; peanuts; oilseed products	Corn; cereal products; seeds; pulses and products	Sorghum; millet; nuts; copra
Cadra cautella	Dried fruits and vegetables; nuts, oilseeds, and products; cereals and product; cocoa	Pulses and products; copra; carobs	Fallen fruits and nuts; spices; citrus pulp
Cadra figulilella	Dried fruits, especially dates, grapes, and figs	Nuts; carobs	Cereal products; oilseed products; cocoa
Ephestia elutella	Cereals; dried fruits and vegetables; tobacco; cocoa	Oilseeds and products; nuts; cereal products	Carobs; pulses and products
Ephestia kuehniella	Cereal products, especially flour	Cereals; nuts; dried fruits and vegetables	Oilseeds and products; dried citrus pulp
Plodia interpunctella	Dried fruits; nuts; cereals; oilseeds and products	Cereal products; cocoa; carobs; confections	Dried vegetables and products; pulses; seeds; citrus pulp; fallen fruits

(From: Gorham, 1991: p. 182)

Nemapogon granella (Linn.) (Corn moth, European grain moth)

Pest status. Of sporadic importance in different parts of the world; usually found in temperate regions, it can survive the winter in England.

Produce. This species is vegetarian, and as such differs from most of the rest of the family, and the larvae damage stored cereal grains. It is recorded to prefer rye, and then wheat; in Europe, oats and barley are seldom damaged, but in North America it is said to infest all types of grain, both in the field and in storage. It is also recorded feeding on nuts, dried mushrooms, and dried fruits.

Damage. The eating of the grain by the larvae is the main damage – this species is a primary pest of stored grain, but only occasionally is serious damage recorded.

Life history. Eggs are laid either on the ripening ear of grain in the field, or on the grains in storage.

The larvae are characterized by having six lateral ocelli on the head, and also the chaetotaxy (Hinton & Corbet, 1955; p. 52); the feeding larvae web the kernels together with silk.

Pupation takes place in the grain inside a silken cocoon.

The adult is a small creamy white moth, 10-14 mm wingspan, heavily mottled brown on the wings, and hindwings grey. On the forewings there are six brown costal spots.

Number of generations per year varies from 1-4, depending on the climate.

Distribution. Thought to have been Palaearctic in origin, but now cosmopolitan in the cooler regions of the world; for example, abundant in the northern states of the USA, but scarce in the southern; also found in parts of Africa.

Nemapogon variatella (Clemens)

This species is said to replace *N. granella* in parts of northern Europe on stored grain, but its precise status seems uncertain.

Most other species of *Nemapogon* are recorded feeding upon bracket fungi in the wild. But within the subfamily Nemapogoninae, the species *Haptotinea ditella* (P., M. and D.) found in Europe and West Asia is recorded feeding on stored grains, rice and groundnuts, and the closely related *H. insectella* (Fab.) is reported to have a similar larval diet.

Tinea pellionella (Linn.) (Case-bearing clothes moth) (Figure 14.107)

Pest status. A widespread and frequently encountered domestic pest that occasionally causes serious damage; but usually more important in houses than produce stores; they may feed on moulds on the walls.

Produce. Larvae feed mostly on animal materials of a keratinous nature (wool, fur, hair of all types, feathers, skin), but stored vegetable products are also recorded.

Damage. Direct damage is done by feeding larvae; on high value commodities such as furs and skins, losses can be very serious; larval cases can contaminate produce.

Life history. Eggs are slightly sticky and may adhere to the substrate; they can be distinguished from those of clothes moths by the sculpturing which consists of distinct longitudinal ridges.

The larvae are distinctive in that they carry a portable silken case, sometimes even up vertical walls – only the head and thorax protrude. The case is constructed of silk, but with other fibres and bits of detritus incorporated, so on infested material it is not at all conspicuous, but on the walls of the building it is very obvious.

Prior to pupation, the larva seals both ends of the case and attaches it to the substrate. At the end of the development period, the pupa forces its way out of the end of the case before the adult moth emerges.

The adult is a small brown moth with darkish forewings on which three faint spots are evident; the hindwings are whitish. Wingspan is 9-16 mm. Males are paler, smaller, and quite active; but females are sluggish and only fly short distances.

Distribution: Widely distributed throughout the Holarctic Region and in Australia.

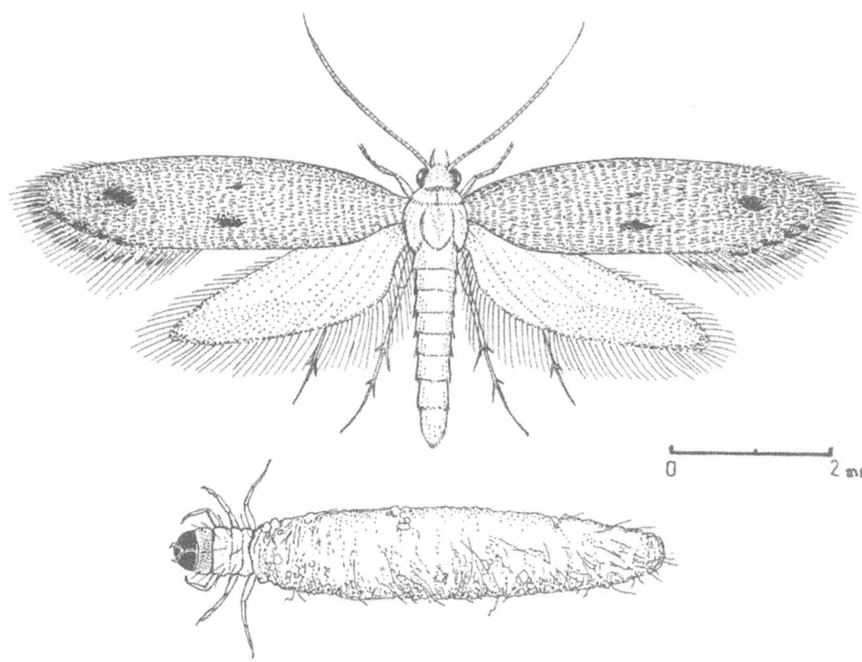

Figure 14.107. Tinea pellionella (Case-bearing clothes moth) (Lepidoptera; Tineidae), adult moth and larva with case.

Other species of *Tinea* recorded as pests of stored products include:

Tinea pallescentella Stainton (Large pale clothes moth). Cosmopolitan; on a wide range of animal materials, especially wool, hair and feathers.

Tinea translucens Meyrick (Tropical case-bearing clothes moth). In tropical countries, this species replaces *T. pellionella*, and together they are totally worldwide in distribution.

Tinea vastella = *Ceratophaga vastella*

Tinea spp. About 6 other species are recorded feeding on woollen materials and other produce of animal origin in storage; in all instances, the larvae live inside a small silken case.

Tineola bisselliella (Hummel) (Common clothes moth) (Figure 14.108)

Pest status. Although this is a very serious pest in domestic dwellings, it is of less importance in produce stores, for it only feeds on material of animal origin.

Produce. It appears to feed only on material of animal origin, preferring keratin, namely fabrics of wool, and other animal fibres. On mixed fibres, such as wool and cotton, the other fibres may be bitten through though not eaten. Woollen fabrics and furs are most at risk. The larvae are also recorded on dried fishmeal and on hides in the tropics.

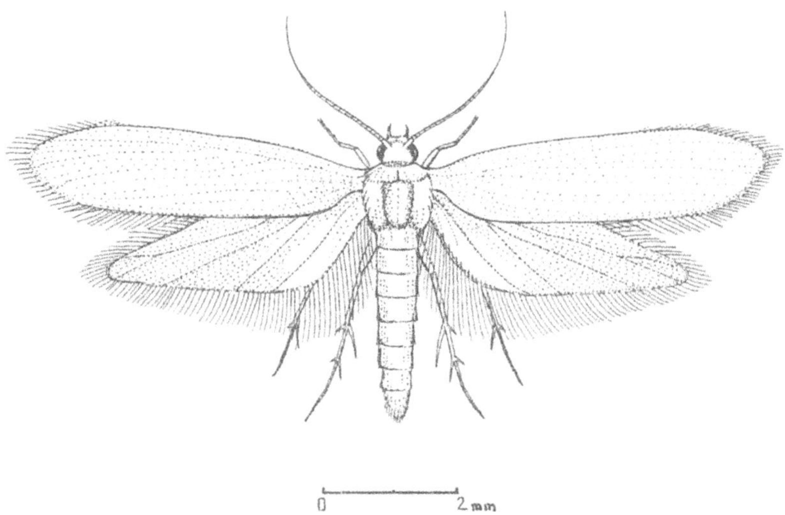

Figure 14.108. Tineola bisselliella (Common clothes moth) (Lepidoptera; Tineidae), adult moth.

Damage. Direct eating makes holes in the fabric, and in mixed fibres the non-animal fibres may be bitten through. Larval frass is usually conspicuous, and the larvae often spin a silken shelter on the substrate in which they live.

Life history. Oval eggs (0.5 x 0.3 mm), slightly sticky, are laid on the substrate; 50-100 eggs per female; they are usually all laid within 3 days. At 30°C the eggs hatch in 6 days (10 days at 20°C).

The caterpillar is creamy-white, without distinct ocelli, the head is brown. After 4 moults, the caterpillar reaches 10 mm in length. The larva spins silk from the modified salivary glands which open under the head – the silk is used to make a loose little gallery or tube in which it lives, and faecal pellets accumulate at the edge of the gallery.

The pupa is shiny reddish-brown, about 7 mm long, and the abdominal segments are quite mobile. Pupation takes place within a spindle-shaped cocoon in the larval gallery.

Figure 14.109. (a) Cow horns covered with larval tubes of Ceratophaga sp. (Lepidoptera, Tineidae); Alemaya, Ethiopia.

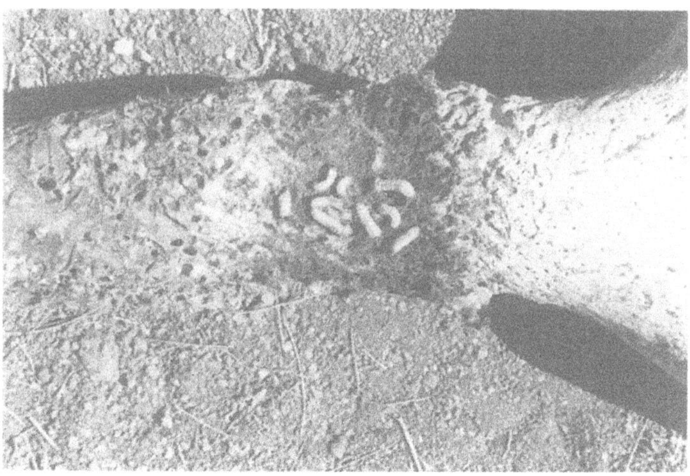

Figure 14.109. (b) Same horn with tubes removed revealing the Ceratophaga larvae and tunneled bone; Alemaya, Ethiopia.

Adults are uniformly buffish (almost golden) in colour, 9-17 mm wingspan, body length (at rest) 6-8 mm; hindwing greyish; both wings with the typical long fringe of setae. Adults are mostly found on the larval food material, but they run swiftly and fly readily, living for 2-3 weeks.

In temperate climates under cool conditions, this species is often univoltine, but in the tropics and in heated premises, breeding may be continuous; larval diapause is known. It can survive a cold winter in England.

Distribution. This species is quite worldwide in distribution, in both tropical and temperate regions, after extensive transportation by human agency and commerce.

Other species of Tineidae (Tineinae) recorded in produce stores include:

Ceratophaga spp. In parts of tropical Africa and Asia; on hooves and horns of cattle and antelopes (Figures 14.109 (a) and (b)).
Monopsis laevigella (D. and S.) (Skin moth). Holarctic; on skins and foodstuffs of animal origin.
Monopsis spp. Four species. Holarctic but now worldwide; on flours, oats, various seeds, woollen fabrics, skins (and birds' nests, owl pellets, fox scats).
Niditinea fuscella (Linn.) (Brown-dotted clothes moth). Holarctic at least; on woollen materials and also on stored vegetable matter.
Setomorpha rutella Zeller. In West Africa; on a wide range of dried plant materials, including tobacco leaf, rice and cocoa.
Trichophaga tapetzella (Linn.) (Tapestry moth). Worldwide; found only on coarse material of animal origin, especially hair, fur and feathers.

A closely related species of *Trichophaga* has been recorded on a cargo of feathers from Central Africa.

14.12.2 Family Heliodinidae (400 species)

A small cosmopolitan group of moths that sit at rest with the hind legs raised or pressed against the side of the abdomen. One species has been recorded associated with grain.

Stathmopoda auriferella (Walker)

Reported to date only from Nigeria on sorghum, first in the standing crop and later in storage. The larvae feed on the outside of the grains, and they live inside small silken tubes. Adults show resemblance to *Sitotroga cerealella* (TDRI, 1984; p. 175).

14.12.3 Family Yponomeutidae (800 species)

A somewhat diverse group, with some small temperate species and large, brightly-coloured tropical ones. Only a few species are of possible concern as far as crop produce in storage is concerned.

Plutella xylostella (Linn.) (Diamond-back moth) (Figure 14.110)

This completely worldwide pest of Cruciferae is a small moth of characteristic appearance, and the caterpillar is a major international pest on vegetable crops. The little caterpillars tend to make 'windows' on cabbage leaves, but sometimes eat right through the lamina to make small holes. Many cabbages and other brassicas are taken into vegetable stores with some holes in the leaves made by this pest. Pupation takes place inside a flimsy silken cocoon firmly attached to the foliage of the plant.

In 1976 a spectacular and costly infestation of broccoli heads was discovered in a refrigerated vegetable store in Hong Kong. Many kilos of this expensive vegetable had been purchased from Taiwan for use in airline meals by a Hong Kong airline, and had been kept refrigerated until needed. But in all the flower heads there were pupae of diamond-back moth inside cocoons clearly visible, but very firmly attached to the stems and flower stalks – the entire consignment had to be discarded. Such infestations could easily occur again elsewhere; even though the damage is really only aesthetic, the produce is rejected.

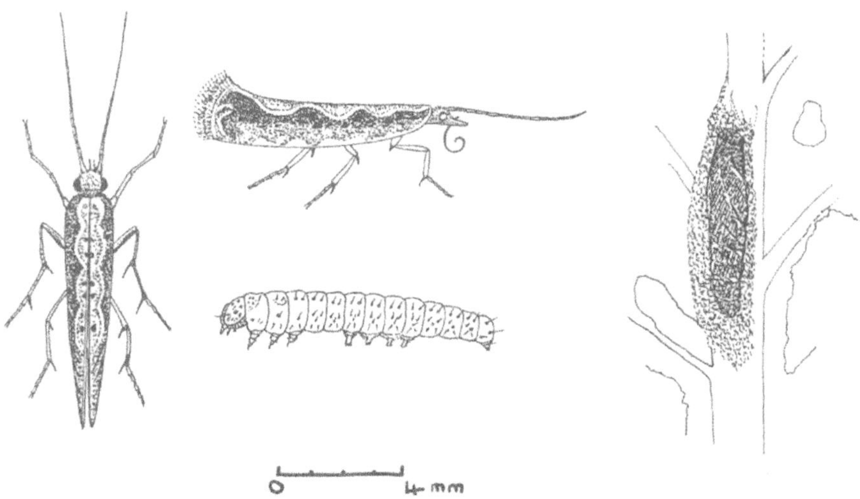

Figure 14.110. Plutella xylostella (Diamond-back moth (Lepidoptera; Yponomeutidae), adult moths and immature stage.

Other species of Yponomeutidae that might occur in fruit stores include:

Argyresthis spp. (Fruit moths). Europe and Asia; several species have caterpillars that bore inside fruits of apple, cherry and plum. The larvae emerge to pupate in soil litter, and so could emerge in storage.

Prays endocarpa Meyrick (Citrus rind borer). S.E. Asia; larvae mine the skin (rind) of citrus fruits.

Prays oleae Fab. (Olive moth). Mediterranean region; larvae tunnel inside the fruits of olive.

14.12.4 Family Oecophoridae (House moths etc.) (3,000 species)

A largish group of small moths with diverse habits; the forewings are not so narrowed at the tips as in the Tineidae, and the hindwings are usually broader, and many species are a little larger (12-22 mm wingspan); the antennae usually bear a basal pecten. Larval habits are varied – some feed in seed heads, or inside spun leaves, or decaying wood, and a few species are known as house moths and live in domestic premises.

The two common domestic species are quite regularly encountered in produce stores, although damage to produce is usually slight.

Endrosis sarcitrella (Linn.) (White-shouldered house moth) (Figure 14.111)

A distinctive little moth, male body length 6 mm, female 10 mm, and wingspan of male about 15 mm and female 17-20 mm; the head, prothorax and base of forewings is white, which gives the moth a characteristic appearance. The white caterpillars grow to about 13 mm, and they feed on a wide range of materials, especially wool, feathers, and a selection of plant materials including meals and damaged grains.

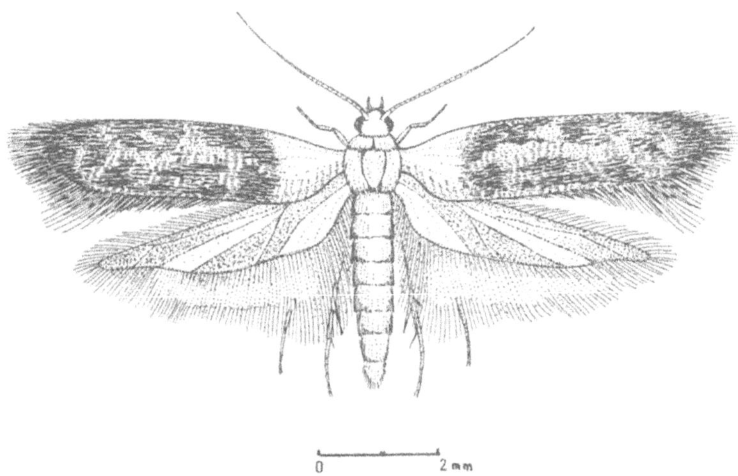

Figure 14.111. Endrosis sarcitrella (White-shouldered house moth) (Lepidoptera; Oecophoridae), adult moth.

It is classed as a general feeder and scavenger; in Europe it is more often found in produce stores than it is in America. The species is cosmopolitan, but more abundant in temperate regions. In the UK, there are usually 2-3 generations per year, but there may be 4. Optimum temperature range is 24-26°C, but will develop at 15°C.

Hofmannophila pseudospretella (Stainton) (Brown house moth) (Figure 14.112)

Pest status. A widespread and versatile pest found in many domestic situations; commonly encountered, but damage levels are not usually high.

Produce. A tremendously wide range of materials is eaten by the larvae, including cereals, cereal products, meals, pulses, seeds, dried fruits, dried plants, wool, fur, feathers, leather, book bindings, cork, and corks in wine bottles, and paper. They can also eat through plastic and insulation materials. In the wild, they live in birds' nests, and feed on corpses; in houses, carpets and Hessian underlays are often infested.

Damage. Direct eating is the main source of damage, but contamination can be a problem; some silk is produced, but not a very conspicuous amount (not as much as by clothes moths). Larvae can digest keratin.

Life history. Eggs are laid in the produce and debris, and development takes 8-110 days according to temperature; they are generally resistant to desiccation.

Larvae are white with a brown head capsule and a small dorsal plate on the prothorax. Fully grown, they measure 18-20 mm. Only a little silk is produced whilst feeding. They require a high humidity for rapid development, and take from 70-140 days according to temperature; larval diapause is quite common, and during this period the caterpillar is resistant to desiccation and insecticide action. They often overwinter as diapausing larvae.

Pupation takes place inside a tough silken cocoon, and can be completed in 10 days; the cocoon usually incorporates particles of produce.

The adult moths show sexual dimorphism, the female being larger, 14 mm long, with 18-25 mm wingspan; the male is 8 mm long with 17-19 mm wingspan. The moths are bronze-brown with darker flecks in the forewings; the hindwings are considerably broader than in *Endrosis*. Males fly more than do the females, who tend to run and hide when disturbed.

The reason for the success of this species appears to be a combination of omnivorous diet, high reproductive capacity, resistant egg and pupal stages, and larvae capable of diapause.

The life cycle in laboratory experiments varies from 190-440 days; in produce stores in the UK, the species is univoltine.

Distribution. Cosmopolitan, but most abundant in the temperate regions.

Natural enemies. Predators include the mite *Cheyletus eruditus*. Larvae of the house moths are parasitized by *Spathius exarator* (Braconidae), and preyed upon by larvae of *Scenopinus fenestralis* (Window fly).

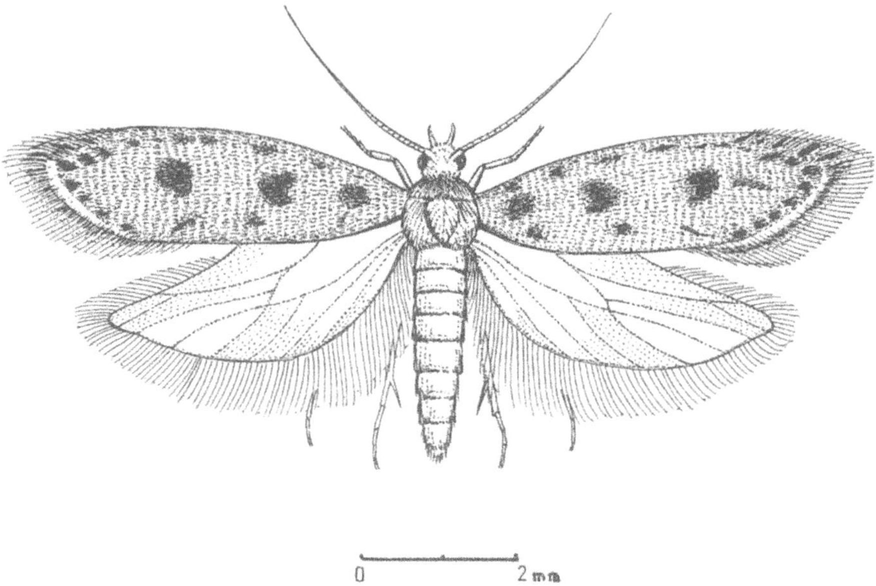

Figure 14.112. Hofmannophila pseudospretella (Brown house moth) (Lepidoptera; Oecophoridae), adult moth.

Three other species of Oecophoridae have been recorded in stored produce, but are of little importance in stores:

Anchonoma xeraula Meyrick (Grain moth). Japan; in grain stores.
Depressaria spp. (Parsnip and carrot moth). Europe, Asia, USA; larvae infest seed heads in the Umbelliferae.
Promalactis inonisema Butler (Cotton seedworm). Japan; on stored cotton seed.

14.12.5 Family Cosmopterygidae (Fringe moths) (1,200 species) (= Momphidae)

There are several species of *Pyroderces* whose larvae infest opened cotton bolls and eat the seeds; coffee berries, bulrush millet and maize cobs in the field are also attacked in parts of Africa and India.

Pyroderces rileyi (Walsingham) (Pink scavenger caterpillar) (= *Sathrobrota rileyi*)

This is a pest of maize in the USA, first in the field and later in storage; damage can be considerable, especially on ripening crops in the field and in crib storage on the farms, though it is seldom serious in commercial shipments (USDA, 1986).

14.12.6 Family Gelechiidae (4,000 species)

A large family with many species, but only a few are associated with stored products. The moths are small, with narrow tapering trapezoidal forewings distally pointed; posterior margin of hindwing is sinuate; both wings have long setal fringes. The proboscis is basally covered with many scales, and the labial palps are mostly long and strongly re-curved (upwards). Larval habits are very varied, but many bore into fruits and seeds or the stems of plants, and a few are leaf miners.

Phthorimaea operculella (Zeller) (Potato tuber moth) (Figure 14.113)

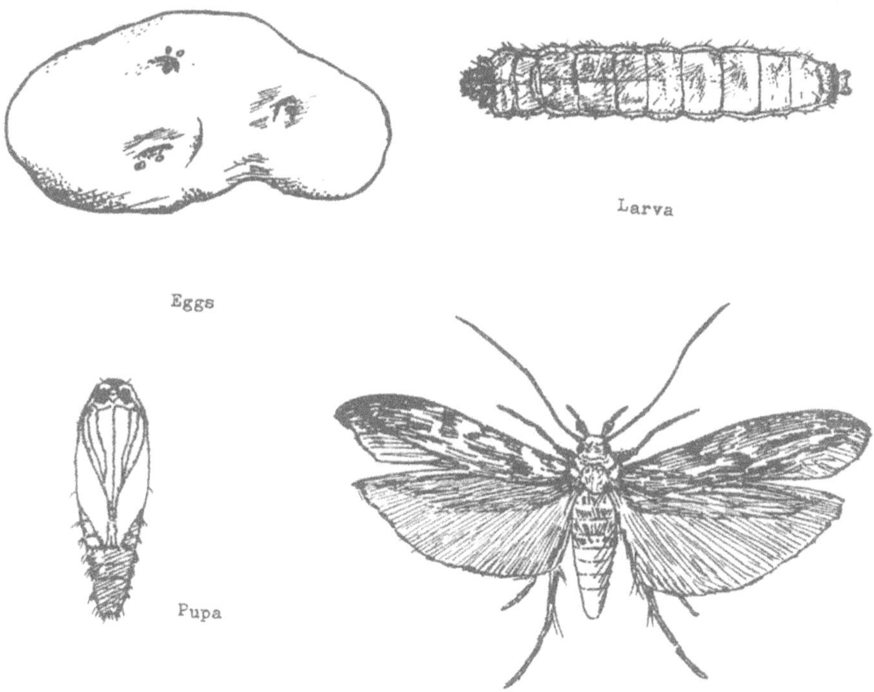

Figure 14.113. Phthorimaea operculella (Potato tuber moth) (Lepidoptera; Gelechiidae), adult moth and immature states; specimens from Kenya.

Pest status. A serious pest of stored potatoes throughout most of the warmer parts of the world. Essentially a pest of the growing crop, where the larvae start as leaf miners before boring down the stems; but it has adapted to survive in stores, and several generations can develop on stored tubers.

Produce. A pest of potato tubers in storage, but also recorded from tobacco, tomato, eggplant and other *Solanaceae*. Leaf mining in tobacco can be of importance, and can be seen in storage.

Damage. In potatoes, the larvae tunnel the tubers, both just under the surface and deep in the tuber; most tunnels become infected with fungi or bacteria, and rots develop (Figure 14.114). Losses of tubers in store can be very high, for tuber moth breeding can be continuous. Leaf mining in tobacco can cause produce rejection.

Figure 14.114. Potato tubers cut to show interior tunneled by larvae of Potato tuber moth, and associated rots, at harvest; Alemaya University Farm, Ethiopia.

Life history. Eggs are laid near the eyes of the tubers in storage; each female lays some 150-250 eggs; hatching requires 3-15 days according to temperature. In the field crop, eggs are laid on the leaves.

Larvae tunnel under the tuber epidermis and then go deeper into the tissues. Larval development takes 9-33 days, and the full grown larva measures 9-11 mm. On the growing plant, the tiny caterpillars start as leaf miners, but gradually move into the petiole, then the stem, and finally down into the roots and tubers.

Pupation usually takes place inside the damaged tuber, within a silken cocoon, and takes from 6-26 days.

The adult is a small grey-brown moth, of wingspan about 15 mm. The moths fly readily, and usually many disperse from the stores into growing crops in the field; they are, however, quite short-lived.

The life cycle can be completed in 3-4 weeks under warm conditions. On stored tubers, there are usually a few generations, causing considerable damage, but then they need to revert to a growing plant for full vigour and development.

Distribution. Almost cosmopolitan throughout the warmer regions of the world, but to date there are no records from West Africa and parts of Asia.

Control. With heavy infestation, fumigation of the tubers may be required. The male moth population can be considerably reduced by the use of sticky traps baited with female pheromones.

Sitotroga cerealella (Oliver) (Angoumois grain moth) (Figure 14.115)

Pest status. A major pest of stored grains throughout the tropics and subtropics; typically starting as a field infestation in the ripening cereal panicles, and the infested grain is carried into the stores. Crop losses can be very high, increasing directly with time.

Produce. Sorghum, maize, wheat, barley, and millets are the main crops attacked, but all cereals are vulnerable, including paddy rice. In produce stores, breeding is usually continuous. It requires whole grains for breeding.

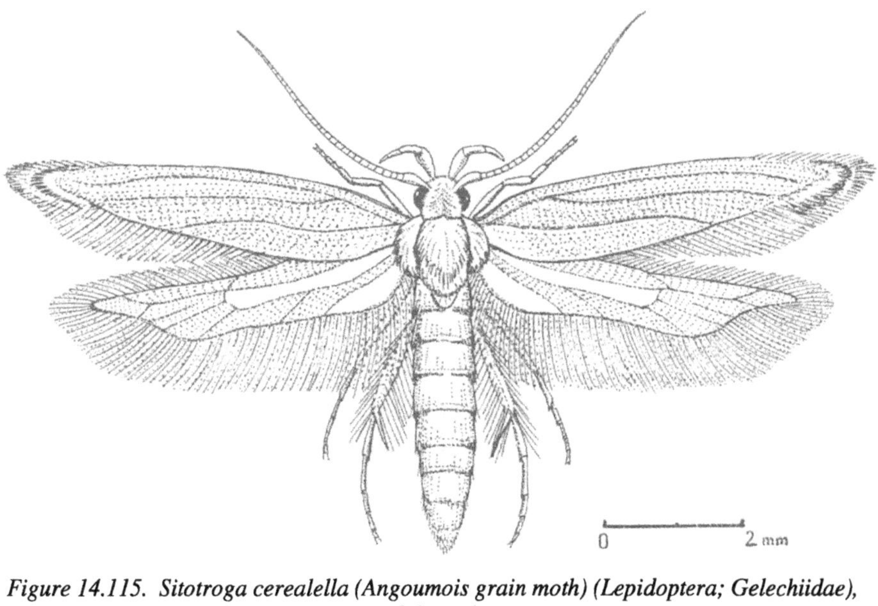

Figure 14.115. Sitotroga cerealella (Angoumois grain moth) (Lepidoptera; Gelechiidae), adult moth.

Damage. Each larva eats out the inside of a single sorghum grain during its development (Figure 14.116); the most serious infestations are in on-farm stores. At the time of harvest, the panicle shows no sign of infestation usually, and the first adults emerge some weeks later in storage.

Life history. Eggs are laid singly or in small groups on the grains; each female lays about 200 eggs over a 5-10 day period.

The newly hatched caterpillar bores directly into the grain, and typically remains inside the grain for both larval and pupal development. Larval mortality is high if the grains are very hard. Larval infestation leaves no visible symptoms on the grains. Total larval development can be completed in 19 days at 30°C; humidities between 50-90% seem to have little effect on the rate of development. Before pupation, the tiny larva constructs a chamber just under the grain seed coat, forming a small circular translucent 'window'.

Pupation takes place within the chamber inside a delicate cocoon, and at 30°C requires about 5 days.

Figure 14.116. Sorghum panicle, about one month after harvest, showing emergence holes in grains by Sitotroga cerealella, initially a field infestation; Alemaya University, Ethiopia.

The adult is a small pale brown moth, 5-7 mm long, and both wings are narrow and quite sharply pointed. The labial palps are long, slender, sharply pointed and upcurved. Adults fly well, and cross-infestation occurs readily; but they are short-lived, and generally survive only 5-12 days.

At 30°C and 80% RH the life cycle is completed in 28 days or a little more; and in suitable stores, breeding may be continuous throughout the year. The intrinsic rate of population increase is estimated at x 50 per month.

It is thought that the larvae compete with those of *Sitophilus*, so this species may be more important under dry conditions – which it tolerates but *Sitophilus* finds less favourable. However, in Ethiopia, at Alemaya, the sorghum was heavily attacked by *Sitotroga*, and the adjacent maize crops equally heavily attacked by *Sitophilus oryzae* – there was clearly a very definite preference being expressed.

Distribution. This species is cosmopolitan throughout the tropical and subtropical parts of the world.

Natural enemies. Parasites include *Trichogramma* spp. attacking the eggs, and both *Habrocytus cerealellae* (Hymenoptera, Pteromalidae) and *Bracon hebtor* (Hymenoptera, Braconidae) attacking the larvae. Eggs are eaten by predacious mites. Populations may be limited by competition from *Sitophilus* weevils and *Rhizopertha dominica*, but population interactions are probably very complex.

Control. Standard insecticide and fumigation treatments are usually effective against this pest. Male moths can be caught in sticky traps baited with synthetic female sex-attractant chemicals.

Keiferia lycopersicella (Wlsm.) (Tomato pinworm)

The tiny caterpillars sometimes bore into the fruits of tomato, and then may be taken into vegetable stores; USA, Central and South America.

Pectinophora gossypiella (Saunders) (Pink bollworm) (Figure 14.117)

A serious pest of cotton worldwide; the larvae inhabit the cotton bolls, and often end up inside the seeds; thus the larvae may be taken into produce stores inside infested cotton seeds. Larval development will continue in storage, and pupation will occur either within the hollowed seed or in a loose cocoon between adjacent seeds, but there will be no breeding of this pest on stored cotton seeds. Some larvae regularly practise diapause when conditions are inclement, and so adult moths may be found in the produce months after harvest.

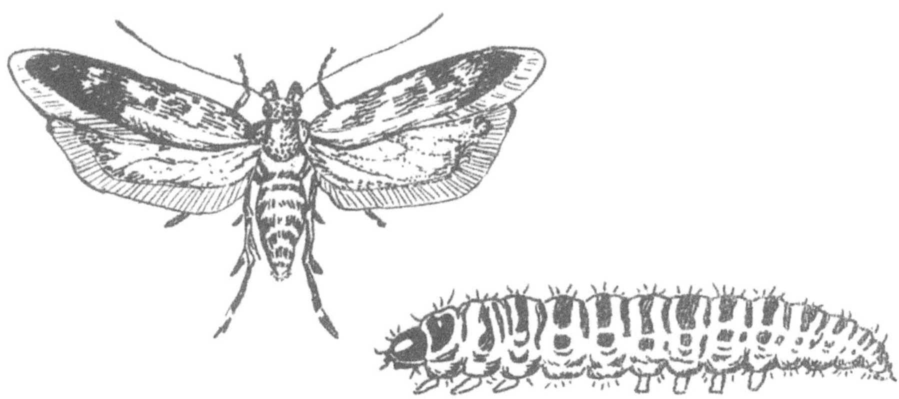

Figure 14.117. Pectinophora gossypiella (Pink bollworm) (Lepidoptera; Gelechiidae), adult and caterpillar; specimens from Kenya.

Scrobipalpopsis solanivora Povolny (Central American potato tuber moth)

A very similar species to *Phthorimaea operculella* in all aspects of its biology, and the damage done to stored tubers; but it is at present confined to Central America, so on a global basis it is regarded as a minor pest.

Anarsis lineatella (Peach twig borer)

In California, the two summer generations (3rd and 4th) attack peach fruits; crops with damage in excess of 4% are rejected by canners, and if more than 5%, the crop fails grading requirements.

14.12.7 Family Tortricidae (4,000 species)

A large group of small dark moths, completely worldwide in distribution but most abundant in temperate regions, where a number are known as fruit tree tortricids, and they are major pests of growing crops. Larval habits are very diversified; some bore into ripening fruits, others are found in legume pods or nuts, eating the seeds inside.

However, there is no one species that is of particular importance in produce stores, but on a worldwide basis there are about 12 species that will occasionally be found in harvested crops, especially on fresh crops in on-farm stores; these are listed below:

Archips argyrospilus (Walker) (Fruit tree leaf roller). USA; Canada; larvae are sometimes found inside harvested walnuts.

Cryptophlebia leucotreta (Meyrick) (False codling moth). Africa, south of the Sahara; polyphagous in a wide range of fruits, including cotton bolls, citrus fruits, and macadamia nuts. Citrus fruits have a small necrotic spot on the rind, and then a small caterpillar eating within. (Figures 14.118, 14.119)

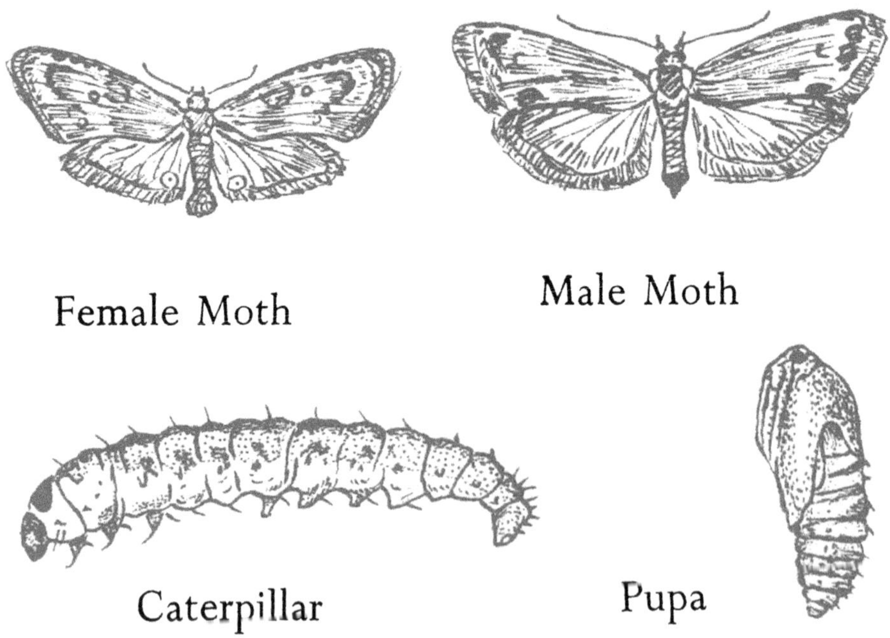

Figure 14.118. Cryptophlebia leucotreta (False codling moth) (Lepidoptera; Tortricidae), adult male and female moths, and immature stages; specimens from Uganda.

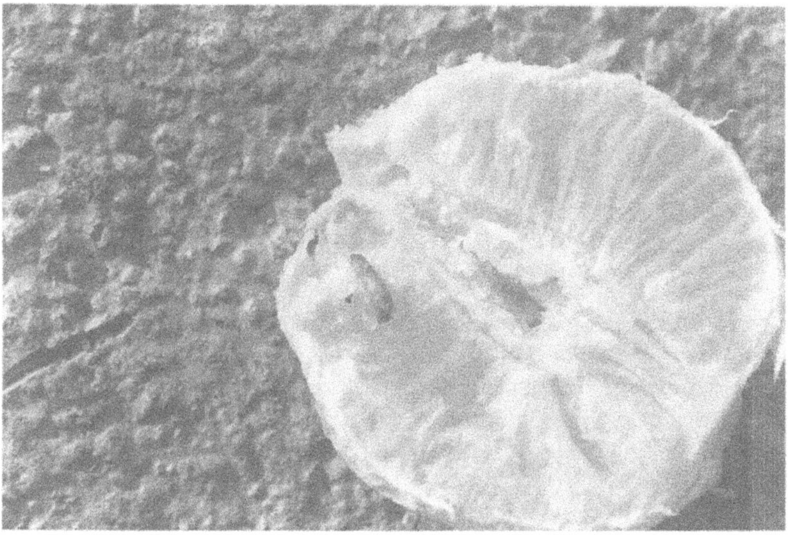

Figure 14.119. Orange fruit cut to show false codling moth pink caterpillar and frass; Dire Dawa, Ethiopia.

Cryptophlebia ombrodelta (Lower) (Macadamia nut borer). Pantropical; a serious pest of macadamia; the larvae bore into the nuts to eat the kernel. Nuts are found in storage with larval emergence holes. (Figure 14.120)

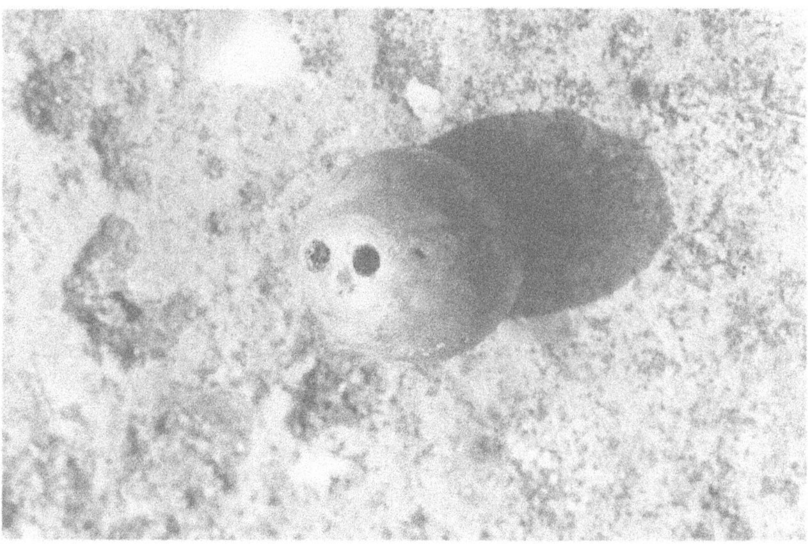

Figure 14.120. Macadamia nut showing two emergence holes of caterpillars of Cryptophlebia ombrodelta (Macadamia nut borer); Malawi.

Cydia funebrana (Treit.) (Plum fruit maggot.) Europe, Asia; the caterpillars live inside the ripening plums.

Cydia molesta (Busck) (Oriental fruit moth). Southern Europe, China, Japan, Australia, USA, South America; caterpillars bore inside fruits (peach, other stone fruits, apple) but emerge to pupate outside. (Figure 14.121)

Cydia nigricana Fab.) (Pea moth). Europe, North America; a serious pest of pea crops – the caterpillars live inside the pod and eat the seeds, and damaged seeds are commonly found in storage. Infested pods give no external sign of damage, so fresh pods may often contain one or more caterpillars with several eaten peas. (Figure 14.122)

Cydia pomonella (Linn.) (Codling moth). A cosmopolitan species attacking apples wherever grown; the small white, later pink, caterpillar bores inside the fruit (all Rosaceae), but emerges to pupate inside a cocoon in packaging etc. Fruit in storage may yield either caterpillars, pupae in cocoons, or later the small dark adult moths; in the USA, walnuts are often infested. (Figures 14.123, 14.124, 14.125)

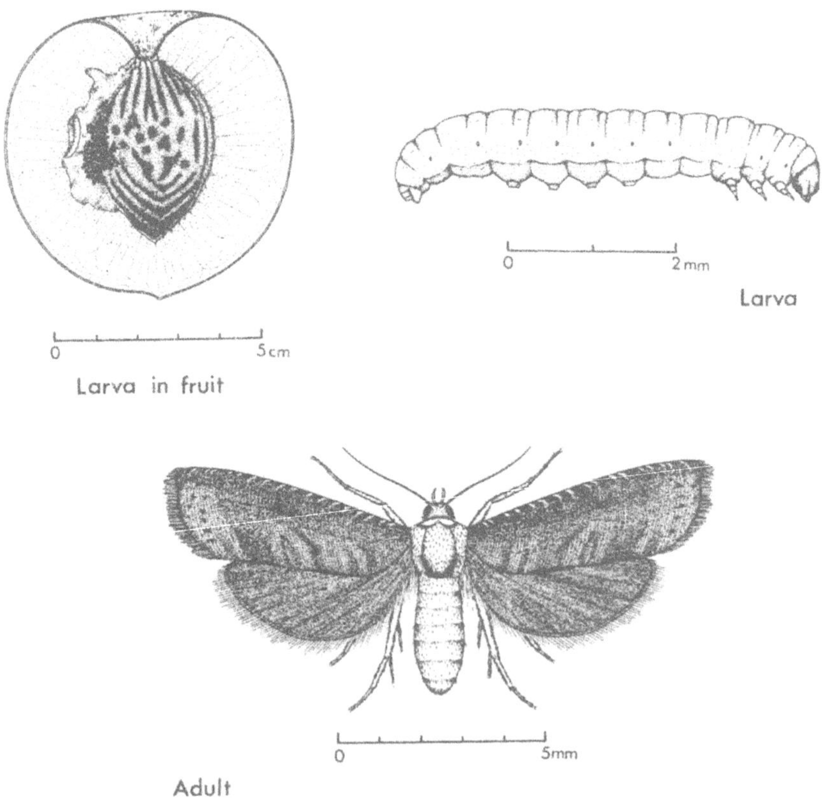

Figure 14.121. Cydia molesta (Oriental fruit moth) (Lepidoptera; Tortricidae), adult moth, caterpillar, and infested peach fruit.

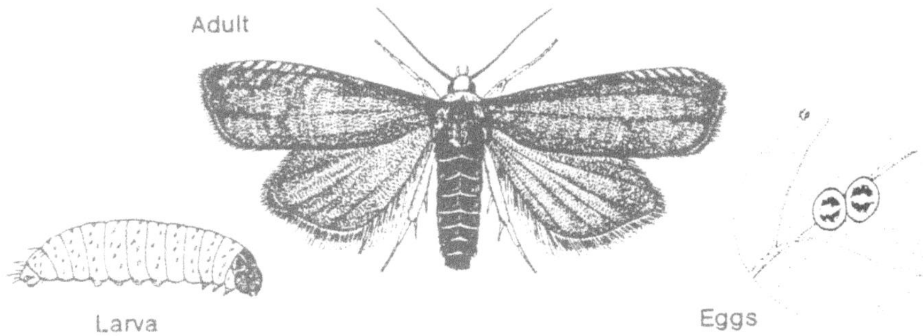

Figure 14.122. (a) Pea moth (Cydia nigricana) (Lepidoptera; Tortricidae) and infested pea pod (b) Peas freshly harvested, showing damage by larvae and frass; Skegness, UK.

Cydia pomonella (Linn.) (Codling moth). A cosmopolitan species attacking apples wherever grown; the small white, later pink, caterpillar bores inside the fruit (all Rosaceae), but emerges to pupate inside a cocoon in packaging etc. Fruit in storage may yield either caterpillars, pupae in cocoons, or later the small dark adult moths; in the USA, walnuts are often infested. (Figures 14.123, 14.124, 14.125)

Cydia prunivora (Walsh) (Lesser appleworm). USA, Canada; on apple, plum, peach, cherry.

Cydia ptychora (Meyrick) (African pea moth). In most of Africa; in the pods of pea.

Cydia pyrivora (Dan.) (Pear tortrix). Eastern Europe, West Asia; on pear fruits.

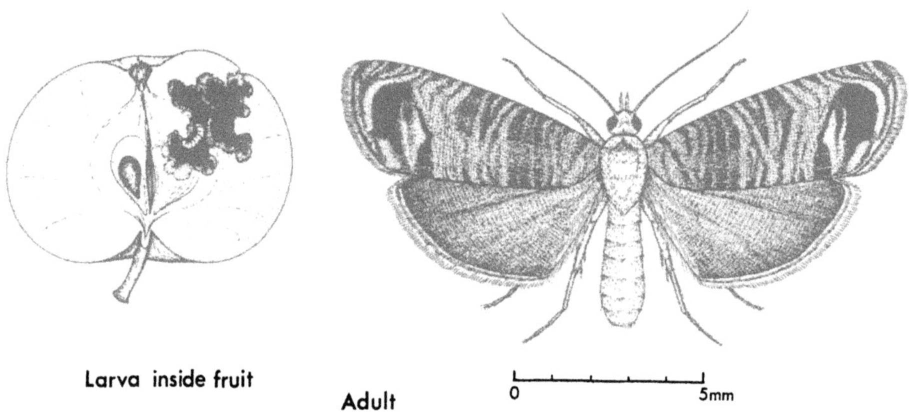

Figure 14.123. Cydia pomonella (Codling moth) (Lepidoptera; Tortricidae), adult moth and infested fruit; specimens from Cambridge, UK.

Figure 14.124. Apple (Golden delicious) showing attack by Codling moth; Skegness, UK.

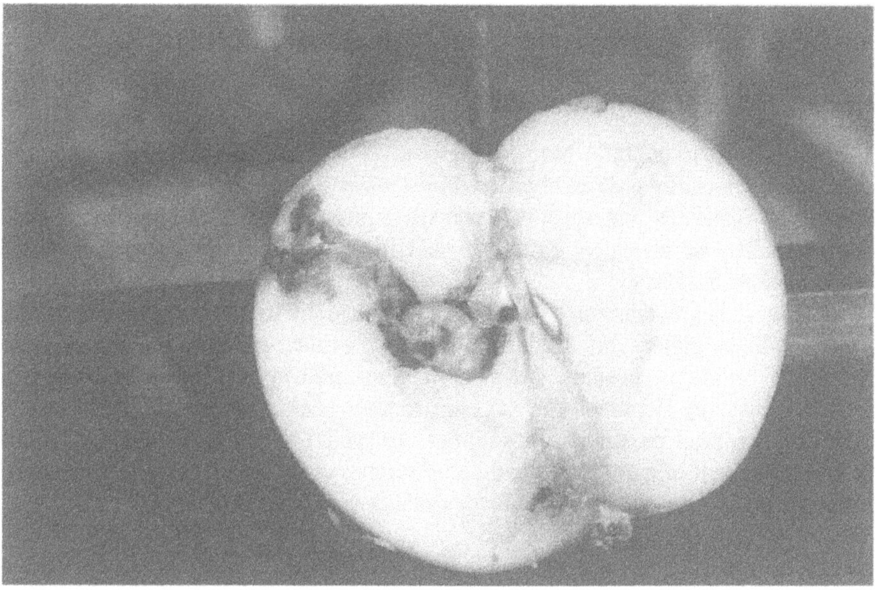

Figure 14.25. Apple cut to show Codling moth larvae in situ; Skegness, UK.

Laspeyresia caryana (Fitch) (Hickory shuckworm). USA; larvae bore inside pecan nuts.

Laspeyresia glycinivorella (Mats.) (Soybean pod borer). S.E. Asia, Japan; larvae bore pods of soybean.

Melissopus latiferreanus (Walsingham) (Filbertworm). USA; larvae bore walnuts and hazelnuts.

14.12.8 Family Pyralidae (Snout moths etc.)

A very large group of moths, worldwide but most abundant in the tropics. In the past, these moths were often regarded as belonging to five separate families, but the present interpretation is of one family and 12 subfamilies. The group is of great importance ecologically and agriculturally; only a few species are found in produce stores with any regularity, but they are of serious economic importance. The moths are smallish with broad and rather rounded wings, and with straight labial palps in most cases, but some have recurved palpi; legs are long and slender. They are nocturnal in habits and use sex pheromones to attract the males. The larvae show tremendous diversity of habits, and it is worthwhile to view the family on the basis of the main subfamilies.

Species of Pyralidae recorded in Britain are illustrated in *British Pyralid Moths* (Goater, 1986). Many of the following species are included having been collected from foreign imports.

Subfamily Pyralinae

Broad-winged moths; nocturnal in habits; if disturbed they tend to scuttle rather than fly; larvae feed on dried plant material (hay, grain, dried dung etc.).

Pyralis farinalis (Linn.) (Meal moth)

A beautifully marked little moth, wingspan 22-30 mm; forewing base and apex reddish-brown, centre pale and bordered by a wavy white line; hindwing pale brown with white markings and several black spots posteriorly. The adult moth has a characteristic pose at rest, with wings held flat, laterally extended, and abdomen curled up over the body.

The larva is greyish with a dark head and shield, up to 25 mm long, and lives in a tough silken gallery attached to a solid substrate; it feeds on stored grain, farinaceous products, peanuts, straw, bran, flour, and even potatoes in storage; in Europe, larval development can take up to two years, but its life cycle can be completed within 2 months in the summer. In the USA, this species is recorded to be most abundant in damp grain and damp refuse, straw etc. It is quite widespread in Europe and North America, and is regarded as being cosmopolitan.

Pyralis manihotalis Guenee (Grey pyralid)

This common name is somewhat misleading, for the moth is basically pale brown with some pale wavy bands, and is a duller and more uniformly coloured moth than the other two species. The female is larger than the male; wingspans from 24-37 mm.

The larvae feed on dried and stored seeds, fruits and tubers, and also on dried meats, hides and bones. It is endemic to South America, but is now found in Africa, tropical Asia and the Pacific Islands; and regularly imported into Europe and the UK.

Pyralis pictalis (Curtis) (Painted meal moth)

Another beautiful moth, wingspan 15-34 mm, female larger than the male; forewings distally reddish-brown and basally blackish with two pale wavy lines; hindwing strongly marked, dark basally, and terminal area paler than forewing. The larvae feed on stored grains. Native to the Far East, but now widely distributed during commerce.

Subfamily Phycitinae (Knot-horn moths)

A large group, mostly tropical in distribution, nocturnal in habits; the adults have the forewings narrowed; most are greyish or brownish, and difficult to identify, but a few are brightly coloured and distinctive.

Larval habits are very diverse, but several are important pests of stored products. In the Phycitinae which have been studied, the major component of the female sex pheromone is the same chemical, so the one synthetic chemical can be used for population monitoring and controlling all the members of this group.

Cadra calidella (Guenee) (Dried fruit moth, Carob moth) (= *Ephestia calidella*)

For positive identification of the species *Ephestia* and *Cadra* mentioned here, it is necessary to make examination of the genitalia, for the moths are very similar, and there is some variation in coloration. This is a small grayish-brown moth of wingspan 19-24 mm; the forewing is grayish-brown or brown, with two cross lines; the hindwing is pale whitish-grey; a Mediterranean species.

The larvae feed on dried fruits (expecially dates and locust beans), nuts, cork, and sometimes other dried plant and animal materials. In the UK, the larval feeding period is September to May, and the adults fly in August and September. On fruits, infestations are most common in the field just before harvest, and this species does not become well-established in stores; but larval diapause increases its pest potential.

Cadra figulilella Gregson (Raisin moth) (= *Ephestia figulilella*)

A smaller species, with wingspan 15-20 mm, but forewing pale yellowish grey with indistinct markings, and the hindwings whitish. The larvae feed on dried fruits of all types, and also freshly harvested carobs (locust beans) and meals. As with the previous species, it is often more abundant on fruits in the field than in stores.

Cadra cautella (Walker) (Dried currant moth, Almond moth, Tropical warehouse moth) (= *Ephestia cautella*) (Figure 14.126)

Pest status. A major pest of stored products throughout the warmer parts of the world, and in heated stores in temperate regions. A penetrator species, but not often recorded as such.

Produce. Dried fruits preferably, but also on a range of stored vegetable materials, including flours, grains, dates, cocoa beans, nuts and seeds.

Damage. Direct eating by caterpillars is the main damage, but the frass-filled silken gallery is a serious contamination of the produce. On seeds, the young larvae often feed preferentially on the seed germ. It is regarded essentially as a secondary pest as far as grain is concerned.

Life history. Eggs are laid in the produce; they are globular, white turning orange; up to 300 eggs are laid per female; at 30°C they hatch in 3 days.

The larva is pale grey with many setae and small dark spots; the head capsule is dark; fully grown, it measures 12-15 mm. The five larval instars can be completed in 22 days at 32°C. The larva lives inside a dense silken gallery filled with frass, amongst the food material. In cooler regions the fully fed larva overwinters inside a silken cocoon. In heavy infestations, the larvae usually leave the produce to pupate on the walls of the store. Pupation takes about 7 days.

The adults are not very distinctive; the forewings are reddish-brown with indistinct whitish cross lines, and the hindwings are pale grey. Wingspan is 14-22 mm, and the wing fringes are quite short. Adults do not feed in the stores, and live for only 1-2 weeks. Adult emergence is markedly prevalent around dusk.

Under optimum conditions (about 32°C and 70% RH), development can be completed in 30-32 days, but development can proceed at temperatures between 15°C and 36°C. Some strains, especially in the southern USA, are able to diapause.

The adults are often not easily distinguished as the wing pattern is usually faint; in addition, the larvae are all very similar in appearance, and a study of chaetotaxy is required for identification.

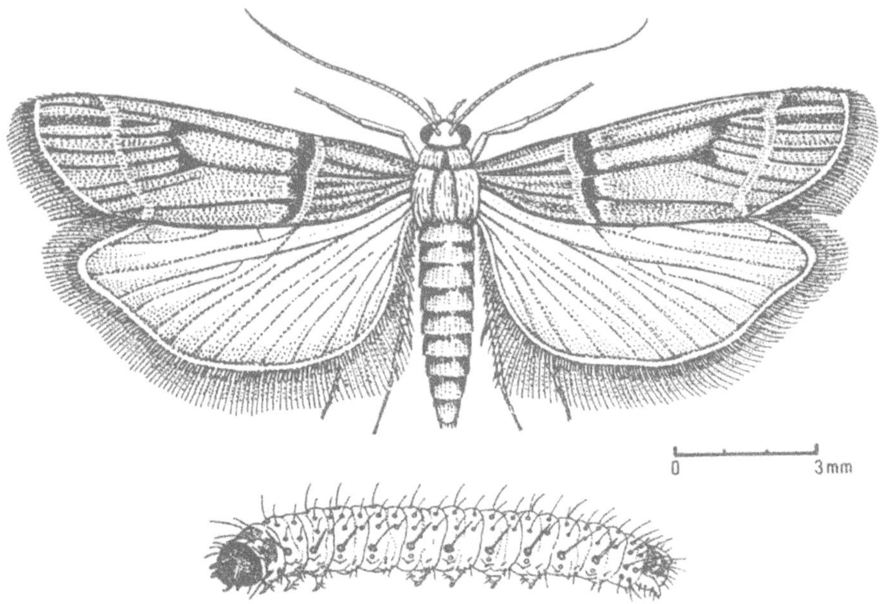

Figure 14.126. Cadra cautella (Lep., Pyralidae), adult moth and caterpillar.

Figure 14.127. Cocoa beans infested by caterpillars of Cadra cautella (courtesy ICI).

Distribution. Cosmopolitan throughout the warmer parts of the world, but less common in arid areas. In temperate countries, it can survive in the summer, but needs heated premises to overwinter.

Natural enemies. Eggs are eaten by the mite *Blattisocius tarsalis*, and larvae are parasitized by *Bracon hebetor* (Braconidae), and attacked by both *Bacillus thuringiensis* and granulosis virus.

Control. Larval populations can be monitored using crevice traps, and adult males using sticky traps baited with synthetic female sex pheromone.

Ephestia eleutella (Hubner) (Cocoa moth, Tobacco moth, Warehouse moth) (Figure 14.128)

Pest status. A serious pest of stored products throughout most of the world; more abundant in the temperate regions, and tends to be replaced in the tropics by *C. cautella.*

Produce. All types of produce of plant origin, including cocoa beans, dried fruits, nuts, seeds, grains, tobacco, flours, meals, and processed flours; also on hay and dried grass.

Damage. A secondary pest that usually feeds on damaged grains and seeds, and often feeds selectively on the germ region, but contamination of high value commodities is often very important and costly. The silk webbing produced by the larvae can be conspicuous.

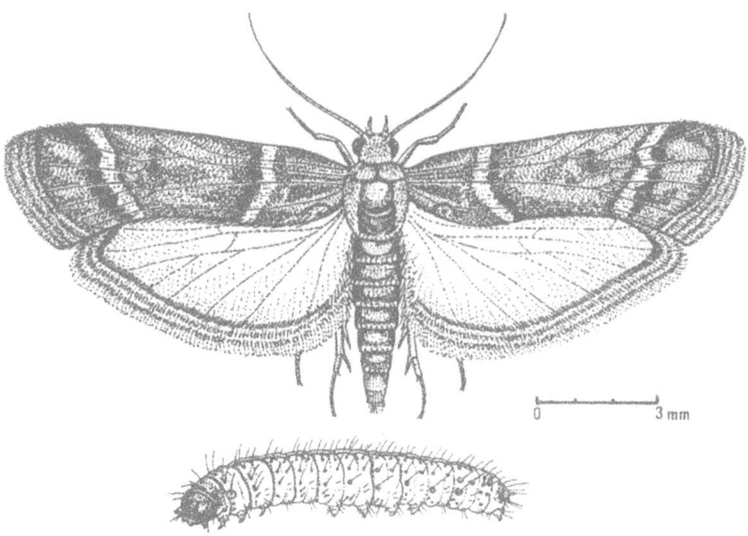

Figure 14.128. Ephestia eleutella (Warehouse moth) (Lepidoptera; Pyralidae), adult moth and caterpillar.

Life history. Eggs are laid in the produce, and take 10-14 days to develop; one female has been recorded laying 500 eggs.

The caterpillars burrow into the produce and spin silk as they feed. Final body size varies from 8-15 mm, and development takes from 20-120 days according to temperature, and at 15°C can take 300 days. The final instar can overwinter in diapause in temperate regions, inside the pupal cocoon.

Pupation usually takes 1-3 weeks.

The adults are small brownish moths, quite pale, about 5-9 mm long, and with a wingspan of 14-20 mm; the forewing cross lines are sometimes quite faint; melanic forms are known. The moths are nocturnal, and fly to lights at night and to flowers for nectar; they live for 1-2 weeks, and may start new infestations at distances up to one kilometer.

Optimum conditions seem to be about 30°C and 70% RH, when development can be completed in 30 days, and in warm climates, breeding can be continuous. Development can proceed at 15°C; in the UK, the species is usually univoltine, but may be bivoltine.

Distribution. Cosmopolitan, but not common in the hot tropics – it is basically a warm temperate species; in the UK and Europe, it probably survives out-of-doors and in unheated buildings.

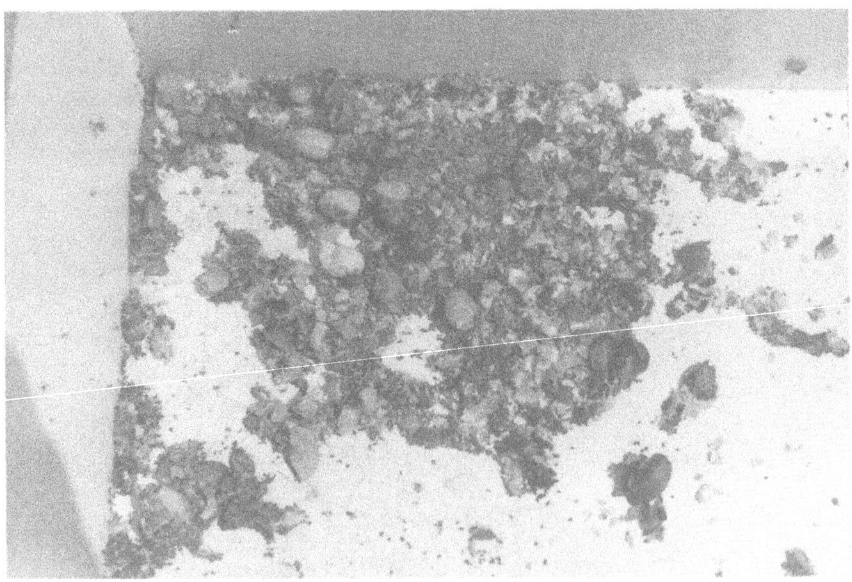

Figure 14.129. Infested peanuts with frass after Ephestia infestation; Skegness, England.

Ephestia kuehniella Zeller (Mediterranean flour moth) (Figure 14.130)

This is the largest member of the genus, with wingspan 20-25 mm; forewing grey, speckled brown and white; hindwing white with fuscous veins. This species is sometimes distinct and recognisable without study of the genitalia.

Larvae feed on wheat flour mostly, but are recorded from a wide range of commodities and from dead insects. Most serious in flour mills, where it can clog machinery with the masses of silken, frass-filled galleries and webs. Optimum conditions seem to be about 20°C and 75% RH, when the life cycle takes some 74 days; but development can proceed at 12°C, although survival rate is poor, and development very protracted. The highest temperature for development is 28°C. Larvae are cold-hardy, and can survive -10° for 3-4 days.

It is found worldwide, but is not abundant in the tropics, and only survives in the UK inside heated buildings – essentially a subtropical species, hence the common name, but a serious pest in flour mills.

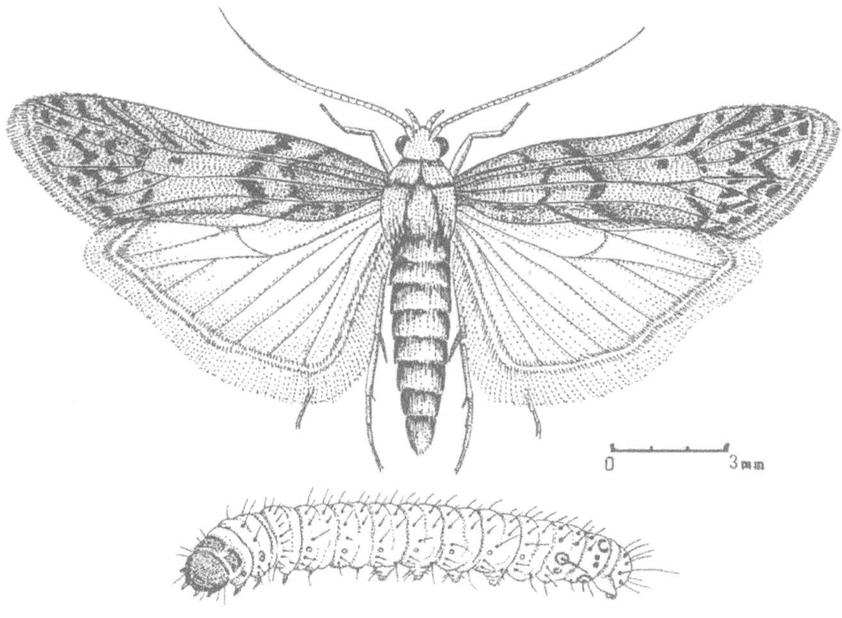

Figure 14.130. Ephestia kuehniella (Mediterranean flour moth) (Lepidoptera; Pyralidae), adult moth and caterpillar.

Plodia interpunctella Hubner) (Indian meal moth) (Figure 14.131)

Pest status. A major pest of stored foodstuffs throughout the warmer parts of the world.

Produce. Larvae feed on stored cereals, dried fruits, flours, chocolate, nuts, dried roots, herbs, some pulses, and dead insects.

Damage. Direct eating of the produce is the main damage, although this is regarded as a secondary pest in grain stores. The larvae spin silk and make a silken web in the food material, which soon becomes contaminated with frass, and very conspicuous. The larvae generally show a preference for the germ region of grains which they manage to attack.

Life history. Sticky eggs are laid in the produce, up to 400 per female; and hatching occurs after 4 days (at 30°C and 70% RH).

The larva is a whitish caterpillar, distinguished from *Ephestia* by not having the dark spots at the bases of the setae (one spot per seta), and the rim of the spiracles is weakly sclerotized and evenly thickened. Larval development proceeds at a very variable rate – there are 4-7 moults – and it can be completed in 2 weeks or slowly, taking as long as 2 years in the UK. The feeding larvae produce silk, and construct a silken web in the produce. Fully fed larvae are 8-10 mm long, and are very active; they leave the produce to find a crevice as a pupation site where a silken cocoon is spun.

Pupation can be completed in 7 days at 30°C and 70% RH.

The adults are small brown moths of distinctive appearance, with forewing yellow basally, and distally bright copper red; hindwings whitish; body length 6-7 mm and wingspan 14-20 mm, and the labial palps are directed forward. They are nocturnal, and fly actively at night.

Under optimum conditions of 30°C and 70% RH, development can be completed in 28 days; in the tropics, there may be 6-8 generations per year, but in Europe, 1-2 is usual. Development ceases at a temperature of 15°C. Some strains can apparently enter diapause as fully grown larvae, induced by short photoperiod, low temperature, or overcrowding.

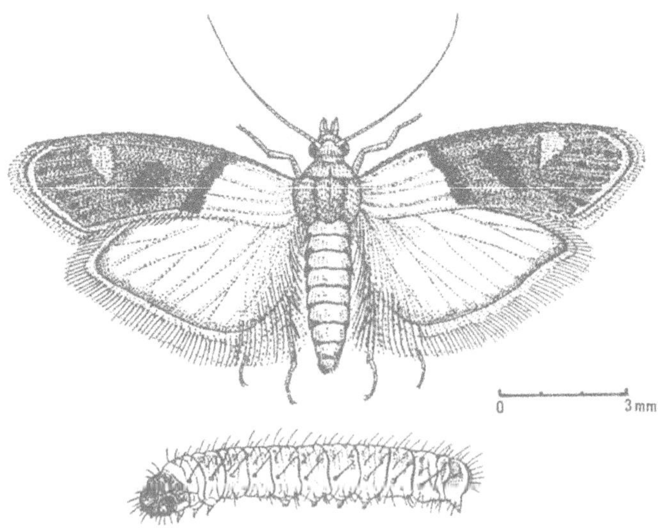

Figure 14.131. Plodia interpunctella (Indian meal moth) (Lepidoptera; Pyralidae), adult moth and caterpillar.

Distribution. Worldwide, but scarce in the hot tropics, where it is to be found mostly in the highland areas – probably most abundant in the warm temperature regions and subtropics.

Natural enemies. Recorded predators include *Xylocoris* bugs, *Tribolium castaneum*, and *Oryzaephilus surinamensis*; parasites *Bracon* spp. (Hymenoptera, Braconidae) attack the larvae, and *Trichogramma evanescens* parasitizes the eggs. Larvae are attacked by *Bacillus thuringiensis*, granulosis virus, and *Nosema* spp. (Protozoa).

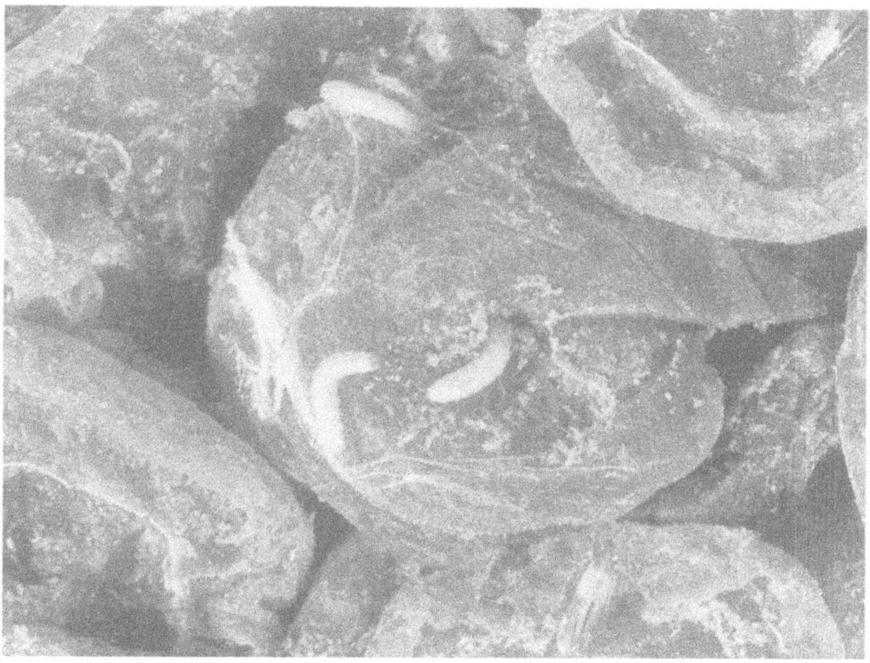

Figure 14.132. Dried apricots infested with Plodia larvae (courtesy ICI).

Other Phycitinae which are recorded in produce stores include:

Citripestis satittiferella (Citrus fruit borer)

The larvae of this moth bore in citrus fruits in S.E. Asia, but there are usually holes in the skin of the fruit to indicate infestation.

Cryptoblabes gnidiella (Milliere)

Native to the Mediterranean region, larvae of this small grey moth have been found in England in imported oranges and pomegranates with great regularity.

Ectomyelois ceratoniae (Zeller) (Locust bean moth)

Another Mediterranean species now established in the Americas, and South Africa; the moth resembles a large *Ephestia kuehniella* in having grayish forewings (wingspan 19-28 mm) and whitish hindwings with fuscous veins. The larvae feed inconspicuously inside dried fruits, seeds, nuts (walnut, almond, chestnut), dates, and beans of *Ceratonia, Robinia* etc. Infestations by this pest are widespread and quite common, and introductions into temperate countries in produce are regular.

A serious pest of carob in Israel (Avidov & Harpaz, 1969) and in citrus, especially grapefruits. Damage to fresh fruits is usually light, but in dried fruits, infestations can be very serious.

Etiella zinckenella (Treitschke) (Pea pod borer)

A cosmopolitan pest of pea and various beans and some other legumes (pulses); the larvae develop inside the pod in the field, but pupation typically takes place in the soil, so this pest is only occasionally found in peas in storage. It is most abundant in warmer countries, although recorded from much of the southern half of Europe, and western USA.

Euzophera bigella (Zeller) (Quince moth)

A Mediterranean species occasionally imported into the UK as larvae inside peaches (or peach stones). In Israel, recorded mainly from quince and apples.

Euzophera osseatella (Treitschke) (Eggplant stem borer)

Another small brownish moth, bred in the UK from larvae in potatoes imported from Egypt. A Mediterranean species found boring in Solanaceae, and quite a serious pest in many locations, but recorded to be relatively scarce in potato tubers in the field.

Mussidia nigrivenella Ragonot

A small brownish-grey moth with white hindwings, reared from larvae eating cocoa beans imported into England. A regular pest of stored maize and cocoa in West Africa; principally a preharvest pest of maize crops, and the infestation in stores usually dwindles as the crop dries.

Paramyelois transitella (Walker) (Navel orangeworm)

This American species is a major citrus pest in California, and now a serious pest of walnuts and almond, although something of a scavenger; usually nearly ripe nuts are attacked, so this pest is carried into produce stores – several caterpillars are often found inside a single nut.

Subfamily Galleriinae (Wax moths, Bee moths)

A small group known collectively as wax moths, for most are associated with the nests of Hymenoptera. Their wings are narrow, and the moths tend to scuttle rather than fly; the larvae live gregariously under silken galleries, and they eat the wax used to make the nest combs.

Corcyra cephalonica (Stainton) (Rice moth)

Pest status. A major pest throughout the humid tropics, especially in S.E. Asia, on a wide range of produce of plant origin. A penetrator species.

Produce. Rice, maize, and other cereals in storage, dried fruits, nuts, copra, spices, some pulses, chocolate and ships' biscuits.

Damage. Direct damage is done by the eating of the caterpillars, and indirect damage is caused by extensive and tough silk webbing – the contamination is often of greater economic importance.

Life history. Eggs are sticky and laid on the produce, and hatching occurs after 4 days at 30°C.

The larva is white, with brown head and prothoracic shield; there is a characteristic seta above each spiracle (starting on the first abdominal segment) arising from a small clear patch of cuticle surrounded by a dark ring of cuticle, and the spiracles are also characteristic in having the posterior rim thickened. The number of larval instars is variable, but usually 7 for males and 8 for females. Optimum conditions are 30-32°C and 70% RH, and diet has a major influence on the rate of development. Silk webbing is extensive, denser and tougher than that produced by other caterpillars, and becomes matted with frass, excreta and pupal cocoons.

Pupation takes place either in the produce or on structures.

The adult is another small brown moth, with pale brown hindwings,and wingspan 15-25 mm. Males have short, blunt labial palps, but the females have long and pointed palps; adults fly at night.

The life cycle can be completed in 26-28 days, under optimum conditions of temperature and humidity (30-32°C and 70% RH respectively) and diet; but this species is recorded to be able to develop at humidities of less than 20%, and on very dry grains.

Distribution. Now a cosmopolitan species throughout the humid tropics, and particularly abundant in S.E. Asia, and there is evidence (TDRI, 1984) that it may be more common in Africa than previously supposed.

Natural enemies. Eggs are eaten by various mites (*Acaropsis, Blattisocius* spp.), larvae eaten by *Sycanus* and *Amphibolus* (Heteroptera), and larval parasites include *Bracon* spp., *Autrocephalus* and *Holepyris*, and *Bacillus thuringiensis.*

Control Population monitoring has employed refuge traps (of corrugated cardboard) for larvae, and sticky, light, and suction traps for adults.

Arenipses sabella Hampson (Greater date moth)

A large moth (wingspan 28-46 mm) with uniformly pale brown forewings and white hindwings. Several specimens have been reared in the UK from caterpillars in imported dried dates.

Doloessa viridis Zeller

A small moth with forewings bright green with grey patches, found in stored rice, maize, cocoa beans, copra, and oil palm kernels in parts of S.E. Asia.

Paralipsa gularis (Zeller) (Stored nut moth)

A small moth (21-33 mm wingspan) with sexual dimorphism in that the female is larger, and in the centre of the pale brown forewing is a distinctive black discal spot; hindwing is whitish.

Larvae are recorded feeding on stored groundnuts, almonds, walnuts, linseed, cocoa, dried fruits, soybean, some cereals, and a wide range of stored foodstuffs. Larvae can undergo diapause, induced by low temperatures or crowding. In England, it is known mostly from dried fruit stores in London, where it apparently prefers to feed on dried nuts. Temperature range 15-33°C for development.

Native to S.E. Asia, now a regular pest in parts of tropical Asia, southern Europe, and the USA, and is imported into cooler temperate regions. In other countries, it is recorded to feed on a wide range of produce. Usually has one or two generations per year.

Subfamily Glaphyriinae

A small group of tropical moths, with characteristic spatulate scales on the upper surface of the hindwings. There are two species that might be found on cabbages and other Cruciferae in vegetable stores, as the caterpillars feed on the foliage of these plants.

Hellula phidilealis (Walk.) (Cabbage webworm)

Recorded throughout Africa, Central and South America.

Hellula undalis (Fab.) (Old World cabbage webworm)

Found throughout the Old World tropics, and now in Hawaii.

Subfamily Evergestinae

A small subfamily erected mainly on the basis of distinctive male genitalia.

Evergestis spp. (Cabbageworms)

At least four species are recorded feeding on Cruciferous crops in Europe, parts of Asia, and North America; and other species feed on wild Cruciferae.

Figure 14.133. Maruca testulalis (Mung moth) (Lepidoptera; Pyralidae), adult moth.

Table 14.2 Bionomics of moth pests of stored products

	Temperature (℃)			*Relative Humidity (%)*		
Species	*Minimum for population increase*	*Optimum for develop-ment*	*Maximum for develop-ment*	*Minumum for Population increase*	*Optimu m for develop -ment*	*Duration of life cycle (days)*[a]
Corcyra cephalonica	18	30	35-37	30	80	25
Cadra cautella	17	30	37	20	75	28
Cadra figuliella	10-12	25	30	20	70	40
Ephestia elutella	15	30	36	30	70	40
Ephestia kuehniella	8-12	25	28	0	70	40
Plodia interpunctella	18	30	35	20	75	25
Sitotroga cerealella	16	30	35-37	30	75	30

[a] At optimum temperature and humidity on good food.

(Adapted from Gorham, 1991: p. 183)

Subfamily Pyraustinae

This is the largest group, most abundant in the tropics; adults are rather broad winged, strongly patterned, with long brittle legs and a jerky flight. Larvae are mostly leaf rollers or web spinners on herbaceous plants, and many are crop pests. A few species might occasionally be encountered with vegetables in storage.

Crocidolomia binotalis (Cabbage cluster caterpillar)

In the Old World tropics, the larvae eat the leaves of cabbage and other Cruciferae.

Dichocrocis spp.

Several species are polyphagous, and in tropical Asia the larvae bore inside the fruits of many different crop plants.

Leucinodes orbonalis (Eggplant fruit borer)

In Africa, India, and S.E. Asia; the larvae bore inside the fruits of cultivated Solanaceae.

Maruca testulalis (Geyer) (Mung moth) (Figure 14.133)

A pantropical species whose larvae bore in legume pods (*Phaseolus* usually), and pupation may take place in the pods; larvae inside the pods of French beans have been imported into England from Malawi (Goater, 1986). The adult moth has a very distinctive appearance.

14.12.9 Family Pieridae (White butterflies etc.)

A large and widespread family of medium sized butterflies, usually either white or yellow in colour, with black or orange markings. The larvae are elongate caterpillars with distinct body segments and setae; in some species, the larvae are gregarious and others are solitary.

Pieris spp. (Cabbage white butterflies) (Figure 14.134)

About six species worldwide are pests of cultivated crucifers, and caterpillars will be taken into vegetable stores in the foliage of cabbages and other crucifers, and in some instances there will probably be pupae also in the foliage.

14.12.10 Family Noctuidae (25,000 species)

An enormous group showing great diversity of larval habits, and many species are very important crop pests. The adults are mostly moderate in size, drab in colour, and nocturnal in habits. The caterpillars are often quite large, and many are striped longitudinally; pupation typically occurs in soil inside an earthen cell. Some caterpillars are distinctive, but in many cases they have to be reared to the adult stage before certain identification can be made, which is always a problem with many different groups of insect pests.

The group referred to as fruitworms are quite large caterpillars, and they attack most fruits from the outside with just the head and thorax inside the fruit whilst feeding – they do likewise to legume pods, so they are not likely to be carried into produce stores. But in larger fruits and tubers, they do penetrate to the interior, and may then be carried into vegetable stores quite easily. Some of the caterpillars which eat fruit of Solanaceae sometimes feed right inside sweet peppers, and can be transported thus.

Some of the leafworms feed on cabbage, lettuce and other leaf vegetables, and are regularly taken into vegetable stores in small numbers on the produce.

Some cutworms spend part of their lives in the soil where they feed on plant roots, but they may tunnel into potato tubers and other root crops (beet etc.). Damaged roots and tubers are common, and sometimes the caterpillars are still inside at the time of harvest, and they may be taken into the stores; cutworms inside potato tubers are a regular occurrence in potato stores.

Figure 14.134. Cabbage leaf infested and damaged by caterpillars of Small cabbage white butterfly (Pieris rapae) (Lepidoptera; Pieridae); Wainfleet, Lincs., UK.

Quite a large number of species of Noctuidae are occasionally taken into fruit and vegetable stores in different parts of the world; but the most likely species to be found thus are relatively few in number, and some are listed below.

Agrotis segetum (Shiff.) (Common cutworm) (Figure 14.135)

An Old World species; bores into potato tubers and various root crops.

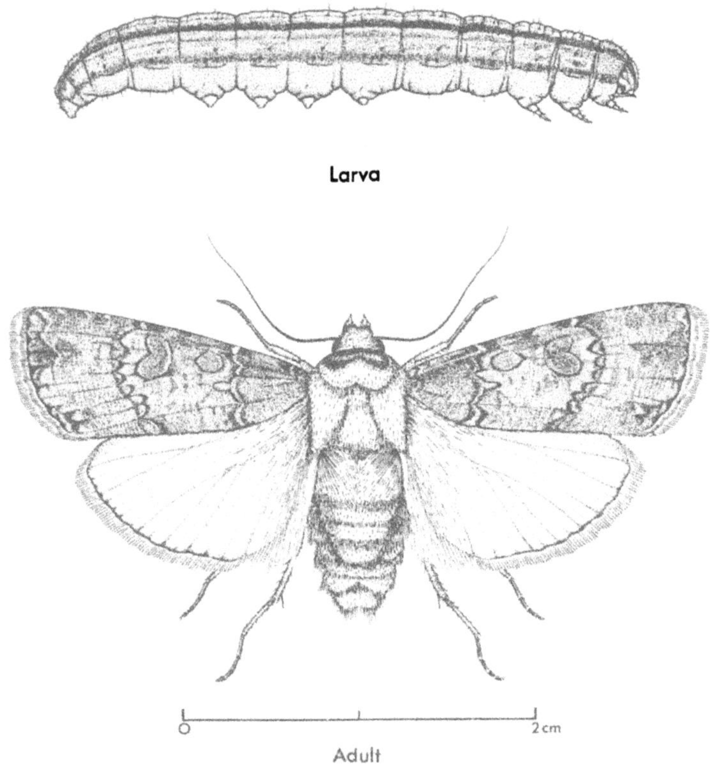

Figure 14.135. Agrotis segetum (Common cutworm) (Lepidoptera; Noctuidae), adult moth and caterpillar, probably the commonest British cutworm; specimens from Cambridge, UK.

Autographa gamma (Linn.) (Silver-Y moth)

Holarctic; polyphagous leafworm on many different vegetable crops; the semi-looper larvae are not easy to identify.

Helicoverpa armigera (Hubner) (Old World bollworm) (Figures 14.136, 14.137)

Old World warmer regions; a polyphagous fruitworm regularly imported into the UK in fruit and vegetable cargoes.

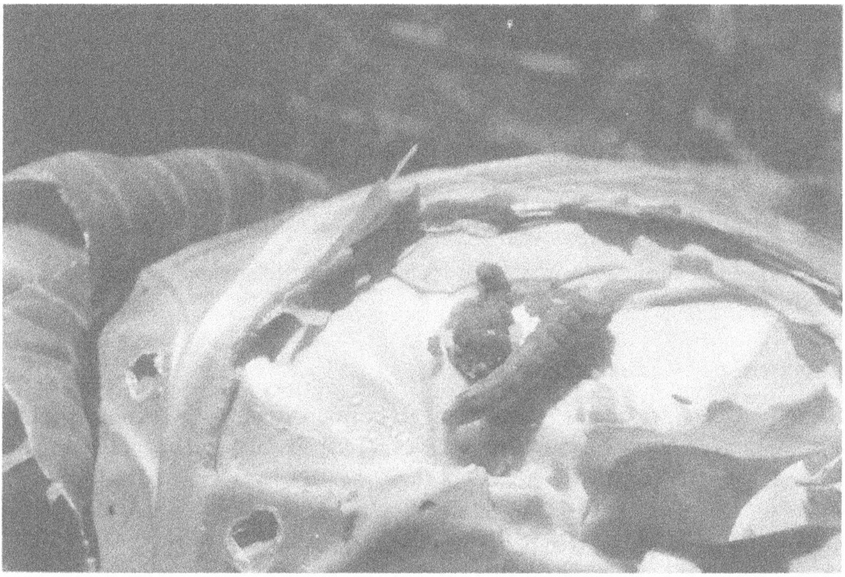

Figure 14.136. Cabbage attacked by caterpillars of Helicoverpa armigera (Old World bollworm) (Lepidoptera; Noctuidae), green form; showing damage and frass; Alemaya University Farm, Ethiopia.

Figure 14.137. Helicoverpa armigera (Old World bollworm) (Lepidoptera; Noctuidae), adult moth from Hong Kong.

Lacanobia spp. (Tomato moth etc.)

Some are Palaearctic, and others Holarctic; polyphagous on a wide range of vegetable and fruit crops.

Mamestra spp. (Cabbage moth etc.)

A north temperate genus (Holarctic); on brassica crops, and other plants.

Noctua pronuba (Linn.) (Large yellow underwing)

A common Palaearctic cutworm, to be found in potato tubers and various root crops in vegetable stores.

Spodoptera spp. (Cotton leafworm etc.)

A large and complex genus, pantropical to cosmopolitan, with several species of great importance; polyphagous fruitworms and leafworms, and regularly imported into the UK in produce, especially in cut flowers.

Trichoplusia ni (Hubner) (Cabbage semi-looper)

Cosmopolitan pest of cabbage and other vegetables; basically a leafworm.

Details of these insects and illustrations can be found in Hill (1987).

14.13 ORDER HYMENOPTERA (Ants, Bees, Wasps) (More than 60 families, 100,000 species)

A very large group of insects and very important ecologically, but only of relatively slight importance in food and produce stores. Adults range in size from minute (0.2 mm) to quite large (4 cm or more); there is great diversity of form and habits within the group, and a range from quite primitive insects to highly evolved and very specialized forms. There are extra veins in the wings, both longitudinal and cross veins, and the small hindwing locks on to the hind edge of the forewing by a series of small hooks. Social life has evolved within the group, and some species live in large communities with polymorphism and division of labour within the colony. Most are predacious and feed on other insects, including many parasitic (parasitoid) species. In the bees, the mouthparts (labium) are modified into a tongue with which they suck nectar from flowers.

Some species of ants are domestic pests, and a few wasps use buildings in which to hang their small nests, but most of the Hymenoptera encountered in food and produce stores are parasitic on the stored products pests (Coleoptera and Lepidoptera mostly), and as such are very beneficial.

14.13.1 Family Braconidae (Bracon wasps)

These wasps are smallish, with long antennae and a distinct pterostigma, and in the female the ovipositor may be evident. They are very important parasites of many insects of cultivated plants. A few species are found in produce stores with some regularity; the two most commonly recorded are probably *Bracon hebetor* Say and *Bracon brevicornis* (Wesmael). They both parasitize lepidopterous larvae (*Plodia, Corcyra, Ephestia* and *Sitotroga* spp. in stores.

14.13.2 Family Evaniidae (Ensign wasps) (300 species)

Most are found in the tropics, but the family is worldwide, and all known species are parasites on the oöthecae of cockroaches. The small black wasps have long flickering antennae, and the small triangular abdomen (gaster) is twitched up and down as the wasp walks. Most of the species are placed in the large genus *Evania* (Figure 14.138).

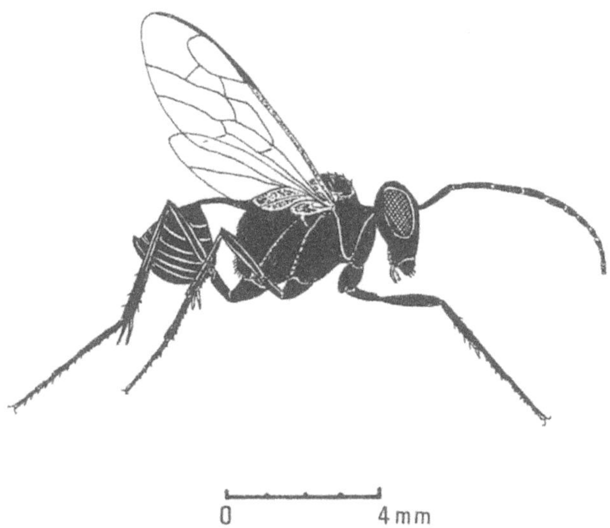

Figure 14.138. Evania appendigaster (Ensign wasp) (Hymenoptera; Evaniidae), adult wasp from Hong Kong.

There are some other families of parasitic wasps which attack the larvae of wood boring beetles, and some of these might occasionally be encountered in produce stores.

Superfamily Chalcidoidea

A large group of small parasitic species usually regarded as containing about 20 different families; the adults have a very reduced wing venation which is quite distinctive.

14.13.3 Family Pteromalidae

Some members of the other families could possibly be found occasionally in produce stores, having emerged from the bodies of other insects, but only these two families are regularly encountered. They are technically termed 'parasitoids' as only the larval stage is parasitic – the adults are free-living. Within the Pteromalidae are many species parasitic on serious pests of cultivated plants, and they are important as part of the natural control complex regulating insect pest populations. There are different levels of host-specificity shown within the family, but frequently each genus of wasp attacks a family of insects, or part of a family.

Three species are of particular importance as natural enemies of stored product beetles, but whether they ever have any real effect on pest populations is not clear.

Anisopteromalus calandrae (Howard), *Chaetospila elegans* Westwood, *Lariophagus distinguendus* (Foerster). The wasp larvae parasitize *Sitophilus* beetles and the larval stage of some other beetles – namely *Lasioderma, Rhizopertha* and *Prostephanus.*

Dinarmus spp. Parasitize larvae of Bruchidae (Coleoptera) in pulse seeds (more usually in the field); also recorded attacking *Trogoderma.*

Habrocytus cerealellae Ashmead. Recorded from the caterpillars of *Sitotroga cerealella.*

14.13.4 Family Trichogrammatidae

These tiny wasps (0.3-1 mm) are egg parasites, that use Lepidoptera and some other insects as hosts, and several species are used widely throughout the world in biocontrol programmes; most of these species belong to the large genus *Trichogramma*, and several species are available commercially. *Trichogramma* spp. are recorded mainly from the eggs of *Sitotroga* and *Plodia.*

14.13.5 Family Bethylidae

A small group of black wasps, usually included in the group Parasitica; they mostly parasitize lepidopterous larvae and Coleoptera, and one genus is recorded on stored produce beetles. The recorded hosts of *Cephalonomia* spp. include *Cryptolestes, Tribolium* and *Oryzaephilus* spp.

14.13.6 Family Formicidae (Ants) (5,000 species)

A very large group of important insects, abundant worldwide; many are polymorphic, and live socially in nests. Most forms are wingless, but the young reproductive forms have wings for their mating and dispersal flights. Mouthparts are of the biting and chewing type, and the workers and soldiers have either a sting or an apparatus to squirt formic acid. An important characteristic of all forms is that on the petiole there are either one or two distinct nodes (swellings). Diet is very varied, according to species and circumstances – many species are opportunistic omnivores, but some are permanently carnivorous, and others equally phytophagous.

Some species are important domestic pests, but in produce stores they are generally less of a nuisance. These ants operate from a large, often underground, nest, and the foraging workers follow pheromone scent trails to food sources; food particles are collected and taken back to the nest to feed the community. Thus, any control measures applied need to destroy the nest and the reproductive queen in order to eliminate the foraging population. In the past, ant control has depended largely on the use of poison baits which are taken by the foraging workers back into the heart of the nest – salts of arsenic in sugar have been extensively used.

A number of different species are regarded as major domestic pests:

Anoplolepis longipes (Long-legged ant)

A S.E. Asia species spread along the trade routes throughout the Old World tropics and to Mexico and Chile. Workers are 4 mm long, pale brown, and they prey on some crop pests, but also cultivate honey-dew producing Homoptera. Newly arrived in the Seychelles, it is causing problems in hotels, hospitals and food and drink processing plants. Nesting takes place almost anywhere – in trees, buildings, leaf litter and soil. Foraging workers gain access to foodstuffs and containers, and harass sleeping humans.

Camponotus spp. (Carpenter ants) (1,000 species)

A widely distributed group – worldwide. Large ants (6+ mm), mostly black in colour; they nest in cavities in wood or underground. Many nest in buildings using existing cavities, but workers enlarge and extend the tunnels. They are omnivorous, and seek sugar and sweet foodstuffs as well as meat and grease.

Iridomyrmex (200 species known) (House and Meat ants) (Figure 14.139)

Many of the species are synanthropic, and domestic pests in Australia, tropical Asia, and South America; some are often important pests in meat processing plants.

Iridomyrmex humilis (Argentine ant)

This originated in Brazil and Argentina, but is a well-known tramp species, and is now worldwide. Workers are tiny (2.2-2.6 mm) and brown; in the tropics, they nest in the soil, rotten wood, and any damp shady place, and in temperate regions they tend to live in heated buildings such as hospitals, restaurants and food manufacturing premises, and houses. Their diet is very varied, but they show a preference for sweets (and meat).

Lasius niger (Black garden ant)

Common throughout Europe and Asia, the black workers are about 5 mm long. They regularly enter houses, warehouses, factories and other buildings. Most nests are actually outside the building under concrete driveways etc., but may be inside cavities in the buildings. Sweet foodstuffs are mostly sought by the foraging workers.

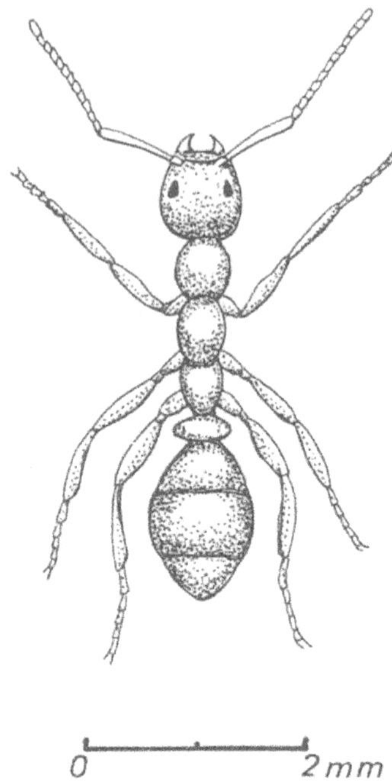

Figure 14.139. Iridomyrmex sp. (House or Meat ant) (Hymenoptera; Formicidae), worker; specimen from Hong Kong.

Monomorium pharaonis (Linn.) (Pharaoh ant) (Figure 14.140)

This tiny warmth-loving domestic pest species is now cosmopolitan in warm situations; it is omnivorous but basically a scavenger. The genus contains 300 species worldwide. Originally from North Africa, they often live in buildings as 'supercolonies' scattered throughout; they can nest almost anywhere. They eat almost anything, especially confectionery and meat products. In the tropics, they have been observed in warehouses feeding on stored grain and a wide range of other commodities.

M. minimum is a house pest in the USA.
M. destructor is a worldwide species recorded destroying rubber insulation round electrical cables in Indonesia.
M. latinode has destroyed rubber products in Java.
M. barbatulum is recorded from granaries in Egypt.

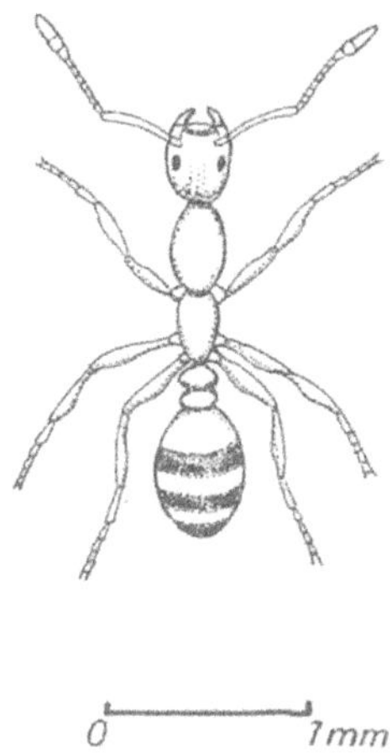

Figure 14.140. Monomorium pharaonis (Pharaoh ant) (Hymenoptera; Formicidae, worker; specimen from Hong Kong.

Ochetomyrmex auropunctatus (Little fire ant)

A Neotropical species, it has become a pest in Florida. Tiny (2 mm) and golden brown in colour, it is restricted by temperature – it is a tropical species. Meat and dairy products are favoured as food.

Paratrechina longicornis (Crazy ant)

Now a pantropical species, spread by commerce. Workers are 2-4 mm long, brown to black in colour. In the UK, they breed successfully in buildings, nesting in wall cavities, under baseboards and the like. The workers travel great distances when foraging, so the nest is often difficult to find. They feed on a wide range of foodstuffs including meats, meat products, seeds, fruits, biscuits and cakes.

P. vividula is a smaller species and a pest in the UK and several other places.

Solenopsis spp. (250 species)

Several species appear to have hybridized in the USA, so identification can be difficult. Workers of the two Fire Ants are 3-6 mm long, and red (*S. invicta)* or black in colour (*S. richteri).* Workers bite and sting when provoked. The Imported Red Fire Ant prefers oily foods, and has been spread by human agency extensively in the USA.

S. molesta (Thief Ant)

A tiny species (2 mm) brownish in colour, found throughout the USA. Food with a high protein content is sought; they do not eat sugar.

Tapinoma (60 species) (House and Sugar ants)

These are urban pests, some of importance, nocturnal in habits, and the genus is worldwide in distribution. All types of food are taken.

Minor ant pests include:

Lasius brunneus (Garden red ant)

In the UK, Europe and West Asia, this is common as a garden ant, often nesting in rotting timbers; they eat meat, cake and jam.

Several species of *Pheidole* and *Crematogaster* and others are regular house pests in the southern USA, but are of little importance in produce stores. *Solenopsis xyloni* damages almond kernels in California while the nuts are drying on the ground.

(From Beatson, in Gorham, 1991; pp. 207-215).

14.13.7 Family Eumenidae (Potter wasps)

This is the first of the families of what are regarded as the true wasps. Potter wasps are so-named because they construct beautiful little urn-shaped nests out of mud, and stick them on to a wall, twig or some other solid substrate. They are solitary wasps, and each nest contains only one larva which feeds on the provisioned prey – paralysed small caterpillars (often Geometridae). Usually, the presence of these wasps inside a produce store is purely fortuitous in that they were are only seeking a sheltered place in which to build the nests. Whether they might use caterpillars from the stored produce to provision their nests is not known at present – the nests that have been investigated usually contained only looper caterpillars (Geometridae).

The two genera found inside buildings in the tropics are as follows:

Eumenes spp. (True potter wasps)

These build the urn-shaped nests of mud stuck on to a solid substrate and provisioned with paralysed caterpillars.

Odynerus spp.

Some dig holes in wood etc., and others use crevices, keyholes, and any type of small space, and the nest is sealed with mud after provisioning the larvae with sawflies or Chrysomelidae (Coleoptera); in Europe, weevil larvae are used as food.

14.13.8 Family Vespidae (True wasps)

Stout-bodied, these insects are yellow and black or reddish coloured; large in body size; many are social, living in nests (some large), either aerial or underground; most sting aggressively if disturbed. There are some thousand species, more or less solitary, in the tropics, and a few hundred social species, some of which are widespread and well-known. They are carnivorous, but also like sugary solutions, and are attracted to ripe fruit and sugar.

The nest is of a papery consistency, being constructed of wood scrapings mixed with saliva, and it may hang from a single pedicle or may be built into a corner or a crevice within a building. The presence of wasp nests inside buildings is again fortuitous, as they are seeking a sheltered location for the nests. The nest may contain up to 12,000 cells in a series of combs and is often about 1 ft^3 in volume. The adult wasps are very aggressive in defence of their nest, and can wreak havoc in any building they inhabit. The colony is annual, and the workers die in the winter; the young fertilized queens overwinter in hibernation, and start new colonies in the spring.

In produce stores, the presence of a wasp nest is usually a cause for concern, but the smallest species (1 cm long) in small colonies of a few dozen are really no threat – their sting is scarcely more painful than a pinprick. The larger species in small colonies (*Balanogaster*) have a painful sting, but there are seldom more than a dozen wasps per nest. But the larger species in the larger nests have to be treated very circumspectly, for if aroused their concerted attack can be very alarming and very painful, and such colonies are best destroyed in produce stores.

In the tropics, there are many solitary and subsocial species that may be found in produce stores. Subsocial species generally live in small colonies, sometimes only 10-20 adults, with a small comb (3-10 cm in diameter), suspended from the ceiling, beams etc., on a thin pedicle. If possible, they should be left alone and tolerated, but if they constitute a threat to humans, they may have to be destroyed. One genus is very widely distributed, as mentioned below. The three genera of Vespidae of importance are as follows:

Balanogaster (Subsocial wasps)

These are slender-waisted, brownish wasps about 2 cm in length; they make single combs attached to ceilings in buildings, and roof overhangs, by a delicate pedicle; there are seldom more than a dozen adult wasps per nest; in many parts of the tropics they are very abundant, and some buildings in Africa often have 10-20 nests, some inside and some outside under the eaves.

Polistes (Paper wasps, Field wasps) (Figure 14.141)

These wasps make a one-comb nest hanging from a pedicle attached to a branch or ceiling beam; the genus is worldwide in warmer regions, but most abundant in the tropics. Most Old World species are yellow with black banding (some have blackish males), but some in South America are totally black. However, they definitely prefer to nest in bushes, and so are only occasionally found in buildings – they are more likely to be found in rural cribs and open on-farm stores than in modern buildings. Each nest contains several dozen to maybe as many as a hundred wasps, and they are very fierce and sting painfully.

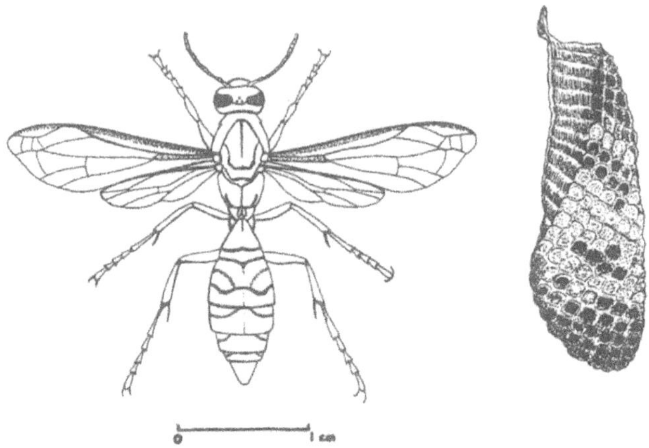

Figure 14.141. Polistes (Paper wasp) (Hymenoptera; Vespidae), adult worker and typical nest; material from Hong Kong.

Vespa (Common wasps) (Figure 14.142)

Some of these wasps are now separated off into other closely related genera, *Vespula* etc., but many are still left in *Vespa*. There are many species, some tropical and some temperate; most colonies are large and may contain 5,000 individuals or more at the height of their activity. Some nests are underground, some hang from tree branches, but quite a few are found in buildings, often constructed in a corner, large crevice or space near the ceiling or roof. *Vespa* colonies are almost invariably a threat to humans, and they have to be destroyed.

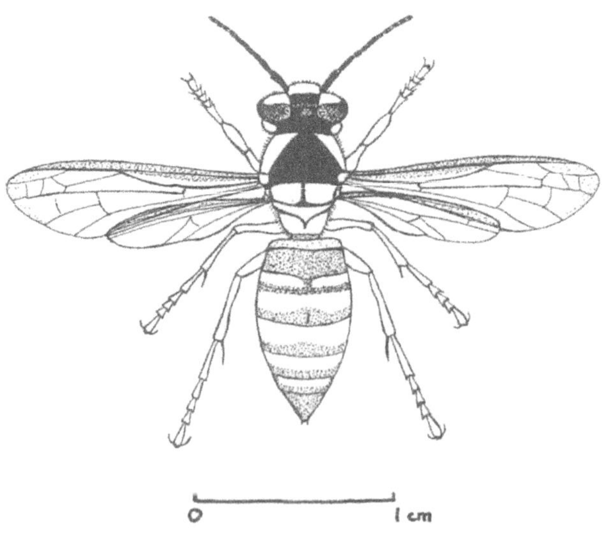

Figure 14.142. Vespa bicolor (Common wasp – in Hong Kong) (Hymenoptera; Vespidae), typical female worker.

14.13.9 Family Sphecidae (Mud-dauber wasps etc.)

A group of fossorial or mud-nest building wasps, with predacious habits, and they provision their small nests with paralysed insects or spiders. The petiole is distinct and sometimes quite long. Only two genera are likely to be encountered in produce stores. The Mud-dauber wasps build their mud nests in small groups on walls of buildings – the nests are provisioned with small paralysed spiders.

Ampullex spp. (Cockroach wasps) (Figure 14.143)

A pantropical genus of metallic coloured wasps (mostly blue or green) with noticeably short wings; they prey on cockroaches, which they paralyse by stinging, as food for the larvae. The paralysed cockroach is dragged into a dark corner, and the wasp lays a single egg on the body.

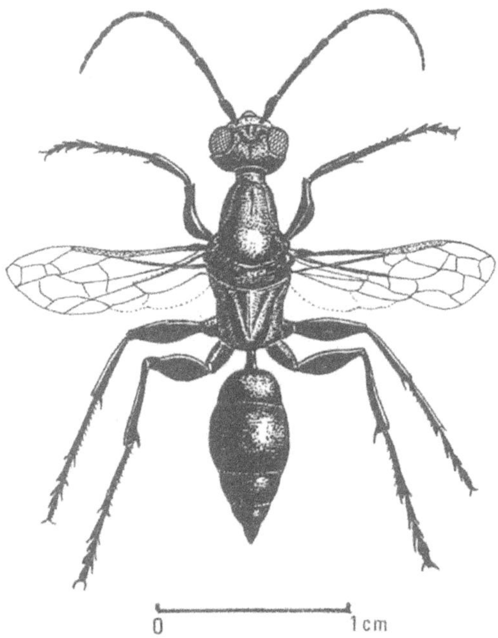

Figure 14.143. Ampullex compressa (Cockroach wasp) (Hymenoptera; Sphecidae), adult female wasp, from Hong Kong.

Sceliphron spp. (Mud-dauber wasps)

These wasps are of no particular importance in the fauna of produce stores, but they are occasionally found in tropical stores; the nest group is usually stuck on to the wall of the building, sometimes inside the building and sometimes on the outer wall.

14.13.10 Family Xylocopidae (Carpenter bees)

These are large, stout-bodied solitary bees, 1-3.5 cm long, which have very large mandibles and dig nest tunnels in dead wood. *Xylocopa* species are found throughout rural Africa and tropical Asia. Several species are regularly found infesting structural timbers accessible from the exterior. The wooden beams are bored with a tunnel of up to 10 mm wide and 20-30 cm deep. When the nest cells are fully provisioned, the tunnel entrance is sealed with a clay plug. Nest holes are conspicuous, but seldom abundant enough to cause any structural weakness. Female bees may sting if provoked sufficiently, and their presence and loud buzzing noise may alarm humans in the store.

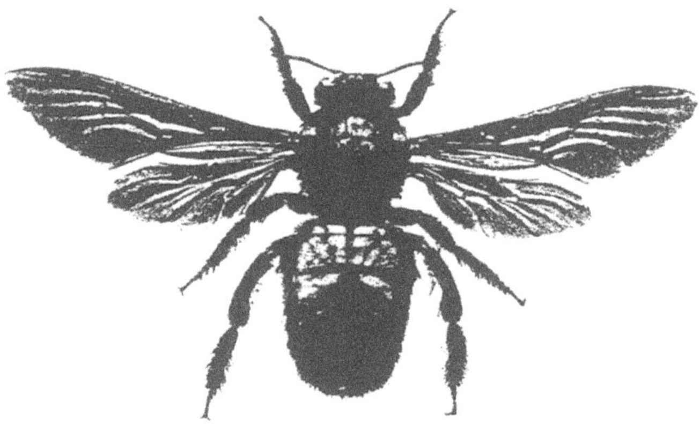

Figure 14.144 (a) Bamboo carpenter bee (Xylocopa iridipennis); body length 26 mm; (Hymenoptera; Xylocopidae); South China.

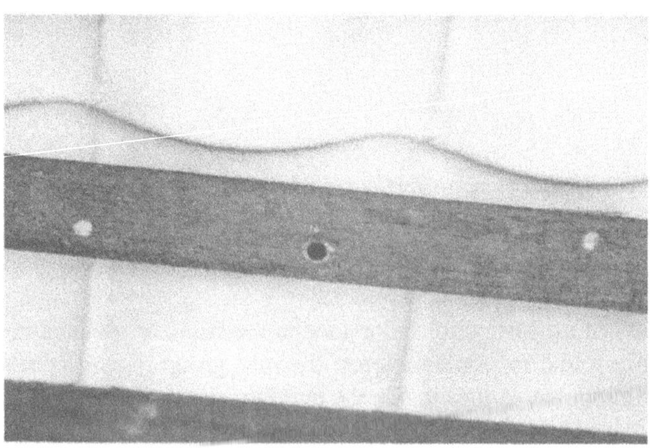

Figure 14.144 (b) Nest tunnels of Xylocopa in process of construction in wooden beam, Sarawak.

14.13.11 Family Apidae (Bees proper)

These are the social bees, and *Apis mellifera* (The honey bee) is of great importance to agriculture being a main pollinator of most entomophilous plants; normally never encountered in produce stores, although in parts of Africa it is common to see a native bee hive fixed under the eaves of dwellings. The genus *Trigona* is sometimes encountered in buildings – the bees are tiny (2-3 mm long) and stingless (sometimes called mosquito bees), and they make their nest in a crevice often with a very conspicuous entrance funnel built of a waxy substance. They cause no harm at all, and if found in a building can be safely ignored.

CHAPTER 15

PESTS: PHYLUM PLATYHELMINTHES
(Tapeworms and Flukes)

The Platyhelminthes are known as Flatworms; they are soft-bodied, flattened animals, many are elongate. There are three classes: the Turbellaria are Planarians and are free-living in water or on land; the thin, flattened body is covered with cilia, and they glide over the surface; they are all regarded as being quite primitive. The Trematoda are Flukes, all parasitic in or on vertebrate hosts, and a few are serious human endoparasites. Class Cestoda are the Tapeworms, a very specialised group of endoparasites, mostly to be found in vertebrate hosts.

15.1 CLASS TREMATODA (Flukes)

15.1.1 Family Fasciolidae

Fasciola hepatica (Liver fluke)

Pest status. A cosmopolitan pest of increasing importance; it is found in sheep- and cattle-raising areas of the world.

Damage/Hosts. Fascioliasis is a serious problem in sheep and cattle, and man becomes infected by accident. Also reported from horses, goats, pigs, rabbits, dogs and squirrels. Host damage includes destruction of liver tissues, damage to bile ducts, atrophy of portal vessels, and often the host is killed.

Life history. The adult is a large fluke, 30 x 13 mm, leaf-shaped, but pointed posteriorly and with an oral cone. It is hermaphroditic, as are most of the Trematoda.

Eggs are large, and pass into the alimentary canal via the bile duct, and eventually leave the host in the faeces. After being in water for 4-15 days at 22°C, the first larval stage, the miracidium escapes. It uses its surface cilia to swim, and seeks out a snail (*Lymnaea* etc.). The larva attaches to and then penetrates the body of the snail, and metamorphoses inside to become a sporocyst. This in turn becomes several mother rediae which give rise to daughter rediae. They turn into cercariae, which escape from the snail and become free-swimming. The cercariae find aquatic vegetation, and encyst (losing their tail) on the foliage as metacercariae. Metacercariae swallowed by the definitive host excyst in the duodenum, and then penetrate the intestinal wall to reach the coelomic cavity. From here, they penetrate the liver, migrating through the parenchyma to settle in the bile ducts where they mature. During migration through the liver, the young flukes actively feed on the host's liver cells. The large leaf-shaped adults develop in the host bile ducts; they can live for 11 years.

Man becomes infected by eating raw or undercooked watercress grown in water downstream from pastures. He can, of course, also be infected by eating raw bovine liver harbouring *Fasciola.*

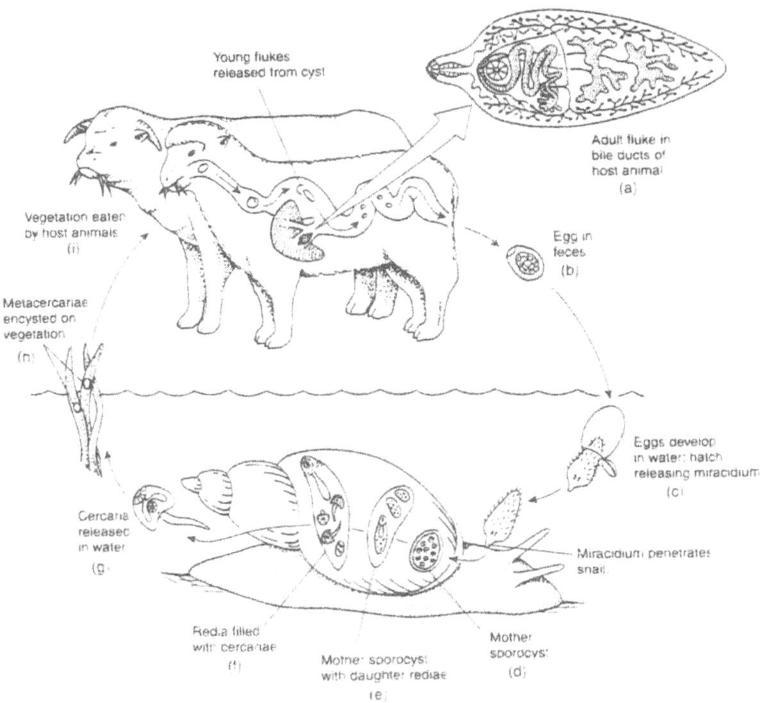

Figure 15.1. Life cycle of Fasciola hepatica (after Schmidt & Roberts, 2000).

Distribution. Worldwide where sheep and cattle are reared; of increasing importance in Cuba, Latin America, Algeria and France. In France between 1956 and 1957, 500 people became infected with *Fasciola*. Different species of snails are used in different regions.

Other species. F. halli is found in sheep and cattle in southern USA; *F. indica* in cattle in India; *F. gigantica* in cattle in Hawaii, where incidence may be up to 80%. *Fascioloides buski* lives in the intestine of man, pig and other mammals, and uses *Planorbis* as intermediate host. In Thailand, the rate of human infection has been as high as 40% (north of Bangkok) – man becomes infected by eating uncooked water plants with metacercariae attached.

F. magna lives in cattle and sheep, and has a life cycle similar to *F. hepatica*, except that a different snail intermediate host is used; this species is thought to be endemic in North America to native deer.

15.1.2 Family Opisthorchiidae

Clonorchis sinensis (Chinese liver fluke)

Pest status. A serious pest in the Orient, where more than 10 million people are infected as a result of eating undercooked fish (carp species).

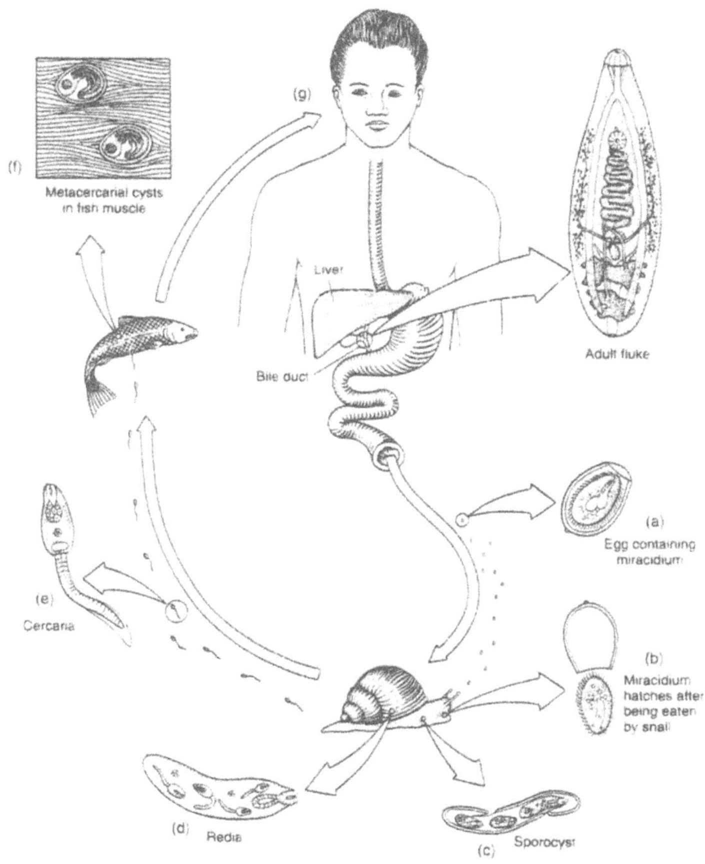

Figure 15.2. Life cycle of Clonarchis sinensis (after Schmidt & Roberts, 2000).

Damage/Hosts. Hosts are freshwater fish mostly in the family Cyprinidae (100 species recorded), and also some Crustacea (*Macrobrachium, Palaeomonetes* and *Caridina*). When the intermediate host is eaten, the metacercariae develop inside the definitive host (man, and other fish-eating mammals). The basic pathogenesis is the erosion of the epithelium in the bile ducts, which thicken and block the duct, and liver function may be impaired; some cases are fatal.

Life history. Adults measure 8-25 mm long by 1.5-5 mm wide, body is elongate leaf-like, with weakly developed suckers. They live in the biliary system, and produce up to 4,000 eggs per day for six months or more; they are hermaphroditic. When the egg is eaten by a snail (*Parafossarulus* etc.) the miracidium hatches and develops quickly into a sporocyst.

After a couple of weeks the sporocyst produces rediae, and each redia produces 5-50 cercariae. The cercaria swims to a host fish and attaches by its suckers. The tail is cast off, and the cercaria bores through the skin and encysts in a muscle. Man is infected by eating raw or undercooked fish or crustaceans. The young flukes excyst in the duodenum and migrate to the liver, and eggs are laid about a month later. The life cycle can be completed in three months, and the adult flukes can live for at least eight years in man.

Distribution. Widely distributed in Japan, Korea, China, Taiwan, and Vietnam, where more than 10 million people are infected. Records outside eastern Asia are from people who visited the area, or ate imported dried, frozen or pickled fish from the area. In many places, up to a quarter of the population are infected; in some places, 80%. Fish farming is a common source of infected fish, as carp are the main species reared.

Other species include:

Opisthorchis felineus is found in eastern Europe and most parts of Asia (from India to Japan), and probably infects a million people.

O. viverrini is common in Thailand, where some 7-8 million people are infected, and they show a high incidence of bile duct cancer.

Both species have a similar life-cycle to *C. sinensis.*

Paragonimus westermani (Oriental lung fluke)

A small (7-12 m by 4-6 mm) fluke found mainly in Asia (also Africa and South America); it lives in the lungs of man and various crab-eating mammals. Eggs are coughed up in sputum, and miracidia are found in the rice and vegetable paddies where they infect snails (*Semisulcospira, Brotia* etc.). Each miracidium gives rise to 144 rediae. The cercariae leave the snail to seek the crustacean second intermediate host, a freshwater crab – in China, the Mitten Crab (*Eriocheir sinensis*) is used; in Korea, a crayfish (*Palaemon nipponensis*); and in Japan, the Philippines and Taiwan, other freshwater crabs are used. Humans become infected by eating undercooked crabs.

In the human, the metacercariae excyst in the intestine, then bore into the coelom through the diaphragm and into the lungs, where they become encapsulated. The wandering flukes may end up in the eye, spleen, liver, muscles or even brain tissue – a heavy infection can be fatal.

Several other species are known from China, and at least one from North America.

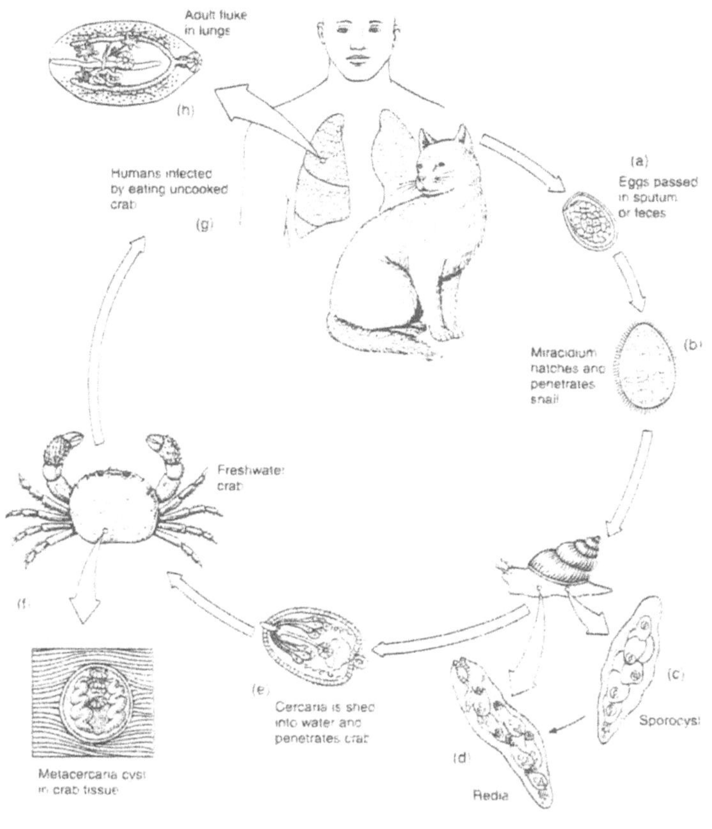

Figure 15.3. Life cycle of Paragonimus westermani (after Schmidt & Roberts, 2000).

15.2 CLASS CESTODA (Tapeworms)

15.2.1 Order Cyclophyllidea

Taenia solium (Pork Tapeworm)

Pest status. A major pest in areas where raw or poorly cooked pork is part of the normal diet.

Damage/Hosts. The adult worms live in the human intestine – man appears to be the only known definitive host. Cysticerci can also develop in humans after self-infection, usually via contaminated fingers. They may form in muscle or subcutaneous tissues where they do little harm, but they can develop in the eye, heart, spinal cord, brain etc. where serious damage can be done – death can result.

Life history. The adult worm measures 180-300 cm, and contains 800-900 proglottids; the small scolex ('head') is only 1 mm in diameter, armed with two circles of hooks for attachment to the intestine wall. Segments of gravid proglottids are excreted every day, each containing thousands of eggs. when the proglottids or eggs are eaten by pigs (dogs, monkeys, camels etc.), or even humans, the liberated oncospheres bore into the intestinal wall from where they are carried in the blood or lymph to viscera, muscles etc., where they form cysticerci. The cysticerci are 6-18 mm long, ovoid and white, and the invaginated scolex can be seen on one side. When the infected pork is eaten by the human host, the scolex evaginates and anchors to the wall of the small intestine. It grows to maturity in 2-3 months. The adult worm can live for up to 25 years.

Distribution. Worldwide, where pigs are reared and pork eaten partially undercooked or raw (sausages etc.).

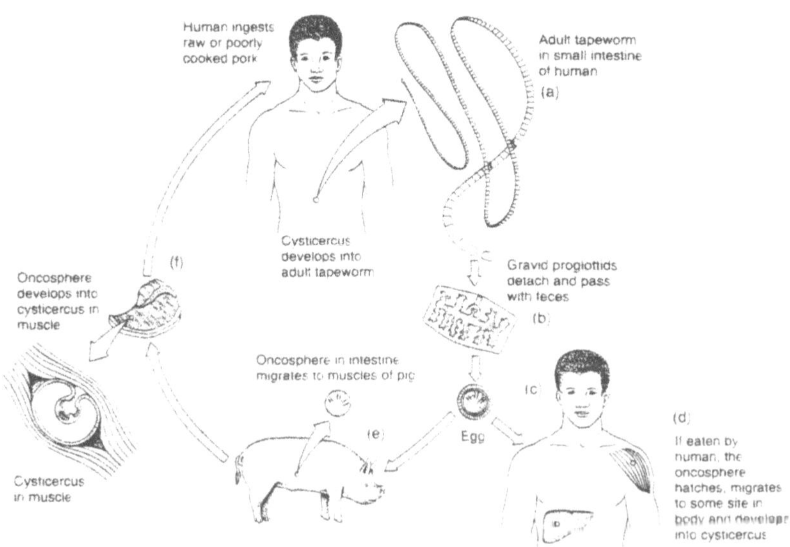

Figure 15.4. Life cycle of Taenia solium (after Schmidt & Roberts, 2000).

Taeniarhynchus saginatus (Beef Tapeworm) (= *Taenia saginata)*

Pest status. This is the most common large tapeworm of man, and is found worldwide. Severe symptoms include physical weakness, abdominal pain, excessive appetite and loss of weight.

Damage/Hosts. Adults live in the human intestine with the scolex firmly attached to the intestine wall, and they produce the symptoms mentioned above.
Life history. The adult worm is usually 1-5 m long (20 m is recorded), and contains 2,000 or more proglottids. The scolex is unarmed (bears no hooks). Proglottids and eggs are passed out daily with the faeces, and the intermediate host is a cow (or some other ungulate), antelope or llama. Human infections start by ingestion of the cysticerci in beef (especially muscles of the head and heart) which is raw or 'rare' (undercooked). The cysticerci are spherical, white and about 10 mm in diameter.

Distribution. Cosmopolitan in distribution, it occurs in small numbers in most countries. But in parts of the world where undercooked or raw beef is eaten, infection rates can be as high as 75% of the whole population. This is the case in Ethiopia, where raw beef is a local delicacy. The educated practitioners of this habit were well aware of their actions, but were willing to tolerate the regular infection in order to keep eating raw beef – on average, they needed anti-helminthic treatment about three times a year (in the 1980s). France is another country where raw beef is regularly eaten and the beef tapeworm is well established as a pest.

15.2.2 Order Pseudophyllidea

Diphyllobothrium latum (Fish tapeworm; Broad tapeworm of man)

Pest status. A major pest in areas where fresh fish are an important part of the local diet. Infection can result from the handling of raw fish, including cleaning, as well as eating the fish. An estimated nine million people worldwide are infected.

Damage/Hosts. Found in fish-eating carnivores (Canidae, Felidae, Mustellidae, Pinnipedia, bears and humans) where it shows little host specificity. In humans, infection results in loss of weight and abdominal pains, and sometimes a type of pernicious anaemia.

Life history. Adult worms measure around 10-20 m long and 10-20 mm broad, and are composed of 3,000-4,000 or more proglottids. The ovoid eggs in water take 8-12 days to develop, and the hexacanth larva is covered with cilia and known as a coracidium. The coracidium, while swimming or crawling, must be ingested by a freshwater copepod (*Cyclops* or *Diaptomus* spp.) within 24 hours, or else it dies. Once in the appropriate copepod host, it loses its ciliated coat, and the naked onchosphere bores through the intestinal wall into the haemocoel. In 14-18 days, the onchosphere turns into the procercoid larva. When the infected copepod is eaten by a freshwater fish such as *Salmo* or *Esox* spp. in North America, or *Onchorhynchus* (an Oriental salmon) in the Far East, the parasite larva bores through the intestine wall and migrates to the body muscles, where it develops into a solid plerocercoid larva with an evaginated scolex at one end. The larva is coiled, and can be either free in the muscle or encapsulated.

Distribution. A widespread pest mostly confined to the northern temperate region, where fish are a major item in the local diet. In Finland, some 20% of the population is infected, and in some other Baltic communities, the incidence approaches 100%. In North America, some 50-70% of northern and walleye pike found in small lakes harbour the plerocercoid larvae. Common in the Great Lakes of North America and the freshwater lakes of Fenno-Scandia, and Russia.

The plerocercoid larvae of these tapeworms found in humans are usually impossible to identify (except for some with a distinctive scolex armature), and they are collectively known as 'spargana'. The infection is known as sparganosis, and is caused by the person drinking water containing an infected copepod. The larva migrates from the intestine and settles in body muscle as a white maggot-like plerocercoid, where it may stay for years with no further development.

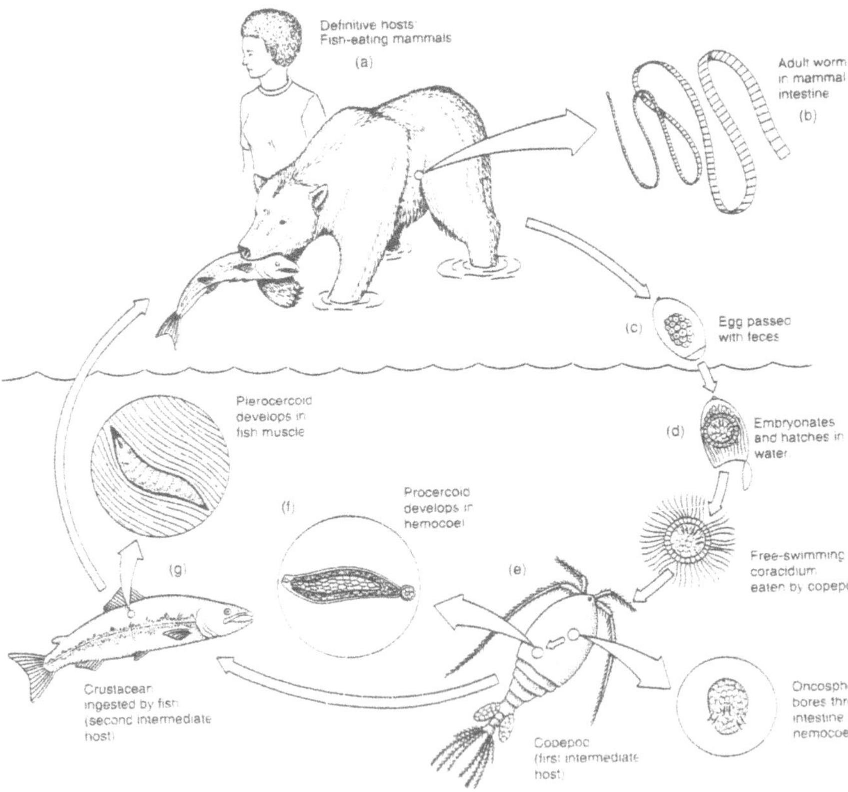

Figure 15.5. Life cycle of Diphyllobothrium latum (after Schmidt & Roberts, 2000).

Control. Modern anti-helminthics are very successful, and often require only a single oral dose – they include: niclosamide and praziquantil as well as albendazole and membendazole.

CHAPTER 16

PESTS: PHYLUM NEMATODA

16.1 INTRODUCTION

These are also called Roundworms; they are elongate, thin, whitish worm-like animals which range in size from a few mm to 40 cm (*Ascaris*). There are many genera known, most are free-living in soil or freshwater, and some in salt water, but the best known forms are internal parasites in animals, or in plants when they are known to agriculturists as 'eelworms'. Several species are important parasites of man, but only a few are associated with foodstuffs.

The classification is basically too esoteric to be referred to here.

16.1.1 Family Trichinellidae

Trichinella spiralis (Pork trichina)

Pest status. A very widely distributed major pest species in man.

Damage/Hosts. Man, and also many other mammals (bears, pigs, rats, raccoons, badgers, moles, porcupines, dogs, rabbits, mice, cats, wolves, ferrets, muskrats), and some birds (owls, crows, pigeons, chickens). Sylvatic trichinosis is the cycling of the infection between wild carnivores and their prey or carrion. Urban trichinosis is the cycling between man, pig and rats. In Alaska, polar bears are the sources of human infection. In man, heavy infections can be very damaging, and death can result. In 1947, it was estimated that 28 million cases of human trichinosis existed, but it is less prevalent today.

Life history. The adults are tiny, 2-3 mm long and very thin. Infection starts by the host (man or another carnivore) eating meat (usually inadequately cooked pork) containing encapsulated larvae. The ingested larvae are released in the duodenum by the action of the host's enzymes. The larvae then penetrate cells lining the host's mucosa, where they develop. On reaching maturity (after 30-36 hours), the male and female worms copulate in the mucosa. The males then die, and the females burrow deeper into the musoca – each ovoviviparous female gives birth to 1,500 larvae over a 9-16 week period, then she dies. Some larvae escape into the intestinal lumen, and thence to the exterior, where they remain infective for a short time. Most larvae in the mucosa (c.0.1 mm long) are carried by the blood vessels into venous blood, which carries them to the heart. From here, they are taken to the peripheral circulation and to skeletal muscle of the jaws, diaphragm, tongue, larynx and eyes, where they develop into the infective stage (on about the 6th day after infection), as intracellular parasites within myofibres. By day 17 they start to coil, and have reached a length of about 1 mm. After 8 weeks, they become infective, and the host cell secretes a capsule around the worm which becomes calcified. The enclosed larva enters developmental arrest, and can live for many months or years. When the encapsulated larva is eaten by a compatible host, the freed larva repeats the life cycle. In urban trichinosis, the parasites cycle between pigs, rats, and man. Infected pork waste is eaten by rats and other pigs feeding on garbage; dead and dying rats are eaten by pigs, and in turn raw or poorly cooked pork is eaten by man, where the infestation cycle ceases and the encapsulated worms are eventually calcified.

Distribution. Worldwide, from the Arctic Ocean (walrus, whales, polar bears etc.) to Africa, Asia, and South America.

Control. Adequate cooking of all pork and pork products, or freezing at -15°C for 20 days, is the most simple and effective method of control of this pest, but the populations in wild hosts cannot be controlled. Chemical control has, to date, not been particularly successful.

16.1.2 Family Ascaridae

Ascaris lumbricoides (Common roundworm)

Pest status. A major international pest species; it is estimated that 600-1200 million people are infected worldwide, the great majority in Asia.

Damage/Hosts. Man is the definitive host, and the migration of larvae through the lungs causes lesions and blood loss. Large numbers of adults can block the intestine. A heavy infection can cause death, but light infections produce few symptoms, if any. Basically, for this parasite, one animal serves as both intermediate and definitive hosts, with the larvae and adults in different organs.

Life history. Adults are large fat white roundworms, males 15-30 cm by 2-4 mm diameter, females 20-40 cm by 3-6 mm. The adults live in the small intestine of the host where they feed on chyme. Each female is capable of laying up to 27 million eggs; these pass out with the host faeces and develop in the soil, where they remain viable for many years, and are extremely resistant to chemicals. After 10-14 days, active larvae are formed in the eggs, and the second larval stage becomes infective. Infective eggs are carried onto food materials, usually by contaminated hands (although some infants will eat soil!).

After the eggs are swallowed, the larvae hatch in the host's small intestine and burrow through the lining into small blood veins where they are carried to the liver, thence to the heart, and then to the lungs. Here they grow for a few days before moving into the bronchioles, continuing up the trachea, through the glottis and down the oesophagus to the small intestine again, where they continue to develop and grow. Occasionally, the adult worms are carried by peristalsis out of the body. In Hong Kong in 1976, a one year-old baby passed a 20 cm *Ascaris* in its stool – the worm was 2/3 the length of the baby!

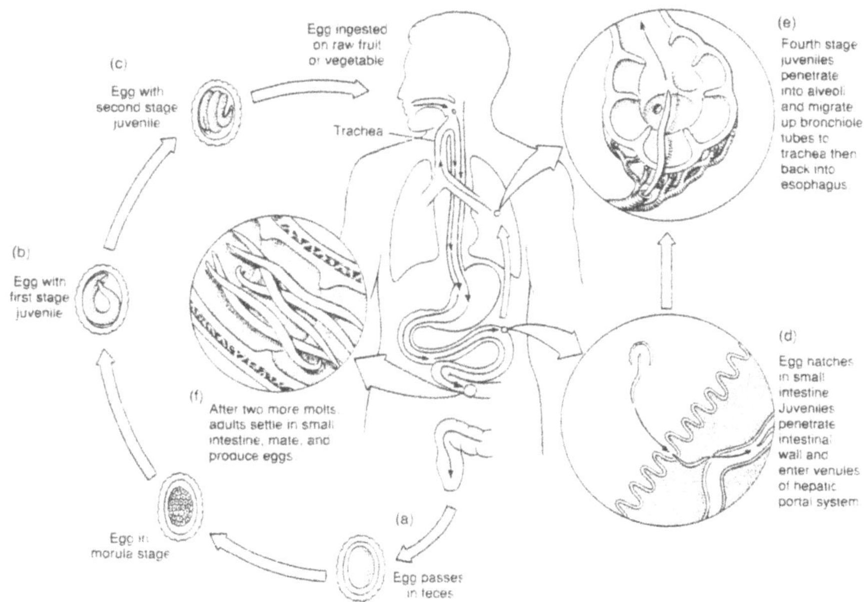

Figure 16.1. Life cycle of Ascaris lumbricoides (after Schmidt & Roberts, 2000).

Distribution. A worldwide distribution, found wherever man resides, most abundant in Asia (500+ million cases of infection), with fewest cases in North America (3 million).

Control. Cleanliness in food preparation, especially the avoidance of contact with faecal matter and soil, is the most effective method of control. Modern ascaricides include ivermectin, albendazole and embendazole, and other benzimidazoles.

16.1.3 Family Anisakidae

Anisakis spp.

Sea fish, such as cod, haddock and whiting (and others), when cleaned will sometimes reveal white nematodes 5-10 cm in length, in parts of the body cavity. Often these are members of the genus *Anisakis* (or closely related genera). The normal definitive host is a marine mammal (seal, whale etc.), but if swallowed by a human, the larvae in the fish flesh can burrow through the stomach or intestine of man, and may cause severe granulomas. The eating of raw fish and pickled fish fillets encourages such infections. Smoking fish or marinating fillets generally does not kill the *Anisakis* larvae, but dry salting usually does. The fish would have to be heated to 50-60°C or cooled to -20°C for 52 hours to kill the parasites. So careful cooking of sea fish is the best method of control.

(For further information concerning helminth and nematode parasites, see Cheng (1986), and Schmidt & Roberts (2000)), and Macrae, *et al.* (1993: pp. 3411-3420).

CONTROL OF HELMINTH PARASITES

The specific chemicals at present recommended for the control of human tapeworms and roundworms include: mebendazole, albendazole, and pyrantel pamoate; also niclosomide and praziquantel.

16.1.4 Family Heterodidae

Meloidogyne spp. (Root knot nematodes) (= Eelworms)

Pest status. Very widely distributed major pests of agriculture, especially in warmer regions.

Damage/Hosts. Some 16 species are known worldwide, but only half of these are likely to be encountered. Field symptoms include plant wilting, loss of vigour and yield, and sometimes death, but the root symptoms are galls of varying size and shape. Inside the galls lie the swollen white female worms, sometimes with a protruding gelatinous egg sac. Some species are very host specific (to tea or coffee bushes), but others are completely polyphagous and recorded from as many as 700-800 different host plants. On various root crops, they make galls either in the actual roots when they look like the illustration (Figure 16.2), or in the swollen root (such as beetroot) when they often make a globular lump on the side of the root. In potatoes and yams, they gall both the fine roots and the tubers, so the tubers have swellings on the surface. These are the vegetable food crops most frequently attacked; they are usually recognised at harvest and so do not often reach the market place.

Life history. The adults are tiny, 1-2 mm long, the males are thin, but females become swollen and pyriform. The egg sac protrudes through the wall of the root into the soil. Each female lays 300-1,000 or more eggs. After hatching, the tiny larvae (first stage) leave the egg sac, and are free-living as they seek a host plant. Second stage larvae are infective, and penetrate plant roots and settle in the tissues where they moult into either vermiform males or saccate females. Development takes about 57 days at 20°C, and 23-30 days at 27°C.

Distribution. Worldwide in warmer regions, up to southern USA and southern Europe, and in greenhouses further north.

Control. Usually control is achieved by a mixture of cultural (crop rotation, use of resistant varieties, and soil sterilization) and chemical methods (soil fumigation, pesticide granules). But in parts of eastern Africa, nematode population levels are so high that control is seldom achieved, partly because of the cost of chemicals.

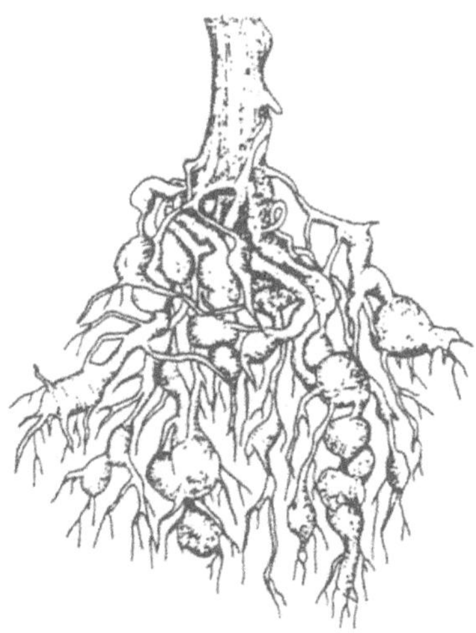

Figure 16.2. Root knot nematode Meloidogyne sp. galls on plant roots.

CHAPTER 17

PESTS: KINGDOM PROTOCTISTA
(Phylum Protozoa)

The 'Protozoa' are minute, mostly unicellular animals that used to be regarded as the most primitive forms of animal life. But there are some unicellular 'Algae' that shared many of the basic characters, and so some forms were difficult to place taxonomically. This presented some problems in the teaching of biology, and so in 1987, the Institute of Biology (UK) commissioned a booklet on *Biological nomenclature* (standard terms and expressions used in the teaching of Biology) – the 3rd Edition was published in 2000, and this redefined the basics of classification. The old Kingdoms Plantae and Animalia were replaced by five new Kingdoms: Prokaryotae (Bacteria), Protoctista (Protozoa and Algae), Fungi, Plantae, and Animalia (Cadogan, 2000).

Kingdom Protoctista now includes both the former 'Protozoa' and the 'Algae' in seven different phyla.

17.1 PHYLUM DISCOMITOCHONDRIA

Giardia lamblia

A tiny flagellate found in the human small intestine, causing the highly contagious disease Giardiasis. A cosmopolitan pest, thought to be present in human populations at levels from 2-68%. Normally water-borne by faecal contamination, there have been many recent outbreaks of epidemic proportions in all parts of the world. In some regions, it is a chronic local complaint due to insanitary conditions. But a good level of control can be achieved by a one-week course of 'Flagyl' (metronidazole) tablets.

Pentatrichomonas hominis

Another tiny flagellate transported by food and drink into the body of man. Natural reservoir hosts include cats, dogs, mice and other rodents.

Sarcocystis hominis and spp.

These protozoan parasites have been known for 100 years, but only in the last 30 have they been correctly identified; it is now known that *S. hominis* is a parasite of man via infected cattle, where it forms a large cyst (sarcocyst) in the skeletal muscle, 1-2 cm in size, and is very obvious in freshly killed cattle. They can form in the heart muscle and sometimes in the brain, when serious damage may result. In man, the sarcocyst wall is digested away, and the escaping bradyzoites penetrate the wall of the small intestine where they develop and reproduce.

17.2 PHYLUM RHIZOPODA

Entamoeba histolytica

This causes amoebic dysentery in man, and is worldwide in distribution. Often thought of as a tropical disease, it actually occurs in both Arctic and Antarctic zones. In the USA, local incidence has varied between 2 and 36% according to location. The cysts responsible for infection are typically carried in water or on food, but have been found in the gut of domestic flies. The trophozoite stage infests the colon and rectum of the host (man, other primates, cats and dogs etc.) and many invade the tissues, but some are carried by the blood circulation to the liver, lungs, brain and other organs, where they become lodged and form abcesses. Liver damage can be serious.

In temperate regions, the disease is usually less pathogenic with minimal symptoms, but in the tropics and subtropics, amoebiasis can be very serious if untreated. Cheng (1986) estimates that worldwide some 100 million people are probably infected. Metronidazole is reputedly effective for treatment of amoebiasis.

CHAPTER 18

PESTS: KINGDOM FUNGI

The Kingdom Fungi includes eukaryotic organisms ranging from microscopic unicellular yeasts to multicellular moulds and macroscopic mushrooms. They are heterotrophic organisms that have to be fed organic compounds for energy and growth, and several are major pest organisms responsible for food spoilage. There is a total of about 100,000 species of fungi, but only relatively few are associated with food deterioration and agricultural products.

The technical classification of fungi is difficult for the non-specialist to follow; this is in part because many fungi have been described from their asexual forms, others from their reproductive forms, and some from both - so some species are recorded in two different forms, and the end result is somewhat confusing.

For the present purpose, it is convenient to consider the stored produce fungi just as 'Moulds' and 'Yeasts', as is the practice in the food industry.

18.1 MOULDS

These are multicellular or multinucleate filamentous organisms that grow as a compact mass of filaments (hyphae), twining and branching and hair-like in appearance. The mycelium is generally clearly visible. Aerial hyphae give rise to the reproductive structures, the conidiophores, which produce millions of spores (conidia). The spores behave like dust particles, and are dispersed aerially. When they land on a suitable substrate, they germinate and start a new mycelium. Other types of spores are produced – some are vegetative (asexual). Many of the moulds important in food spoilage reproduce mainly asexually (also called Fungi Imperfecti).

Moulds are ubiquitous, and mostly live in the soil where they break down decaying vegetable material. Cereals and other crops generally acquire colonies of mould in the field, and then bring them into the stores where they may or may not continue to develop. They are thus common contaminants of foodstuffs of all types.

Moulds can grow in quite harsh environments, and can withstand more severe stresses than most micro-organisms – they require less moisture than bacteria and yeasts, and can grow on drier substrates than bacteria. Once mould growth starts, it will probably continue, being fed moisture released by its own respiration. They are highly aerobic, and can be completely inhibited by carbon dioxide. Moulds grow over a wide range of temperatures, but most have their optimum between 25 and 35°C. Some can grow at temperatures from 0-5°C, a few can withstand freezing, and a few can tolerate 60°C +. They can utilize a wide range of carbohydrates, from simple sugars to starch and cellulose, as well as proteins and various inorganic salts.

Thus they can attack all known types of foodstuffs and produce in storage. The biochemical activities of the mould result in deterioration and spoilage of foodstuffs and animal feeds, also causing tainting; some secondary metabolites are compounds toxic to humans and livestock – these are collectively known as mycotoxins. A few of these secondary metabolites are beneficial, specifically those responsible for the flavour (and colour) of certain cheeses (Camembert, Stilton, Roquefort, and some cream cheeses).

Mycotoxins refer to the diverse, naturally occurring chemical substances produced by certain moulds (microfungi). These secondary metabolites produce a number of adverse effects on a range of biological systems, including death of livestock and humans if large doses are ingested. Three of the most widely encountered mycotoxins are the aflatoxins that *Aspergillus flavus* produces on groundnuts, and zearalenone produced by *Fusarium graminearum* on mouldy cereal grains. *Claviceps purpurea* and other species infect a wide range of cereal crops and grasses (rye grass especially), and the end result is a long (5 mm) dark grey or black sclerotium or ergot which develops in place of a grain, and it projects visibly out of the side of the cereal head.

Table 18.1 Examples of mycotoxin-producing moulds and associated diseases

Mould	*Mycotoxin*	*Disease type*
Aspergillus flavus	Aflatoxins	Cancer
Aspergillus parasiticus		Liver damage
Claviceps purpurea	Ergot alkaloids	'St Anthony's fire'
		Gangrene
		Convulsions
Aspergillus ochraceus	Ochratoxin A	Nephropathy
Penicillium viridicatum		Teratogenic effects
Penicillium crustosum	Penitrem A	Tremors
Penicillium cyclopium		Neurological disorders
Aspergillus versicolor	Sterigmatocystin	Cancer
		Liver damage
Fusarium spp.	Trichothecenes	Dermatitis
		Digestive disorders
		Blood disorders
Fusarium graminearum	Zearalenone	Oestrogenic effects
		Abortion

(From: Macrae, *et al.* 1993: p. 3,214)

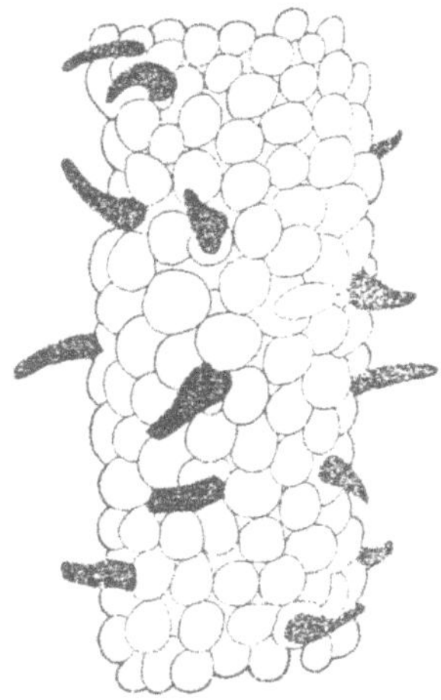

Figure 18.1. Ergot (Claviceps sp.) on millet ear; Africa.

Ergots are found worldwide, with different species in different continents and on different crops. The sclerotia are harvested with the grain, and if eaten by man or cattle can cause serious poisoning. On rye grass, the infected head may be grazed by livestock.

Some of the commonest moulds to be found in foods are briefly reviewed below:

Aspergillus

Many species known, but 4 or 5 are widely encountered on grain. *A. flavus* and *A. parasiticus* produce aflatoxins on mouldy peanuts (groundnuts), which can cause extensive liver damage. The genus is common on grains, nuts, oil seeds, and some dry-cured meats. *A. oryzae* is non-toxic and used in oriental food fermentation to make soya sauce and miso. The conidia may be green, black, brown or yellow.

Penicillium

Another widespread genus with many species recorded from a wide range of food products (grains, cheeses, fruits etc.). Several species are used in the production of blue-veined cheeses, others for Camembert and Brie cheeses. The conidia are mostly coloured green, grey or bluish. Several mycotoxins are produced by *Penicillium* spp. – the famous penicillin used to control bacterial infections is produced by *P. notatum.*

Fusarium

These are mostly field fungi where they infect grain plants worldwide, in both tropical and temperate regions. They live in the soil and cause rots (root and stem), wilts and blights. But some affect the grain head in the field, and may produce mycotoxins. *Fusarium* species can invade fresh fruits and vegetables in storage, causing rots and spoilage.

Alternaria

Several species are found on plant materials and foodstuffs worldwide, including fruits, nuts, grains (wheat, sorghum), some vegetables (tomatoes, bell peppers), and in cold-stored meat and spices. They can grow at low temperatures on refrigerated foods. Some species are capable of producing mycotoxins, and so present both food spoilage and potential health hazards.

Other genera recorded as food pests, but of lesser importance generally, include: *Trichothecium, Trichoderma, Cladosporium, Geotrichum, Mucor, Absidia, Thermomyces, Rhizopus etc.*

CONTROL OF MOULDS

As with other micro-organisms which produce aerial diaspores (such as spores), they are so widely distributed in the atmosphere that for most storage and food processing facilities it is pointless to try to prevent access to the produce, for the spores are already there. However, some production and packing lines can be enclosed with high efficiency air filters which can remove the vast majority of particles and spores from the working atmosphere. And container cleanliness can be achieved if required. But unless a food has been previously heated and subsequently aseptically place in a hermetically sealed container, the presence of mould spores has to be expected. The main aim is to keep conditions such that spore germination and mycelium development are prevented, or at least kept to a minimum. Atmosphere, temperature and moisture levels are three aspects that can be successfully manipulated. Reduction of oxygen level, preferably its complete absence, will inhibit most (but not all) moulds, and similarly raising the level of carbon dioxide to about 70-80% greatly reduces the growth rate of most moulds. Most moulds possess little heat resistance, especially in the absence of sugar and salt, and pasteurization can be effective, but a salt concentration of 3% can prevent this. The majority of moulds are mesophilic, and so grow at moderate temperatures in the 10-40°C range. Lowering the temperature decreases the rate of metabolism and slows the rate of food deterioration. Fungi show a wide range of behaviour in relation to water availability, but generally are more tolerant of water stress than other organisms (including yeasts). Stored grain with a moisture content at or below 14% has very little mould development, and for many species 15% is low enough to arrest development. The critical moisture level for flour, for mould development, is 16-17% m.c., and so flour is usually stored with a slightly lower moisture content, but, of course, will carry spores into the ensuing cereal products.

Some chemical additives to suppress moulds are permitted in foodstuffs, at very low levels (between 0.1 and 0.2%), and these include sorbic acid, benzoic and propionic acid and salts, which can be added to bakery products, some cheeses, fruit products and carbonated drinks.

18.2 YEASTS

About 500 species of yeasts have been identified, and a number have been used by man in the production of foods (bread etc.), and fermented beverages, and nowadays yeast in tablet form is often prescribed medically as a source of B-complex vitamins. But these beneficial uses not withstanding, many yeasts are responsible for food spoilage. About one quarter of all known species can be isolated from foods, but only about a dozen species play a significant role in food alteration (known as spoilage yeasts). Two of the most important genera are *Saccharomyces* and *Zygosaccharomyces.* Food yeasts are mesophilic with a preferred temperature range of 25-27°C, but are capable of growing in the range 0-40°C. The main damage done by yeasts to foodstuffs and beverages is the modification of the organoleptic (i.e. odour/taste) characteristics. The economic losses consequent to this are thought to be very high. Modern technologies in food processing have apparently exacerbated this situation by using less severe processing conditions, so as to preserve natural flavours, tastes and colours, and also using lower concentrations of salt and sugar. Most yeasts are facultatively aerobic, but do grow more vigorously in the presence of oxygen. The genus *Candida* is renowned for its role as a human skin parasite (thrush, athlete's foot etc.), and does also have some food spoilage species.

Control of yeasts can be achieved in some cases by pasteurization, by γ-radiation, or by freezing. Preservation of foods by addition of benzoic acid (and other weak organic acids) or sulphur dioxide treatment is achievable, as well as a high level of cleanliness of processing equipment, and disinfection. Packaging to include inert gases, and the storage of finished food products at low temperatures, will also help to reduce yeast activity.

CHAPTER 19

PESTS: KINGDOM PROKARYOTAE
(Bacteria)

The contamination of foods by micro-organisms is usually a mixed population of bacteria, moulds and yeasts, and they interact and compete according to which are better adapted to the prevailing conditions.

Under optimal conditions, the bacteria usually dominate, for they have a generation time of less than an hour (in some cases 10-15 minutes), whereas yeast and mould populations take several hours or days to duplicate. Thus, moulds and yeasts only compete successfully with bacteria when conditions are unfavourable to bacteria, usually a low pH and low water activity, and also the presence of chemical preservatives.

There is a wide range of different types of bacteria, some that can develop at low temperatures (0-7°C) called psychrotrophic bacteria; others can withstand pasteurization (called thermoduric; also the spore-forming bacteria); some are anaerobic or facultatively aerobic. So each food product has to be given preservative treatment appropriate to its particular situation, and against the types of bacteria it is likely to encounter.

Spoilage of food by bacteria is preventable if the food is processed and handled correctly, but the more process stages, the greater the risk of infection.

A few of the most important bacteria involved in food spoilage are listed below:

Campylobacter

Now the biggest cause of food-poisoning worldwide, with an estimated 500,000 cases a year in the UK alone. Extreme cases result in intensive care treatment, and can lead to death. The total number of deaths annually in the UK from food poisoning is about 200, with millions falling sick. About half of the poultry produced for human consumption carries this organism. The level of incidence appears to be rising annually. Under-cooked meat seems to be the main source of infection. (See Chapter 20).

The importance of this group was not realised until almost 1980, and now it is recognised as being the main causal agent for food poisoning. Two particular species are involved in most cases, *C. jejuni* and *C. coli*. They live naturally in the intestine of wild birds and some mammals as commensals, and the faecal shedding of bacteria by wild and domesticated hosts means that virtually all surface water is contaminated. Meats and milk are the main sources for human infection, and refrigeration tends to prolong their survival. The ever-increasing production of broiler chickens is probably a major reason for the increase in *Campylobacter* infections.

(For more details, see Macrae *et al.* (1993) pp. 594-606)

Clostridium

Clostridium perfringens

This is probably the most widespread of all the pathogenic bacteria; there are five toxigenic types, some causing human illness, others affecting domestic animals. It is a spore-producing form and is commonly found in fresh meat and poultry products. Spores can survive many food processing procedures, and it can grow over a wide temperature range, so is often implicated in human food poisoning. Heat-resistant strains are common. It forms part of the normal intestinal flora in humans and animals, and also occurs widely in the soil. Because of its abundance in faeces, it is also found widely in sewage-polluted water.

Vegetative cells can grow over the temperature range of 15-50°C, and even up to 70°C, but are rapidly inactivated at 75°C. If cooked meat is kept above 63°C or below 10°C, safety should be ensured. Most of the spores are heat-sensitive, and killed in a few minutes at 100°C, but the heat-resistant ones (which are far less abundant) survive this. Refrigeration is the usual method of control – the spores are not killed but merely de-activated, and they will germinate and grow when conditions are suitable, so the time factor (i.e. the time after cooking and before consumption) is often critical.

Clostridium botulinum

This causes human botulism – many different strains are known, some confined to animals. Spores are found in soil and on produce associated with soil, in fish, and (for some reason) also in honey. Most outbreaks are associated with home canning of vegetables, sometimes canned fish or meats, especially pickles fish and home-cured ham. Nitrites and other chemicals have an inhibiting effect on these bacteria. (Macrae *et al.* (1993) pp. 1043-1061)

Escherichia coli

This bacterium lives naturally in the large intestine of warm-blooded animals, including man; it does not survive long outside the gastro-intestinal tract, and its presence in water has long been used as an indicator of faecal pollution. Many strains are known, and they are divided into four main groups: the enterotoxigenic, the enteroinvasive, the enteropathogenic, and the enterohaemorrhagic (including E. coli 0157). Humans and cattle appear to be the main sources of *E. coli*, and the ways in which their faeces give rise to infections include:

- Contamination from food handlers with poor personal hygiene.
- Contamination of animal products with faeces during slaughter and dressing. Also milk contamination from dairy cattle.

Table 19.1 The main bacteria responsible for food poisoning

Organism	*Incubation period (hrs)*	*Duration*
Salmonella (infection)	6-48 (usually 12-36)	1-7 days
Staphylococcus aureus (toxin in food)	2-6	6-24 h
Clostridium perfringens (toxin in intestine)	8-22	24-48 h
Cl. botulinum (toxin in food)	12-96 (usually 18-36)	Death in 1-8 days or slow recovery 6-8 months
Bacillus cereus (toxin in food)	Vomiting 1-6 Diarrhoea 8-16	6-24 h 18-24 h
B. subtilis (toxin in food)	<1-14 (median 2.5)	1.5-8 h
B. licheniformis (toxin in food)	2-14 (median 8)	6-24 h
Escherichia coli (infection and toxin)	12-72	1-7 days
Vibrio parahaemolyticus (infection)	2-48 (usually 12-18)	2-5 days
Yersinia enterocolitica (infection)	24-36 (3-5 days)	3-5 days
Streptococcus (toxin in food)	3-22	24-48 h
Campylobacter jejuni (infection)	3-5 days	Days or weeks
Listeria monocytogenes	1-70 days	Days to years, mortality rate 30%

(From: Macrae, *et al.* 1993: p. 2013)

- Direct contamination of sea food (shellfish and others).
- Contamination of plants by irrigation or fertilization (especially the use of 'night soil' in the Orient).
- Contamination of food processing equipment and food contact surfaces.

The foods most frequently associated with *E. coli* infections include meat, meat products, fish, poultry, milk, cheese, vegetables, shellfish and water.

These bacteria are sensitive to heat and are killed by proper cooking, but freezing may have little effect on some strains. A high level of personal and physical hygiene will minimize infections.

E. coli 0157 caused a food poisoning outbreak in Scotland in 1997, with 500 people affected and 20 killed; the source was traced to a local butcher.

(Macrae *et al.* (1993) pp. 1613-1628)

Listeria monocytogenes

Only recently (1980's) recognised as a major source of human food poisoning. It is cosmopolitan (together with four other species) and found in soil, vegetation and water, and reported from many mammals and birds. The occurrence in ruminants is probably associated with bulk silage as a food, and it is dispersed through the use of animal slurries as field fertilizer. Generally, pasteurisation is effective in controlling this pest, but refrigeration is not.

The incidence of listeriosis is low (about 3-8 cases per million population in North America and Western Europe); but it has a mortality rate of 30%.

Salmonella

These are mobile, rod-shaped bacteria, active in both air and its absence; more than 2,200 types are known, and some are quite host-specific, such as *S. typhi* in man. But the non-specific ones are more often associated with food poisoning, and they are widespread. It is possible that all these types actually belong to a single species, *S. enterica*, and they represent subspecies and serotypes. They are primarily intestinal parasites of vertebrate animals. Poultry, pigs and cattle are the main domestic reservoirs, and modern conditions of intensive rearing are encouraging a population build-up of the bacteria, especially in the case of poultry. About 50% of poultry carcasses carry *Salmonella*, as also do eggs (both externally and internally), and the mechanical preparation of chicken carcasses encourages the spread of the bacteria. In recent years, it appears that the outbreaks of this pest have become more numerous, in both Europe and North America. In the UK, it is estimated that there are 100,000 cases per year. Heat pasteurisation at temperatures around 70°C is effective against most strains, as is freezing and frozen storage; and ionizing radiation treatment would be very effective if widely used.

Salmonella typhi causes typhoid (enteric) fever, and is still a major problem in many parts of the world; and vaccination continues to be the best preventative measure against this disease.

CHAPTER 20

FOOD POISONING

This refers to a disturbance of the gastrointestinal tract following the consumption of contaminated food or water. Mostly, it refers to bacterial infection, but it can be applied to a wide range of organisms, as indicated below:

Table 20.1 Agents of food poisoning

	Examples
Bacteria	*Salmonella, Staphylococcus aureus, Clostridium perfringens, Cl. botulinum, Bacillus cereus*
Viruses	Parvovirus; Norwalk agent; small, round, featureless viruses
Moulds	*Claviceps purpurea* (ergotism) *Aspergillus flavus* (aflatoxicosis)
Protein prion	BSE in cattle, vCJD in humans
Marine protozoa	*Gonyaulax tamarensis* (paralytic shellfish poisoning) *Gambierdiscus toxicus* (ciguatera poisoning) *Prorocentrum lima* and several species of *Dinophysis* (diarrhoeic shellfish poisoning)
Parasites (helminthes protozoa, amoebae)	*Trichinella spiralis, Taenia solium, Giardia lamblia, Cryptosporidium, Entamoeba histolytica*
Chemicals	Heavy metals, e.g. lead, zinc, copper, cadmium mercury, arsenic. Insecticides, herbicides, fungicides, cleaning agents, disinfectants
Toxic plants	Toadstools, hemlock, deadly nightshade Raw or undercooked red kidney beans (Phaseolus vulgaris lectin) Apricot kernels (glucoside-amygdalin), green potatoes (solanine)
Toxic food animals	Puffer fish (tetrodotoxin), red whelk poisoning (tetramine) Scombroid fish, poorly processed (histamine)

(From: Macrae, *et al.* 1993: pp. 2013 & 2014)

But the organisms which are mainly responsible for food poisoning are micro-organisms as defined by WHO, listed below:

- *Aeromonas hydrophila* and *sobria*
- *Bacillus cereus*
- *B. licheniformis*
- *B. subtilis*
- *Brucella* spp.
- *Camplobacter jejuni* and *coli*
- *Clostridium botulinum*
- *Cl. perfringens*
- *Cryptosporidium* spp.
- Enteroviruses, e.g. hepatitis A virus (HAV), small round structured virus (SRSV)
- *Escherichia coli*
- *Listeria monocytogenes*
- *Mycobacterium bovis*
- *Plesiomonas shigelloides*
- *Salmonella* spp.
- *Staphylococcus aureus*
- *Streptococcus zooepidemicus*
- *Vibrio cholerae*
- *V. parahaemolyticus*
- *V. fluvialis*
- *V. vulnificus*
- *Yersinia enterocolitica*

The main sources of these micro-organisms are animals (including fish), but enteroviruses, *Salmonella typhi* and *V. cholerae* are of human origin, and *Cl. botulinum, L. monocytogenes, Bacillus* spp. and *A. hydrophila* are general environmental contaminants. Milk and vegetables, spices and prepared foods of many types are also sources (see Table 20.2).

Bacterial food poisoning is classed into three basic types according to the mode of action of the bacteria involved.

- Infection type – this is when the bacteria are consumed with food and they multiply *in vivo* (*Salmonella, Vibrio* and viruses).
- Toxin type – this is when the ingested food contains a toxin produced previously by microbial growth. Examples include *Staphylococcus aureus* (enterotoxin), *Bacillus cereus* (emetic toxin), and *Clostridium botulinum* (neurotoxin).
- Intermediate type – where the organism (e.g. *Clostridium perfringens*) releases an enterotoxin, but in the bowel, causing fluid production and diarrhoea.

In a recent article (Guardian, 7 April 2001), a quote from the Centres for Disease Control and Prevention (CDC) in America stated that every day in the USA, roughly 200,000 people contract food poisoning, of whom 900 are hospitalized and 14 die (= 5,000 deaths per year). It is estimated that more than a quarter of the American population suffer a bout of food poisoning each year, although most of these cases are never reported to the authorities or diagnosed. Figures of a similar proportion are quoted for the UK, as seen in *New Scientist* (10 March 2001, p. 18), with an annual death rate of about 200. In 1999, 90,000 cases of food poisoning were reported, but because most cases go unreported, some experts believe the true number could be 30 times higher!

20.1 PREVENTION OF FOOD POISONING

Clearly, certain foodstuffs are more at risk of bacterial infection than others (see Table 20.2 below), but many of these bacteria are part of our normal intestinal micro-flora, and as such they must be expected to be present all of the time in human domestic situations – the aim must be to attempt to keep the populations low enough for damage to be minimal. Similarly, the soil-dwelling species may be difficult to avoid.

Table 20.2 Main food sources of the common microorganisms causing food poisoning

Agent	*Food source*
Salmonella spp.	Raw meat and poultry, raw milk, eggs
Staphylococcus aureus	Cold meats and desserts (much handled during preparation): dairy products, especially if prepared from raw milk
Clostridium perfringens	Raw meat, poultry, dried foods, herbs, spices, vegetables
Cl. botulinum	Improperly preserved meat, fish and vegetables
Bacillus cereus and other Bacillus spp.	Cereals, dried foods, dairy products, meat and meat products, herbs, spices, vegetables
Escherichia coli	Many raw foods, particularly of animal origin
Vibrio parahaemolyticus	Raw and cooked fish, shellfish and other seafoods
Yersinia enterocolitica	Raw meat and poultry, meat products, milk and milk products, vegetables
Streptococcus spp.	Raw milk, raw meat and poultry, foods prepared by infected food handlers
Campylobacter jejuni	Raw poultry meat, raw or inadequately heat-treated milk, untreated water
Viruses	Raw shellfish, cold foods prepared by infected food handlers
Listeria monocytogenes (a widespread environmental organism)	Meat, poultry, soft cheese and other dairy products, vegetables, shellfish

(From: Macrae, *et al.* 1993: pp. 2014 & 2016)

However, there is a list of things about which many food-handlers are very careless, as illustrated in Table 20.3, mainly in connection with the cooking process; in addition, there is the ever present problem of hygiene and cleanliness, both personal, to avoid faecal or soil contamination of food, and the sterilization of production machinery and equipment to remove food debris.

Table 20.3 The most common factors contributing to outbreaks of food poisoning in England and Wales from 1970-1982

Contributing factor	*Percentage of outbreaks in which the factor was recorded (1479 outbreaks studied)*
Preparation too far in advance	57
Storage at ambient temperature	38
Inadequate cooling	32
Inadequate reheating	26
Contaminated processed food	17
Undercooking	15
Contaminated canned food	7
Inadequate thawing	6
Cross-contamination	6
Raw food consumed	6
Improper warm-holding	5
Infected food handlers	4
Use of left-overs	4
Extra-large quantities prepared	3

(From: Roberts, 1986)

But many researchers believe the basic problem lies with the slaughterhouses and farms, and their poor hygiene. In the summer weather, the food quickly heats up, and even half an hour of heat is enough to make a chicken carcass dangerous, for the bacteria grow very rapidly given the slightest opportunity. Cattle usually arrive at the abattoir splashed with faeces, and the chances of meat contamination are very high.

CHAPTER 21

CONTROL OF PESTS OF STORED FOODSTUFFS

21.1 BACKGROUND AND HACCP

With many storage facilities, such as the silos for bulk grain and the like, by having cleaning and drying facilities as part of the complex, the grain can be stored dry enough to prevent serious pest problems, and should a pest outbreak occur, fumigation facilities can be utilised.

But since the advent of Supermarkets on such a large scale, there has developed simultaneously a very extensive food industry to supply these retail giants, the annual value of which must be billions of pounds/dollars, in each of the major countries (UK, USA etc). And all the different companies are fiercely competitive, and very secretive, about the details of their programmes. But, clearly, the food production business is not only dealing with vast sums of money and making huge profits, it is also having to face many hazards with regard to public health and food safety. The single greatest possible threat comes from microbial infection of the food products, because serious infections could put people's lives at risk.

In the USA, this awareness led to the widespread adoption of the hazard analysis critical control point (HACCP) programme. This was originally developed for NASA in order to ensure the safety of astronauts, but the principles can be just as easily applied to safety in the modern food industry. A flow diagram is used to define the process in an accurate and sequential form.

21.1.1 Definition of Hazard Analysis and Critical Control Points

When applied to the food industry, an HACCP programme involves the systematic analysis of food production or food handling processes from the purchasing of the raw materials through to the end use of the product by the consumer. An HACCP audit seeks to identify the source and severity of potential hazards that may threaten the safety or integrity of the product. It involves the identification of critical control points (CCPs), i.e. location, stages, operations or raw materials which, if not properly controlled, pose a threat to product acceptability and consumer safety. In some cases, implementation of a CCP can virtually eliminate a hazard; in other cases, the CCPs identify those points where rigorous quality testing is necessary.

21.1.2 Implementation in Food Production

Different food processes are vulnerable to different hazards. There is no single hazard analysis which covers all cases, but the general principles will still apply. Each HACCP programme must be highly specific and take into account every factor that may affect the safety of a particular product. Although details will vary, there is a generally recognised sequence of events involved in completing a HACCP programme and implementing the findings.

21.1.3 Role in Food Production

One of the main difficulties faced by the food industry in its efforts to maintain and control the microbiological status of products is the time taken for conventional laboratory methods to produce useful results. During the last 20 years, the food industry has developed a different approach to the problem. Understanding the sources of contamination, and the way in which different micro-organisms respond to processing and storage conditions, enables prediction of their behaviour and the design processes and products to eliminate them or prevent them from developing. This concept forms a basis of the HACCP approach to the control of microbiological hazards. It involves a detailed analysis of a process, coupled with an in-depth understanding of the sources and behaviour of any micro-organisms likely to gain access to the food. It leads to the design of safe processes which effectively eliminate food poisoning bacteria, and indicates the points in the process where critical conditions must be monitored to ensure safety.

The HACCP approach is applicable not only to food manufacturing, but also to all parts of the food supply chain. A properly designed and operated HACCP system is essential to food handlers or processors intent on ensuring the safety of their operation in the food supply chain.

21.1.4 Defining the Process

(See Figure 21.1)

21.1.5 Identifying the Hazards

Identifying hazards or potential sources of hazards is not easy, but acknowledgement of that fact is no defence should a processor be found to have compromised consumer safety. Obvious sources of hazards are as follows:

- Raw materials: micro-organisms, trace contaminants such as heavy metals and pesticides.
- Product formulation: pH, water activity, preservative level.
- Processing stages: time and temperature of cooking, re-contamination.
- Product shelf life: packaging, storage, storage temperature, instructions for use.

21.1.6 Assessing Severity of Risk

Once the potential hazards have been identified, the associated risk of each should be quantified, usually into high-, medium- or low-risk categories.

21.1.7 Determining Critical Control Points

Control of hazards is exercised at CCPs. These can be locations, procedures, practices or processes, such as maintaining staff hygiene standards, correct operation of thermal processes, control of product storage temperatures etc. A CCP

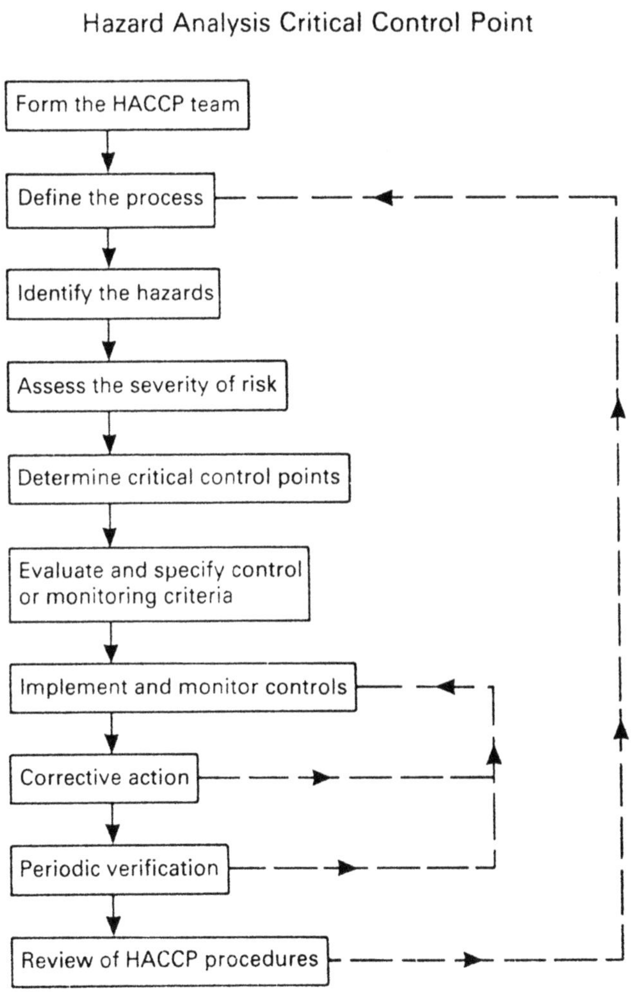

Figure 21.1. Summary of the stages of a HACCP programme and feedback systems (From: Macrae et al., 1993: p. 2290)

can be classified as a CCP1 or a CCP2. Where control of a hazard can be virtually ensured, the control point is designated as a CCP1. If the control point can minimize, but not completely control a hazard, it is designated as a CCP2. The distinction between a CCP1 and a CCP2 is not always clearly defined, but all CCPs are control points where control can and must be exercised. Other control points may exist in the process, which are not labelled CCPs because of the low risks attached to the hazard, or because a better control point to that hazard exists. Appreciating the difference can save money by preventing waste of effort in exercising control at a point of little consequence compared with others in the process.

21.1.8 Hazard Analysis and International Quality Standards

Increasingly, food businesses are using standards or specifications, such as British Standard BS 5750 or International Standards Organisation ISO 9000, as a quality system backbone on which to build an effective means of quality assurance.

British Standard BS 5750 (Part 0) or ISO 9004 (Sections 0.2 Guide to Quality Management and Quality System Elements) requires the following:

> "Consideration has to be given to risks (of hazards) such as those pertaining to the health and safety of people, dissatisfaction with goods and services, availability, marketing claims and loss of confidence."

Within the 'requirements of society' are:

> "Requirements including laws, statutes, rules and regulations, codes, environmental considerations, health and safety factors and conservation of energy and materials."

In order to meet the requirement of such specifications, potential hazards to the product and process must be assessed, identified and monitored according to a documented system.

Hazard analysis and quality management systems are therefore inextricably mixed in the quest for safe food, and the compilation of adequate proof of such safety.

In the UK, certificated courses, administered under the auspices of both the Royal Society of Health (RSH) and the Institution of Environmental Health Officers (IEHO), and run by training organizations registered by the above bodies, provide basic, intermediate and advanced level training in food hygiene to all facets of the industry. Similar schemes may be found in other European countries and in North America.

The potential scope of the application of HACCP is immense, as are the potential benefits. A systematic stepwise evaluation of a process from start to finish will highlight certain areas which may improve not only product safety, but also product quality and efficiency of production. Although the HACCP system is primarily used to ensure the microbial safety of food, it is equally applicable to the control of sources of potential physical or chemical contamination.

Apparently, the World Trade Organisation (WTO) is concerned about the quality of food products exported internationally, and their Codex Alimentarius Commission (Codex) laying down guidelines for food producers/exporters, emphasizes application of HACCP and the associated Good Manufacturing Process

(GMP) and Code of Practice (COP) to ensure that foods produced are safe, nutritious and correctly labelled. In Malaysia, 36 companies are now HACCP certified, and 19 of these are in the prawn aquaculture business (Sarawak Tribune, 21/11/2001). From January 2000 to July 2001, 18 cases of Malaysian frozen seafood products were rejected by European Union (EU) countries, and now it is clear that HACCP certification is a mandatory requirement for companies where food is produced for export to the USA and EU countries. The companies in Sarawak that have not yet been certified are working towards this end. HACCP auditors visit the premises to check on the production practices both before and after HACCP certification.

A major attraction of this system is that it brings about a change in regulatory procedure by consideration of the process as well as the end product. Previously, the procedure was to detect the problems only at the end of the production line, but with HACCP the food safety problems are identified early, and often prevented.

21.2 LOSS ASSESSMENT AND PEST SAMPLING METHODS

Before contemplating control measures, it is first necessary to establish that there is in fact an infestation problem, and that damage is either being done or is likely to be done. The next major point is to identify the pests concerned, and to establish precisely which species are of prime importance. Then some sort of probable damage level assessment needs to be determined. These are aspects that have been neglected in the past, yet they are the basis for a rational control programme. It is known of course that, in the tropics especially, produce infestations are often so heavy that no scientific assessment is really required to know that control measures are needed. But ideally some assessment of cost : potential benefit ratio should be made as the basis of a scientific control programme. The empirical approach will show a correlation of pest population size to a particular damage level over a period of time.

21.2.1 Pest Infestation

With field crops, it is very important to make an assessment of either pest infestation levels or damage levels, or both, when planning an IPM project. In the past, too often pesticides were applied to a crop when they were either not really needed or not effective.

Pest infestations are assessed in two basic ways: the *incidence* of a pest (or damage symptoms) is the proportion of plants (or plant parts) in a sample which are host to the pest (or which show damage symptoms), and is usually expressed as a percentage. The *severity* of the infestation is a measure of the size of the pest population, or the extent of the damage done, and is usually measured as so many insects per plant, or plant part. With stored grain, the incidence could be the number of insects per sample, say, 100 grains or one spearful, and the severity the proportion of the 100 grains actually damaged. For unthreshed maize or sorghum, it could be the number of insects per cob (or panicle), and the proportion of grains damaged per cob. For fruits and vegetables, similar scales can apply. And, of course, the total damage caused to a crop is a combination of severity of infestation together with *duration* (time). But with many foodstuffs, the tolerance levels for infestation or damage are far lower than for field crops. Although damage detected early in the processing chain of events could be tolerated and the items rejected, at the end of the production line no infestation/damage is acceptable. Contamination is often the major factor causing food products to be rejected; there does not have to be any actual damage done – the dead pest body's presence is sufficient.

In ecological studies of populations of both plants and animals, several methods of assessment are employed. These are based upon the proportion of area covered within the habitat (for plants), or the number of animals seen or sampled in relation to area, or proportion of plants examined (sampled). Botanists are able to use more precise systems for plant population assessment because of the immobility of the organisms, and the three most widely used methods employ between 4 (Raunkiaer. 1934), 6 (Braun-Blanquet. 1927) and 11 (Domin) abundance categories. But for small, highly mobile insects, the latter level of precision is not feasible, especially when different recorders are being used. In the UK, the *Biological sites recording scheme* advocates the use of four abundance/frequency categories for population size assessment, without the use of lengthy or detailed sampling procedures, and this approach would seem to be appropriate for assessing field populations of insect pests on crops or stored products. These categories of abundance are as follows:

Abundant	(A)	=	Very Common	(VC)
Frequent	(F)	=	Common	(C)
Occasional	(O)	=	Uncommon	(U)
Rare	(R)	=	Rare	(R)

The alternative categories in the right-hand column would appear preferable as their designation is somewhat more obvious. Further details on pest population assessment can be found in the publication by Southwood and Henderson (2000), Bardner and Fletcher (1974), and FAO/CAB (1981).

21.2.2 Loss/Damage Assessment Methods

For many years, a basic problem in field crop protection was the lack of suitable methodology for assessing probable crop losses, and it was not until 1971 that the FAO and the Commonwealth Agricultural Bureau (CAB) published their manual, *Crop loss assessment methods*. Supplements to this manual have also been published. Nowadays, enough research has been done and published so that workers involved with any of the world's major crops should have no difficulty in defining a suitable methodology for infestation assessment and probable yield loss estimation.

With stored produce infestation studies, there was a greater delay, and it was not until 1975 that FAO organised a special session devoted to postharvest food losses (basically, stored grain), and in the following year organised two international workshops on methodology. The proceedings of the two workshops were almalgamated into a single manual (Harris & Lindblad, 1978). There are also a number of papers dealing with aspects of basic methodology in BCPC Monograph No. 37 (Lawson, 1978). At least, with stored grain there is not the diversity of methodology that is required for field crops, and the total number of major pests involved is far fewer. But with food products generally, there is an enormous diversity of material, and a broad range of pest organisms and destructive agents to contend with.

21.2.3 Damage Assessment

It was suggested by Bardner and Fletcher (1974) that the term 'injury' be used for slight (i.e. non-damaging) effects of insects feeding, or other activities on the growth or appearance of crop plants; and that 'damage' is injury resulting in a measurable loss of yield or reduction in quality. This distinction would be most useful, but to date in the literature, these two terms are used more or less synonomously, and, of course, the term 'economic injury level' is well established.

The extent of crop damage is usually proportional to the numbers of insects present, and would accordingly be rated as follows:

					6
Very Severe	(VS)	=	4	or	5
Severe	(S)	=	3		4
Mild	(M)	=	2		3
Very mild	(VM)	=	1		2
				or	1

In some systems of recording, a numerical categorisation is used, and a six-point scale seems to be quite popular; presumably in this case, number 1 (and 'very mild') come under the 'injury' category of Bardner and Fletcher (1974), and would be detactable but not of any economic importance. With simple damage, such as leaf lamina being eaten or apples infested with Codling Moth larvae, damage can be expressed easily as proportion of lamina destroyed or percentage of fruits infested per tree. In some systems, numbers of pests present are correlated empirically against expected loss of yield (percentage), on a scale of no loss (0%) to total loss (100%).

Because each crop and food product has its own characteristics, and the vast diversity of pests and types of pest damage and all the other factors involved with crop and produce production, it is not possible to generalise extensively. Damage assessment will remain different for each crop and type of food product, and sometimes also for each major locality, as the pest complex usually varies regionally. It is, however, generally agreed that for most purposes a damage assessment scale of not more than six levels is preferable, for easily recognition by non-experts, and it is recommended that a large number of small (easily categorizable) samples be taken rather than a small number of large samples; this also caters better for the usual uneven pest distribution within the crop or stored produce.

21.2.4 Produce Sampling (Monitoring)

This usually refers to grain. Stored grain has to be monitored to check the physical condition of the grain, to look for signs of damage, to check for attack by micro-organisms, to look for insects, and to determine the extent of any contamination. Traditionally, most grain was stored and shipped in bags, and the spear probe was developed as a simple sampling tool. However, the results obtained were seldom representative of the consignment as a whole, and it is now known that this system of sampling is really quite inefficient.

The United States Department of Agriculture (USDA) developed a more sophisticated tool for use with bagged grain. This is a compartmented bag probe – a pointed metal tube 54 cm long (2.5 cm diameter) with three elongate slots at different levels which can be opened and closed using a terminal screw-nut (Harris & Lindblad, 1978; p. 150). This probe will take samples at three different levels in the bag simultaneously. Longer versions are available for sampling bulk grain in storage bins, wagons etc.

Sometimes a whole bag (sack) is taken at random for testing, and then a machine called a produce flow sampler may be used, which will remove a specified amount of grain at random from the bag for examination.

21.2.5 Physical Parameters

Measurement of the major physical parameters is of considerable importance, for temperature and moisture levels etc. will be the key factors in deciding when to apply control measures, and these measurements must be able to be done quickly and easily, yet with a high level of accuracy. In Europe and North America, there is now available a vast range of equipment that will make almost immediate recordings; sources of equipment are listed in Harris and Lindblad (1978; Appendix C, pp. 167-86).

The main factors are obviously temperature, atmospheric moisture (in bulk grain etc.), grain moisture content, and sometimes grain weight, protein and oil content of grain and other seeds.

Temperature is traditionally measured with a wide range of thermometers and thermocouples; bulk grain stores usually employ long probes (thermocouples) inserted into the grain to different depths, and many are constructed to give continuous temperature recording as part of the general store monitoring. In the larger grain stores, the whole environment monitoring programme is computer controlled.

Atmospheric moisture (as relative humidity) can now be measured very easily using one of many different probes available commercially. These probes operate on several different basic principles, but typically they give a direct reading on a digital display unit, and also give a direct temperature reading simultaneously. The length of the probe can vary from 0.5-1.0 m for use in bags and bulk grain respectively. A new range of models recently developed is the ELE Tempeau – Insertion Moisture Meters (Hemel Hempstead, UK).

Grain moisture content is now also equally easily measured using a wide range of equipment. Figure 8.15 shows an apparatus made by Dickey-John, Inc. (USA); a small sample of grain is inserted into the pan, and a direct reading of the moisture content as per cent, grain weight and temperature results – an attached printer will give a paper print-out for each sample for easy recording. Similar machines are programmed to measure and record grain protein content, oil content of some seeds, and moisture (Figure 8.16).

21.2.6 Pest Population Sampling

The first thing that has to be done when confronted with an infested store or damaged produce is to determine which pest species are involved, so that their relative importance can be assessed. Then it is necessary to establish the extent of the population of each of the major pests. And sometimes the precise extent of the damage to the grain or other produce has to be assessed to complete the infestation picture. Sometimes, in the tropics, the grain or produce is swarming with insects, so it is very obvious which species are involved. Otherwise, it is necessary to sample the produce or to collect the pests using traps or other sampling devices. The behaviour and the life history of the pests are of paramount importance when endeavouring to sample the pest population. Even for determining the most abundant pests, it may be necessary to use several different but complementary methods simultaneously. With high population densities, sampling is much easier; the problems arise when confronted with low pest densities. Few pests are uniformly distributed throughout the stored bulk grain; often they eventually concentrate in the surface layer (10-20 cm), but when the grain is mixed or redistributed, the insects are also redistributed. Sampling has to be done on a random basis, bearing in mind possible areas of concentration. Most adult insects are mobile and can be attracted to a trap, and so can some larvae, but some larvae are sluggish, and some such as *Sitophilus* are confined within the grain and quite immobile. With adults, some can fly and will fly to traps; some are nocturnal or crepuscular, and times of activity will vary according to species; a few species have winged adults, but they are reluctant to fly. It is important that very low pest densities be recognised, for some species have a great reproductive potential, and a small number can give rise to a large population in a couple of months under warm conditions. But very low population densities are difficult to detect, and even more difficult to assess. Pest detection methods fall into two main categories, the first being produce sampling, and the second some form of trapping.

21.2.7 Sampling

Produce sampling depends upon the removal of a sample which is then sieved and inspected visually for signs of pests. Sometimes a method of flotation is used to separate rodent hairs or droppings, or sometimes to separate mites from the produce. In the USA, insect frass and contaminant detritus, faeces etc., is collectively referred to as filth by the Food and Drug Administration, and some of their techniques for filth detection in foodstuffs involve fluorescence and a range of chemical tests. The physical size of the pest in relation to the particles of produce is important. Whatever method of sampling is used, it should follow a predetermined pattern according to a carefully considered methodology. Clearly, it would be preferable to use a method widely adopted or standardised throughout the country or the world as a whole. But the sampling pattern must be closely followed and carefully replicated and recorded in order to ensure that results are comparable, and so that results can be analysed. As with field sampling, it is generally recognised that a large number of small and easily categorised samples are preferable to a few large and carefully counted samples.

21.2.8 Spear Probe Sampling

This is the traditional method developed for use with bagged grain. The spear probe (also called spear, probe, trier, or thief) was often a piece of hollow bamboo stem (internode) about 15-20 cm long and 15-20 mm diameter, cut diagonally at the tip. This was pushed into the bag (between the sacking strands) and grain flows out through the hollow tube. More recent spears are made of metal, some have a handle, some are quite long. The collected sample is sieved and examined for pests. It is a very convenient and easy method, but unless a large number of samples is taken, it may be quite inaccurate. As already mentioned, the use of a compartmented grain spear (bag probe) will enable simultaneous grain samples to be taken at different depths in the produce. The length of the probe can be determined by the storage procedure – a 54 cm probe with three collection slots for use with bags, or a 5 m probe for bulk grain.

21.2.9 Vacuum sampling

With bulk grain, vacuum samplers with a long nozzle are used to take samples at different depths, and the collected samples are sieved.

21.2.10 Random Sieving

Small quantities of grain can be taken by any suitable method and sieved. For some species, this method may be effective if a large number of samples is taken.

21.2.11 Sample Dividing

Often it is necessary to take a large number of samples in order that they are truly representative of the whole consignment, and the final collected sample can be quite disconcertingly large. There are, however, machines available for sample reduction, known generally as dividers. If a divider is not available, the traditional farm method, called coning and quartering, can be employed, whereby a pile of produce is mixed by shovelling material from the periphery of the cone (pile) to the apex, and then the pile is divided using a flat piece of wood. The process can be repeated until a suitable final sample size is reached.

21.2.12 Trapping

These methods all rely on the mobility of the insect or pest to take it into the trap receptacle. With the most attractive traps, there will clearly be a considerable concentration of insects in the trap. The basic problem with trapping as a method of population sampling is that of correlating the trap catch with the actual population present. Some traps catch insects more or less at random (pitfall traps), but others attract the prey and may bring them some distance to the trap. Trapping in produce stores is often useful for pest population monitoring, and in some cases for determining which species are present. Trapping-out projects using light and pheromone traps are seldom feasible in produce stores (and are only rarely successful in field crops), but the pest population is contained within the store, and the use of a good light trap (or Insect-O-Cutor) can often be recommended as part of an integrated pest management programme, for it is quite effective and can be conveniently situated high in the store and out of the way. For flying insects it is an advantage to be able to use aerial traps, for they can be hung from rafters and beams or wall-mounted well above the produce and out of the way of people.

If the top 20 or more important stored produce pests are considered, it appears from the literature that there are fundamental differences between the different insect species with regard to success in trapping them. A few species are relatively easy to trap, others almost impossible, and in the vast majority of cases there is no reliable relation between trap catches and produce infestation levels. Rodent trapping, particularly of rats, is even more of a problem, for some populations show remarkable learning processes, and trap-shy rats are very difficult to catch. The three main basic types of insect traps commonly used are:

21.2.13 Light Traps

Normally, these are mercury vapour lamps that emit ultraviolet radiation, for these are very effective at attracting most moths and some beetles. Such traps are many times more effective (about 20 times) at attracting flying insects than are ordinary tungsten lamps emitting white light. In produce stores, the catch container would normally contain a killing agent. Occasionally, a light trap is used in conjunction with sex pheromones, to give an enhanced catch, if one particular species is of concern.

21.2.14 Sticky Traps

These range from the old-fashioned sticky flypaper to pieces of waterproofed card, and now a flat piece of plastic covered with a slow-drying adhesive. If coloured yellow, these traps will catch a wide range of flying insects. Yellow plastic strips (about 20 x 10 cm) with a sticky covering are being used in greenhouses in Europe with some success. With some sticky traps, pheromones are added so that particular insect species can be caught in extra numbers.

21.2.15 Pheromone Traps

Some traps are based solely on the use of a particular sex pheromone, and have a trapping chamber where the insects are caught. Several traps are constructed with a plastic funnel leading into the reception chamber containing a strip of dichlorvos. Recent work on micro-encapsulation of pheromones enables a slow release of chemical over a long period of time. Many insect pest species have had their sex pheromones analysed, and now quite a number have been synthesized and are available commercially. *Ephestia* moths apparently all react to the same basic major component chemical, so the one chemical in the trap will entice males of all the *Ephestia* species in the store. *Ephestia* funnel pheromone traps are recorded as being generally very effective. Chemical attractants have been discovered for a number of pest species, and some very effective ones are in use with fruit flies (Tephritidae), and can be useful in tropical fruit stores to monitor fly populations. Apart from the *Ephestia* traps, it does appear (as mentioned below) that in produce stores pheromones are often less useful than the use of food aroma attractants.

21.2.16 Suction Traps and Water Traps

Although commonly used with field crops, these traps are not appropriate for storage situations, for the insects in grain and other produce are non-flying and have to walk or crawl into the traps. Some of the adult beetles which have functional wings will, of course, be caught in both types of trap; at the other extreme, there are some larvae which have very limited powers of locomotion and their dispersal will be minimal, and these species are very difficult to trap.

The following are the three main types of trap to put into the grain, with a number of variations of each.

21.2.17 Pitfall Traps

These are quite simple, and are usually glass or plastic jars with a wide mouth, buried in the substrate up to the rim, so that walking insects fall into the jar. For field use, such jars usually contain a little water with added detergent and preservative (formalin), but in produce stores the jars are often used dry. Some beetles are able to climb up a smooth glass surface and so can escape from such jars, but most are trapped inside. Some jars have a slippery coating of plastic painted along the inner rim so as to prevent insect escape. Sometimes insects are enticed into the trap by the inclusion of a food attractant (see below).

21.2.18 Crevice or Refuge Traps

Most stored products pests are cryptozoic and often nocturnal, and they prefer to hide away in crevices. This habit can be exploited, and simple traps can be made of a folded piece of corrugated cardboard buried in the grain. Several different versions are in use, and some are given proprietary names (e.g. Storgard trap). These traps need to be removed daily (or at regular intervals) and the insects collected. Moth larvae and both adult and larval beetles will come to these traps, but they are often not very effective unless baited.

21.2.19 Bait Bag Traps

These are envelopes of plastic mesh, about 10-20 cm x 5-10 cm, half filled with an attractive food such as broken groundnuts; the mesh holes are large enough to allow the insects to crawl through to the interior of the bag. The bags are pushed into the produce at intervals and left for a few days or weeks before collection and examination. (Figure 21.2).

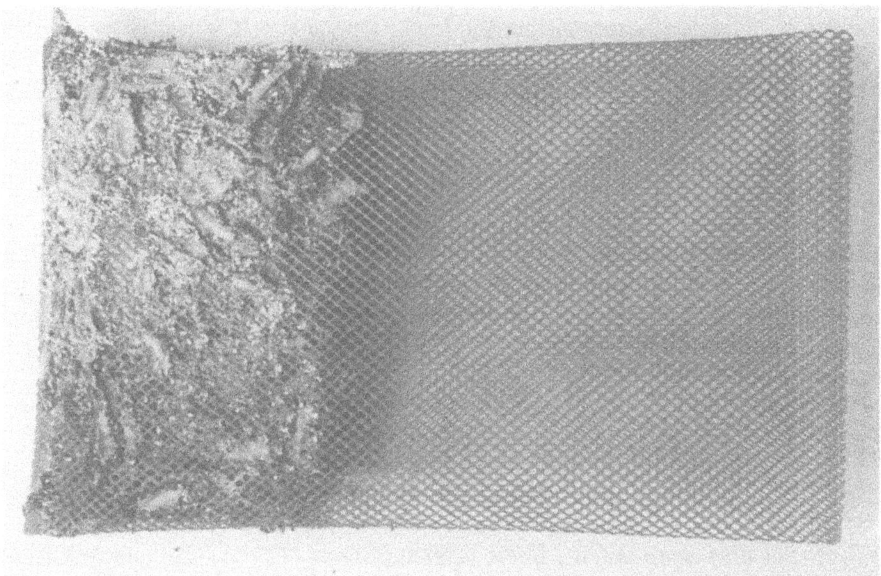

Figure 21 .2. Bait bag made of red plastic mesh, containing broken groundnuts filled with Ephestia larvae.

21.2.20 Probe Traps

Because of the difficulty in finding a good trap for insects in stored grain, there has been much research effort over recent years, and the USDA developed a probe trap which has recently been tested in the UK by staff at the MAFF (now DEFRA) Stored Products Laboratory at Slough. It consists of a plastic tube 30 cm long, perforated with downwardly pointing holes, and usually baited with some type of food attractant. The tube is pushed into the grain to a prescribed depth and left in position for a set time. The inside of the tube is painted with a film of plastic to make it slippery so that once the insects fall into the tube they cannot climb out. It was reported that in experiments these probe traps caught about 5% of insects introduced into bulk grain, but this was at least ten times more effective than the use of spears or random sieving.

21.2.21 Food Attractants

It is thought that the reaction from stored produce insects to foods is just as complicated as the seeking of the host plant by phytophagous field insects. There have been a number of laboratory experiments reporting the response of several major stored grain insect pests to various food components from wheat and oats, but the situation is far from clear. Of course, for trapping purposes it is necessary to use a food component more attractive than the surrounding commodity. It has been discovered that crushed (kibbled) carobs (pods of *Ceratonia siliqua*) are very attractive to stored grain insects, and their addition to traps and bait bags greatly increases the numbers of insects caught. Research is being directed at analysis of foodstuffs using electro-antennagram (EAG) techniques in an endeavour to isolate the active components responsible for insect attraction. If it can be established precisely how the insects are attracted to a particular commodity, then firstly it will be possible to bait probe traps more effectively, and secondly it might be possible to prevent infestation from occurring.

21.2.22 Rodent Traps

These come into a separate category, for they are sufficiently large and specialized to catch only larger creatures. Although rats are often notoriously trap shy, often these simple and cheap traps are worth using as part of an integrated pest management programme; certainly for rodent control, a multi-pronged attack is always recommended.

21.3 INTEGRATED PEST MANAGEMENT (IPM)

The idea of IPM arose initially in the context of pests of field crops and fruit trees. It was basically concerned with achieving a balance between the use of chemical poisons to kill the insects and avoidance of destruction of natural enemies (predators and parasites) of the pests. In field populations, the importance of natural control of insect pests cannot be over-emphasised, for most insect populations are kept in check most of the time by the actions of the natural predators and parasites. In the past, the accidental destruction of natural enemies in many crop situations (especially fruit orchards and plantations), which has occurred with alarming regularity in most parts of the world, almost invariably precipitated a pest population upsurge. In this situation, the concept of integrated control, and later IPM, was developed, and it came at a time when insect resistance to pesticides was becoming extensive, so it was a timely development.

With the practice of produce storage, man has created what is basically an artificial situation. Crop cultivation and agriculture in general is to be regarded as a natural situation, as it only differs essentially from a natural vegetation community in a quantitative sense. Thus, whereas natural control of field pests is of paramount importance, it is of little significance in produce stores. The one exception would be rats, because the destruction of their natural predators (carnivores and snakes) has led to a general increase in their numbers; but they are in the category of visiting (temporary) pests, as they usually live outside the stores and enter in their quest for food.

However, the idea of the integrated approach is applicable to any pest situation, and it is otherwise suitable for use in the context of the produce stores. The biology of the pests needs to be known in detail, especially the influence of climate on development, food preference, behaviour, competition, reproductive potential and the like. Then an overall view of the pest situation can be taken in relation to a particular store (or group of local on-farm stores) in order to decide which are the major controlling factors. Ways in which the pest population can be kept to a minimum can then be evaluated. Sometimes a pest eradication programme is necessary, but otherwise it is often sufficient to keep the pest population to a low level where damage is insignificant. Decisions can be made when all the relevant data are accumulated and assessed relative to each other. A simultaneous use of several different methods of control in a carefully balanced and integrated way is certainly more likely to lead to success.

One point of overwhelming importance is, however, that if the store is well-constructed and can be rendered air-tight, regular fumigation with a toxicant such as methyl bromide will kill more or less everything in the store. So this is another reason for regarding a stored produce pest situation as unique in comparison with others (crop pests, medical or veterinary pests). But of course in many situations throughout the world, especially in the tropics, such a simple answer is not possible, and crops are infested at the time of harvest, and infested grains and seeds are taken into the local stores, so that cross-infestation soon occurs.

The major factors which indicate where efforts should be directed are probably:

- Store construction - to be weatherproof and dry; airtight if possible; to be as pest-proof as possible.
- Use of clean produce - infested grain should be treated before storage, if possible.
- Store hygiene - store to be cleaned and fumigated regularly; produce to be inspected regularly (pest monitoring).
- Physical control - grain to be dried before storage; cool ventilation; use of 'Insect-O-Cutors' if applicable; store valuable or very susceptible materials in containers.
- Produce protection - fumigation if feasible, although some of the most effective fumigants are now banned in the USA and Europe; insecticides to be added if necessary; plant materials with deterrent effects.
- Suitable crop varieties - use naturally resistant varieties or land races for preference; soft-grained maize varieties are just too vulnerable for on-farm storage.

The evaluation of the different aspects of the pest/produce/store situation is vitally important, especially relative evaluation so that control input can be concentrated on the most likely aspects, and those most likely to lead to success. Some measures are far more cost effective than others. In general, store construction, although relatively expensive, would be an aspect of a control programme that should always be worthwhile in the long term. Similarly, store hygiene is usually quite inexpensive, and thus is always worthwhile.

Rodent control has been discussed separately (Chapter 10), and it has been stressed by several workers that for effective population reduction (or possible eradication), rat control requires a concerted effort using several different methods simultaneously.

In Europe and North America, it appears that emphasis is placed upon pest and disease control in the growing crop, and then store hygiene and the drying and cooling of the grain. And the not-so-high summer temperatures together with the cold winter period help in minimizing insect pest populations, and field infestations of ripening grain are few.

In the tropics, climate conditions are far more conducive to pest population development, and field infestations of maize and sorghum by *Sitophilus* and *Sitotroga* are both common and extensive. So the end result is that most grain crops are already infested when taken into storage, and postharvest fumigation almost always needed (but seldom available). Failing that, the addition of contact insecticide to stored grains is the usual alternative when money and facilities permit. In these situations, where infestation levels are often so high and losses serious, it is to be hoped that the extension and outreach programmes can convince the farmers that a broad integrated (holistic) approach to the problem of grain and foodstuff storage is needed. Also, it is sometimes necessary to convince the government authorities to ensure that supplies of the required materials be made available. Basically, the more multi-pronged the approach to the storage problem, the better. Total control can seldom be anticipated or even hoped for , but a major reduction in damage levels can be the aim.

A recent IPM project conducted in England proved to be very successful, as reported below:

> "Long term storage of grain with minimum pesticides proved feasible
>
> "Grain remained free of major pest infestation and was stored successfully for up to two years without chemical treatment of the bulk using an integrated pest control strategy developed with Home-Grown Cereals Authority funding. The Central Science Laboratory (CSL), Slough, conducted trials on wheat deliberately infested with insects and mites, in 20 tonne bins. Pest control in the bulk of the grain was achieved through cooling with ambient air blown by low volume fans. Thermostats turned the fans on when outside temperatures were below those of the grain. Insects – particularly saw-tooth grain beetles – and mites were detected at times during the long storage period at the surface of the grain. These infestations were brought under control by raking pirimiphos-methyl into the surface of the grain. CSL has also now demonstrated that the strategy can be effective in a 2,500 ft commercial grain store. " (Source: HGCA)

According to project leader David Armitage, use of the integrated strategy will reduce costs for growers and storekeepers, and also minimize pesticide usage. He estimates that storage costs for grain stored 3 m deep in farm stores can be cut to just 3p a tonne with this technique. Full details of the work are contained in HGCA Project Report No 57, *Integrated pest control strategy for stored grain: surface pesticide treatments of aerated commercial and farm stores to control insects and mites,* price £10 inc p & p, available from HGCA, Hamlyn House, Highgate Hill, London N19 5PR. Tel: 071-263 3391; Fax: 071-281 0884.

A useful publication, titled 'Guidelines for pest management research to reduce stored food losses caused by insects and mites', was published by ODNRI as Bulletin No.22 in 1989 (McFarlane, 1989).

21.4 LEGISLATIVE CONTROL

There are three aspects of legislation of relevance here:

21.4.1 International Phytosanitary Regulations

In 1953, FAO established an International Plant Protection Convention, and the member countries agreed to the setting up of eight plant protection zones and an International Phytosanitary Certificate whose use is designed to curtail accidental or careless distribution of pests and pathogens around the world with agricultural produce. Each country has quarantine facilities, so that if there is any doubt about a consignment it will be put into quarantine for a specified time to ensure that it is pest free. Some of the fruit flies (Tephritidae) and some fruit tree scale insects (Diaspididae) are very serious pests that are not established in certain countries, or particular states within countries, and concerted efforts are made to restrict the dispersal of these pests. A recent development of importance was the banning of the use of ethylene dibromide in 1984 in the USA, following demonstration of it having carcinogenic properties – for years it had been used to fumigate fruit and some vegetables, especially against the fruit flies (Tephritidae); as yet, no equally suitable alternative fumigant has been found.

21.4.2 National Legislation (Pests and Diseases)

Within each country, there are laws and regulations referring to the prohibition of certain noxious pest organisms that are to be kept out, or if accidentally introduced, then certain specific measures must be taken to ensure their eradication. These are mostly medical and field crop pests, but a few stored produce pests are involved (Khapra beetle etc.). Presumably the recent accidental introduction of *Prostephanus truncatus* into Tanzania resulted from the failure to notice the infestation of imported maize; this new species differs only very slightly from the quite common *Rhizopertha.*

21.4.3 Health Legislation

Relating to foodstuffs, there are health regulations in each country that are concerned with the quality of food materials, and they specifically refer to contamination (adulteration), damage by pest organisms, and the presence of pest organisms. In the UK, there has been a series of Acts and Regulations over the last 50 years designed to safeguard the health of the public.

In the UK, the Ministry of Agriculture, Fisheries and Food (MAFF) had responsibility for foodstuffs regarding public health, in conjunction with the Department of Health. MAFF comprised three separate directorates dealing with agriculture, fisheries and the Food Safety Directorate dealt with all aspects of food safety and consumer protection. Latterly, the work of MAFF was heavily influenced by the EEC, of which the UK is a member. In relation to food safety, several committees were established to deal with pesticides, food additives, microbiology, veterinary medicines, labelling etc. Eventually, all the existing legislation was brought together, strengthened, updated and made more flexible to form the Food Safety Act 1990. A nice summary of existing legislation (up to 1993) is given in Macrae, *et al.* (1993; pp. 3132-3136). In 2001, MAFF was reorganised into the Department for Environment, Food and Rural Affairs (DEFRA).

Also in 2001, the UK government established the Food Standards Agency, a quasi-governmental body, to monitor the UK food industry and conduct research in many aspects of food-related topics, including diseases, chemical contaminants, meat hygiene, safety of novel foods etc., microbiology and various types of risk assessment.

Very recently, the UK Public Health Inspectorate has suffered a name change, to Environmental Health (Institute of Environmental Health Officers). In the USA, the Federal Food, Drug and Cosmetic Act (1944) has given the Food and Drug Administration broad and far-reaching powers referring to pesticide residues in foods and edible materials, and contamination, insect pest and rodent filth, damage etc. Under this legislation, cockroaches in food stores, the presence of rats, contaminated food for sale, are all offences, and the Health Inspectorate and FDA staff are responsible for advice for legal prosecution of offenders.

The Environmental Protection Agency (EPA) based in Washington DC (20460) regulates pesticides and their use under the authority of the Federal Insecticide, Fungicide and Rodenticide Act (1980). Pesticides approved for use are registered in the Federal Register, and their use is carefully proscribed, especially for application in food-handling establishments.

21.5 STORE HYGIENE (SANITATION)

There are several aspects to this approach, and mostly they are not expensive, special equipment is not needed, nor extra facilities required. But sanitation methods will help to reduce pest populations, although often only slightly, but they can be a meaningful part of an IPM programme. The main aspects are listed below.

21.5.1 Sound Structure

The building or structure should be intact, with tight-fitting doors, and, as far as possible, it should be rodent and bird-proof; insect-proofing would be very desirable if feasible. Wooden structures can be creosoted or otherwise treated to ensure longevity, and to resist insect and fungal attack. Clearly, modern concrete and sheet metal structures are expensive, although worthwhile in the long term; recognising the desirability of such stores, and acknowledging the construction cost problem for most farmers in many countries, government grants have been made available for this purpose. In tropical countries, the peasant farmers being assisted in the various appropriate technology programmes should aim at building sound storage structures; often, for a relatively small financial outlay, a sound structure lasting many years can be constructed.

21.5.2 Cleanliness

This involves sweeping up, removal and destruction of produce residues, spillage, and rubbish generally. Careful inspection of nooks and crannies in the structure is important, for insects lurk in such places. Vacuum cleaning is preferable if possible, for ordinary sweeping usually leaves behind mites and insect eggs; all possible foci of infestation should be removed. Fumigation or spraying of empty premises is always desirable. In parts of Africa, the underground storage pits when empty have a fire built in the base for a few hours.

21.5.3 Clean Produce

As far as possible, the produce being taken into the stores should be clean (i.e. uninfested) and dry. It has been estimated that in many parts of Africa field infestations of maize by *Sitophilus* weevils are often at a level of 10-20%, and similar levels have been observed on sorghum by *Sitotroga cerealella*.

21.5.4 Clean Sacks and Containers

Sacks are vulnerable to insect and mite infestations, and they provide sites for eggs to be hidden and for pupal cocoon attachment. Sacks, bags and reusable containers must be thoroughly cleaned between uses; if insecticide treatment is not possible, they should at least be placed on an ant nest for a few hours.

21.5.5 Use of Sealable Containers

The use of small containers which can be securely sealed is strongly recommended, for this does minimize cross-infestation. If the containers are air-tight, insect pests can be asphyxiated and killed.

21.6 PHYSICAL METHODS OF CONTROL

Under this heading come the main environmental factors of temperature, moisture, atmosphere etc., and the use of traps.

Most of the time, the single most important factor in the longterm storage of grain and other foodstuffs is moisture. This is in part because of the ubiquitous nature of fungal spores and bacteria – they are literally everywhere! In addition, insects only develop at a particular level of moisture availability – under dry conditions they do not develop. There is an exception, as usual, in that the Museum beetle (*Anthrenus museorum*) only thrives under dry conditions, and moist air can actually cause its demise.

21.6.1 Drying

At the time of harvest, wheat in the UK usually has a moisture content of 18-20% (occasionally lower, but more often higher), and field drying is no longer practiced, but for long term storage it must not exceed 16%, and should ideally be less (13%

or less would be perfect). At a mc level of 14%, fungal development is minimal, and insect development is retarded. *Sitophilus* weevils are reported to stop feeding at a grain moisture level of 9.5%. In the UK, wheat is usually dried to a mc of 14-15% for long term storage, and large continuous–flow driers are used.

In the tropics, air drying is the usual method, and the unthreshed maize and sorghum is stacked in the field, or else the maize cobs are gathered and stored in open cribs in which to dry. In wetter regions, there is a problem in that field drying is often impossible. In S.E. Asia, with rice it is the practice to harvest by hand and thresh soon after; the grain is then dried in the sun in small quantities which can be easily handled.

Pulses are typically dried to a lower mc, and 9-10% is usual. But it has been shown in California that kidney beans at moisture levels below 11.5-12.0% become very susceptible to mechanical damage (testa cracking especially). In the UK, it is also known that peas and field beans (*Vicia faba*) dried at high temperature (above 60°C) will suffer fine cracks in the seed testa.

With dried fish and some dried meats, it is vitally important that the initial drying is rapid, for in the early stages the dividing line between drying and putrefaction is sometimes fine.

A final point about moist grain storage is that most contact insecticides are markedly less effective on moist grain.

21.6.2 Temperature

All living organisms have a range of temperature in which they thrive, and critical temperatures above which they die of heatstroke, and below which they die of cold. Animals and plants are basically divided into tropical, temperate and boreal (arctic) species according to where they normally live, and to which temperature regime they have evolved and adapted. Animals are in addition regarded as eurythermal if they can tolerate a wide range of temperature, and stenothermal if their tolerance range is narrow. This categorization applies essentially to poikilothermous animals such as insects and reptiles; for homoiotherms such as mammals (and most birds) are able to adapt to a wide range of temperature and physiological adjustments. Thus the common rat and the house mouse are now almost totally worldwide in distribution, and *Mus* is recorded living in deep frozen beef carcasses at -20°C or lower. According to the basic nature of the organism concerned, there is much variation in the reaction of stored produce pests to temperature, and generalisation has to be limited.

21.6.3 Heating

This is usually not too successful, for a slight rise in temperature will just increase the metabolic rate of the organisms, and a temperature high enough to kill the pests will invariably cause deterioration of the produce. But kiln treatment of timber has been very successful for killing timber insects and pests. Cotton seed has been heat treated to kill larvae and pupae of pink bollworm. And hot water treatment (dipping) is a standard method of killing nematodes, mites and insect larvae in flower bulbs and onions, and has been used in Hawaii with unripe papayas to kill fruit fly larvae. But the process has to be done very carefully otherwise the bulbs or fruit are damaged.

Because of the withdrawal of ethylene dibromide and methyl bromide as fruit fumigants, there is renewed interest in the use of hot water treatment on fruit in Hawaii for export, also Thailand and Taiwan are successfully following suit. Another method is a high temperature, forced air treatment. Vapour heat is the controlled application of humidified air at between 40 and 60°C to kill the pest or ameliorate the disease, without damaging the fruit, and for some fruits it works well.

21.6.4 Cold Treatment

This tends to be effective, especially against tropical pests; in general, a reduction in temperature of a few degrees makes a considerable reduction in metabolic rate (development and activity) of insects and mites. Many stored produce insect pests are actually tropical, and only survive in the UK for a short time in the summer, being killed by the winter temperatures in unheated stores. Some of the cosmopolitan insect pests are multivoltine in the tropics, but only univoltine in the cooler climate of the UK. The effects of cold may be misleading, though, for many insects will become comatose at low temperatures, but will revive as the temperature rises. Bluebottles trapped accidentally in a domestic refrigerator are still moving (but sluggish) on release, and shortly after will revive and fly away. Cold is an excellent treatment to reduce infestation or to suppress a population – for example, *Oryzaephilus* cease development at 19°C and *Lasioderma* at 18°C, but to kill the insects requires exposure to very low temperatures, and even then only the more tropical species will die. In general, animals of all types are far more susceptible to a rise in temperature than to a fall of the same magnitude as far as temperature death is concerned.

Some grain stores, and also many fruit and vegetable stores, have facilities for chilled air ventilation, and there is often use of mobile refrigeration units, but many stores in the UK use ambient air ventilation which cools sufficiently (Figure 8.12).

It should be remembered that temperature effects are a result of the combination of actual temperature and duration of exposure.

A combination of temperature and humidity control can be very effective, and is a far better aim than using either factor alone. In Israel at the Volcani Institute for Agricultural Research, there was a three year project (1982-85) in the Negev Desert for long term grain storage. The grain was placed in a large plastic bag with sealed openings each end, and buried in a pit in the desert. The desert air is very dry and, at night over the 'winter' period, very cold, and in the cool season at night the bag was opened and cold dry air was pumped through the container, which was resealed during the day; after three years, storage damage to the grain was negligible.

In modern food production systems, chilling and refrigeration play very important roles, and the great majority of prepared foodstuffs are sold from supermarkets either chilled or frozen. This, of course, extends the shelf-life of most products (meat, fish etc.), and suppresses or even kills many of the micro-organisms responsible for food spoilage. Chapter 4.10 looks at cold storage of produce.

21.6.5 Suffocation (Atmosphere Modification)

If a container is filled with grain and then hermetically sealed (made airtight), the living organisms inside soon use up the available oxygen and they then suffocate and die. Air with only 2% oxygen will asphyxiate *Sitophilus* weevils; but the store or container does have to be air-tight. More effective is to fill the sealed container with an inert and preferably heavy gas such as CO_2 or N_2 – both are used with great success (Chapter 4.11). The modern practice of sealing foodstuffs inside plastic bags will also have this effect if the bag is packed tight with produce so that only a little air is included. Replacing the air with an inert gas is becoming more commonplace, and is very successful.

21.6.6 Light Traps

To be effective at attracting insects, the radiation has to be mainly ultraviolet, for, in general, light trap catches are some 20 times more when using ultraviolet from a mercury vapour lamp as opposed to the white light from a tungsten lamp. It is now known that black light peaking at around 3,574 Angstroms (357 nanometres) is the most attractive for most nocturnal flying insects. Most moths and many beetles are attracted to these traps which are used both for faunal identification and for population monitoring. For use in produce stores, it is customary for the collection chamber to contain a potent insecticide or other killing agent.

Light traps are only occasionally used in trapping-out programmes for population depletion; it is now more usual for an electrocutor to be used for this purpose.

21.6.7 Electrified Light Traps (Electrocutors)

A recent development is the commercial production of light traps with an electrified grid over the tubes, although an experimental model was built in the USA as long ago as 1928. The light emitted is a mixture of ultraviolet and visible radiation, so that it appears as a faint blue glow to the human eye. The trap is wall-mounted, high and conveniently out of reach, and will kill many flying insects daily. Latest models are not too expensive to buy, and not expensive to run, and are widely used on domestic premises of all types. As a means of insect pest population reduction, these traps are very useful and widely employed; the best known models are probably those made by 'Insect-O-Cutor' in the USA, and now in the UK.

One drawback with the use of these electrified light traps is that apparently the ultraviolet tubes deteriorate with age, and there is a change in the spectrum of radiation with a reduction in the attractive wavelength range. This change is not evident at all to the human eye.

21.6.8 Sticky Traps

In the 21st Century, it is quite surprising to observe the number of old-fashioned flypapers in use in food stores and domestic premises, the advantages presumably being that they are very cheap and convenient to use, hanging high above people's heads, and they do trap urban flies and other insects.

In greenhouses, yellow plastic sticky boards are being used to trap adult whiteflies (Aleyrodidae) and other insects. There is no insecticide being used – purely a sticky film to physically trap any insect that lands, and the yellow colour is attractive to most flying insects and will to some extent induce alighting.

21.6.9 Irradiation

A Joint European Committee for Food Irradiation (1980) concluded that up to a level of 10 kGy, there was no risk of toxicity. Equipment is now being constructed in various countries for bulk treatment of grain, and also some smaller facilities for use with various processed foods. Two different systems are being used for grain protection; the most widely used is a cobalt irratiator emitting gamma radiation, and the other is an electron accelerator, which is still largely experimental for this purpose, but seems to be very effective. Insects are killed directly if the ionizing radiation is intense enough, and at lower levels insect sterilization results. It would appear that this is one of the methods of food protection (preservation) likely to become widely used in the future.

A useful publication, although now out of date, is the joint FAO/IAEA book (1975) on *Requirements for the irradiation of food on a commercial scale.*

21.6.10 Thermal Disinfestations with Microwaves

Radio frequency energy in microwaves has been used to kill insects in stored grain, but it is thought that death results from selective heating of the insect bodies in the stored produce. Since this is a very convenient way to disinfest food, it is likely to become more important as the technique is further developed.

21.6.11 Miscellaneous Methods

Other physical methods that have been used to try to control insect pests in stored produce include the use of sound energy, which was successful in experiments but not as yet recorded on an industrial scale. Inert abrasive powder added to grain has resulted in the removal of the waxy epicuticle so that the insects die of desiccation. Diatomaceous earth is a natural siliceous powder sometimes used for this purpose.

Hygroscopic powders added to the grain are being tested, and it appears that this method of attacking insect pests shows promise as a method of control.

Commercial preparations of silica-gel (silicic acid) are being used with success, but they are slow in action and require several weeks to work, and are only effective as long as the grain is quite dry.

As mentioned later, in many farmers' stores, it has been a tradition to add sand or wood ash to the grain as a method of protection – and often this does seem to have a beneficial effect.

For physical methods to manage stored food pests, see Gorham (1993) pp. 399-418.

21.7 BIOLOGICAL CONTROL

The produce store is basically an unnatural habitat, and so it is not surprising to discover that natural control of stored products pests by existing local predators and parasites amounts to very little. There is one exception, and that is concerning the abundance of rats in the tropics – the widespread and extensive destruction of predatory snakes and various carnivores such as foxes, civet cats etc. has undoubtedly contributed quite seriously to their general upsurgence.

There is some interest in trying to find natural enemies of a few pests of on-farm stores. At the present time, for example, there is a search being conducted in Mexico and Central America (as well as in East Africa) for natural enemies of the larger grain borer (*Prostephanus truncatus*). The Greater Grain Borer in East Africa has proved difficult to control in the smallholder farm situation with pesticides, so a major concerted effort is being made to obtain natural enemies from back in Central America. To date, some insect predators and some parasites have been collected and introduced into East Africa. Whereas it may not be feasible to try to establish natural enemies in a large warehouse filled with grain, it may well be possible in a smallholder farmer's plot.

However, there is a number of different animals that are to be found in produce stores and domestic premises that are undoubtedly beneficial in that they are feeding on the pests found there. For example, in parts of S.E. Asia in domestic premises, geckos and sometimes skinks are abundant and quite tame, and their daily intake of domestic insect pests is very considerable. Predators such as these should be encouraged whenever possible.

The main groups of natural enemies that are found widespread in many parts of the world include the following. It should be remembered that in a few cases the species are cosmopolitan; in other cases, the genus is worldwide by with different species in different geographical regions, otherwise there are different genera in each family in different regions, but with similar biology.

Insecta -	Heteroptera	-	Anthocoridae (Flower bugs)
			Reduviidae (Assassin bugs)
	Coleoptera	-	Carabidae (Ground beetles)
			Staphylinidae (Rove beetles)
			Histeridae
	Diptera	-	A few species only
	Hymenoptera	-	Braconidae
			Ichneumonidae
			Chalcidoidea
			Formicidae (Ants)
Arachnida -	Acarina	-	Some mites
	Aranaeida	-	Arachnidae etc. (Spiders)
Amphibia -	Anura	-	Ranidae etc. (Frogs)
Reptilia -	Lacertilia	-	Gekkonidae (Geckos)
			Scincidae (Skinks)
	Serpentes		Elapidae (Cobra etc.)
			Colubridae (Rat snakes)

Mammalia -	Carnivora	-	Felidae (Cats)
			Canidae (Foxes, dogs)
			Mustellidae (Weasels etc.)
			Viverridae (Civets)

Biological control (*sensu stricta*) refers to the deliberate introduction of predators or parasites, often exotic, in an attempt to control pest numbers, but little has been attempted to date, for the situation is usually not appropriate.

In Ethiopia, most households in the rural villages kept a cat, and these were very successful in suppressing the local rat population – the very hungry cats ate the rats (*Rattus rattus*), and the less hungry just killed them.

A recent review article dealing with the sterile male release treatment (SMRT) for stored products pests after γ-radiation is by Hasan and Khan (1998).

21.8 PESTICIDES (Chemical Control)

Pesticides are chemical poisons that kill pest organisms. Some of the very toxic fumigants kill almost all biological organisms, but most poisons are only effective against certain groups. For example, insecticides kill insects; herbicides kill weeds; fungicides kill fungi; bactericides, bacteria; rodenticides, rodents etc. Some major pesticides in current (or recent) use are listed below, but fungicides and bactericides are not included. Rodenticides are referred to in Chapter 10. For further information, see Appendix 1, pp. 415-437.

SOME MAJOR PESTICIDES IN CURRENT OR RECENT USE

Chlorinated Hydrocarbons (= Organochlorines) (= OCs)

The first of the 'modern' insecticides, whose development started with DDT in 1940. Broad-spectrum; persistent; stomach and contact poison.

- DDT group - DDT
- BHC group - BHC
- Cyclodienes - Aldrin, Dieldrin, Endosulfan etc.

Most are now banned in many countries because of environmental damage, and pest resistance is very widespread.

Organophosphorus Compounds (= OPs)

Developed as nerve gases (World War II); very poisonous; contact and systemic poisons; high mammalian toxicity; short-lived; some quite specific; action is inhibition of acetyl-cholinesterase – thus a nerve poison.

- Bromophos Dimethoate Malathion
- Chlorpyrifos-methyl Etrimfos Pirimiphos-methyl
- Diazinon Fenitrothion Tetrachlorvinphos
- Dichlorvos

About 50 compounds have been used as insecticides.

Carbamates

Also act as anti-cholinesterases; more broad-spectrum of activity; developed because not so toxic to vertebrates.

- Bendiocarb
- Carbaryl Methiocarb Propoxur

About 12+ compounds being used; most are fungicides and herbicides.

Miscellaneous Compounds

Inorganic salts and organic fumigants. Amongst earliest compounds used.

- Copper acetoarsenite (Paris Green) Aluminium phosphide
- Lead arsenate Carbon tetrachloride
- Lime sulphur Ethylene dibromide
- Mercurous chloride Methyl bromide

Natural Organic Compounds

Mostly from plants with compounds toxic to insects.

- Nicotine - from Tobacco; very poisonous; fumigant and contact action.
- Neem - foliage and seeds have insecticidal properties.
- Rotenone - from roots of *Derris* spp.; powerful fish poison.
- Pyrethrins - from flowers of *Pyrethrum*; rapid knock-down; used with synergists; only poisonous to insects; very safe to use; rapid degradation in sunlight.
- Pyrethroids - new synthetic insecticides; broad spectrum, persistent, but some vertebrate toxicity. Eight or more compounds regularly used (Permethrin, Cypermethrin, Bioallethrin etc.).

Organic Oils (= Hydrocarbon Oils)

- Tar oils - heavy, dark, toxic, phytotoxic; 'winter washes'; creosote; wood preservatives.
- Kerosene - of great antiquity; mostly used as fungicide and herbicide.
- Petroleum oils - distilled, lighter (mineral and white oils).

Biological Compounds

- Antibiotics and entomophagous fungi.
- Bacteria - *Bacillus thuringiensis* (Bth) – caterpillars killed.
  - *B. popilliae* – Japanese Beetle Milky Disease.
- Viruses - most effective against caterpillars and beetles.

Insect Growth Regulators (IGR) (including *juvenile hormone analogues* (JHA) and *chitin synthesis inhibitors* (CSI))

Chemicals with hormonal properties that disrupt insect physiological development.

- Methoprene

- Diflubenzuron etc.

Most pesticides are either crystalline solids or oily liquids in the pure state or the technical product, and they are usually effective in quite small quantities (i.e. a litre of two per hectare). In order to apply a pesticide to a crop, it is usually necessary for the chemical to be contained in a carrier fluid or powder. The earliest carrier (or diluent) used was water, and the chemical was either dissolved in it, or, if it was water-insoluble, presented as a suspension or emulsion. In the case of emulsions and suspensions, it is important that the mixture be stable enough to ensure the application, under practical conditions, of a solution of uniform and known concentration. The physical stability of a spray solution may be conferred by the addition of supplementary materials. Thus, the sedimentation of a suspension can be delayed by the addition of protective colloids or dispersing agents. The coalescence of the scattered droplets in an emulsion may be retarded by the addition of an emulsifier. Some chemicals either occur as a powder, or else they can be formulated as such for application to grain in storage.

21.8.1 Formulations

As previously mentioned, most pesticides are crystalline solids or oily liquids, and as such are usually not suitable for direct spraying. In a few cases, the technical product is soluble in water; then the pesticides can be prepared as a very *concentrated solution* (c.s.) which only requires dilution to the appropriate strength for spraying. Usually, these concentrated solutions have to have wetting agents or detergents added (see later in the chapter). This type of spray solution is typically very homogeneous, and spreads a very even level of pesticide over the target.

Many solid substances that will not dissolve in water can be ground and formulated as *wettable powders* (w.p.). Wettable powders are powders which can easily be wetted, and do not resist the penetration of water, and are miscible in water in which they pass into suspension. Others are more accurately termed *water-dispersible powders*, in that when mixed with water, they remain as individual particles in suspension for a considerable period of time. Various additives (dispersants) can be included in the formulation of wettable powders to delay the process of sedimentation.

Oils and other water-immiscible liquids, if agitated with water, break up into tiny droplets which, on standing, rapidly coalesce to form a separate layer. This coalescence may be retarded or prevented by the addition of auxiliary materials known as surfactants or emulsifiers. With mixtures of oil and water, two types of *emulsion* are possible. The oil may be dispersed as fine droplets suspended in water, which is then the continuous phase, giving an oil-in-water (o/w) emulsion, or the water may be the disperse phase, giving water-in-oil (w/o) emulsion. The type of emulsion generally required for crop spraying practice is the o/w emulsion which, as water is the continuous phase, is readily dilutable with water. Thus, pesticides insoluble in water may be dissolved in various organic solvents forming an *emulsifiable concentrate* (e.c.), which can be diluted in water to an appropriate spray strength. The 'breaking' of an emulsion is the usual way in which the toxic dispersed phase comes into play, breaking occurring after the evaporation of most of the water. Various chemical substances can cause the 'inversion' of an emulsion, which then becomes useless for spray purposes. Sometimes, emulsions are caused to 'cream' (named from the analogous creaming of milk, resulting from differences in specific gravity between the dispersed and continuous phases.

Some pesticides are more suitably formulated as *miscible liquids* (m.l.). In this case, the technical product is usually a liquid and is mixed (dissolved) with an organic solvent, which is then, on dilution, dissolved in the water carrier.

At times it will be necessary to know precisely the quantity or proportion of the pure chemical in any formulation; sometimes this may be shown on the pesticide container, but more often it is not. However, it will be available on the company data sheets, and in such sources as *The pesticide manual.* The usual method of expression of the proportion is as grams of *active ingredient* per kilogram of formulation (g a.i./kg) for powders, and grams per litre (g/l) for liquids.

21.8.2 Spraying

When water is the carrier, the usual method of application of the spray is by passage under pressure through special nozzles which distribute the chemical in a fine spray over the crop. In general, the type of spray application can, for convenience, be expressed as high-volume, low-volume, or ultra-low-volume according to the amount of carrier liquid.

During the last 40 years, especially when viewed worldwide, there have been many different definitions published for high- and low-volume spraying, with very little overall agreement as to rates. There has tended to be variation, not only from country to country, but sometimes according to the crop concerned, and also, it appears, according to the availability of local water supplies in some cases.

In the UK recently, there has been a concerted effort to promote the following definitions:

		Rate (l/ha) for trees and bushes	Rate (l/ha) for ground crops
High-volume	(h.v.)	>1,000	> 700 (600)
Medium-volume	(m.v.)	500-1,000	200-700
Low-volume	(l.v.)	200-500	50-200
Very-low-volume	(v.l.v.)	50-200	5-50
Ultra-low-volume	(u.l.v.)	< 50	< 5

21.8.3 Droplet Size

This is very important in crop protection – very small droplets give the best coverage of the host plant foliage, but are subject to drift on air currents. But for stored grain etc., droplet size is of far less importance, and it is usually easier to obtain a uniform distribution of the spray over the grain.

21.8.4 Controlled droplet application (CDA)

This term was coined in 1975, in discussions on herbicide application, in order to emphasise the importance of having a known and uniform size of droplet in sprays of liquid pesticides, but the basic concept had, of course, been known for a long time. However, awareness of the precise importance of droplet size in pesticide sprays, and methods of achieving both uniform droplet size and pattern, are relatively recent. (Matthews, 1979).

21.8.5 Solvents

Anyone with an elementary knowledge of applied mathematics will be aware that the volume of a sphere increases according to the cube of its linear measurement (radius or diameter), whereas the surface area increases according to the square. Thus, smaller droplets have relatively large surface areas, which implies that the rate of evaporation of spray droplets is higher with smaller droplets. Consequently, solvents used in u.l.v. formulations must have low evaporation rates, and water should never be used. The more volatile liquids would also lead to evaporation of the solvent in the atomizer, and might cause crystallization of the pesticide there. Liquid pesticides are sometimes sprayed undiluted in u.l.v. application, but more often some solvent is required.

The solvent must be non-phytotoxic, of low volatility, of high dissolving power for the pesticide, of low viscosity, and compatible with the pesticide. Not many of the solvents generally available will fit all these categories, and so clearly good u.l.v. solvents are difficult to find. Xylene is too volatile, other aromatic hydrocarbons of lower volatility are often highly phytotoxic. Alcohols and ketones show higher phytotoxic effects at lower volatility, and other common solvents show the same variations in properties. Sometimes a mixture of various solvents can be used with reasonable success. In the laboratories of Phillips-Duphar have been developed several compounds called *adjuvants*, which dramatically decrease the phytotoxicity of many low-volatile solvents, thus the range of solvents that can be used for u.l.v. formulations has now increased. Conventional emulsifiable concentrates usually have a low flash point, but this seldom represents a fire hazard since the concentrate is mixed with a large volume of water before spraying. However, with u.l.v. techniques, the solvent must be of low volatility, for with rotary atomizers with electrical systems and many rotating parts, the possibility of electrical discharge is always present.

With the use of these adjuvants, a wider range of low-volatile solvents can be used for u.l.v. formulations, and many of these formulations are now regarded as being special ultra-low-volume (s.u.l.v.) formulations. Their characteristics are high concentration, low volatility, low phytotoxicity, with a much lower viscosity than most u.l.v. preparations, and with a flash point above 75°C.

21.8.6 Spray Residues

In conventional spraying, the spray liquid contains a large amount of water, containing various wetting agents, dispersants and emulsifiers. The pesticide is generally present as a finely dispersed phase, solid in the case of wettable powders, or liquid when using emulsions. In u.l.v. sprays, the pesticide is generally present as a true solution in an oil carrier, or sometimes the technical material is sprayed as such. Such differences in the spray make-up will have a profound effect on the behaviour of the spray droplets on the biological target.

After deposition of a spray droplet, the droplet will assume a certain shape by spreading over the target surface. With conventional aqueous spray liquids, this spreading depends greatly on the properties of the target: on hydrophilic surfaces, the droplet will spread to a thin film, but on hydrophobic surfaces, the droplets usually retain their spherical shape. The spreading of oils on most smooth surfaces is much better than that of aqueous solutions. The s.u.l.v. formulations spread to a thin film on most surfaces, even hydrophobic leaves, with very little run-off. However, there are times when run-off is wanted, and in these cases u.l.v. application is not the most suitable method of application.

The formation of a residue from an emulsion droplet is very complicated; evaporation of the water, breaking of the emulsion, and crystallization of the pesticide from the oil phase can occur simultaneously. Often from u.l.v. sprays, the crystalline residue on the surface is particularly coherent and very resistant to dislocation, and also the rain-fastness of u.l.v. residues seems to be better than that of conventional formulation.

21.8.7 Dusting

Sometimes it is more convenient to use a dust instead of a spray: the need for water is obviated; the dust may be bought ready for use and is easier to handle than spray concentrate; the dusting appliances are generally lighter and easier to manipulate than sprayers. For dusting, the active ingredient is diluted with a suitable finely divided 'carrier' powder, such as talc. The dust is usually applied by introduction into the air-stream of a fan or turbine blower. However, in practice, dusts are often not easy to apply; frequently the powder 'cakes', usually through absorption of atmospheric moisture, or 'balls' in the hopper through static electrification. Also, it is difficult to ensure that the dust is homogeneously mixed.

It is generally found that crop dusting is only practicable in the calmest weather, and that best results are obtained when the dust is applied to wet or dew-covered plants. When dusts are applied to dry foliage, usually not more than 10-15% of the applied material sticks to the foliage.

But dust/powder application to stored grain and pulses is usually much easier, is very effective, and avoids the use of water.

21.8.8 Baits

The use of poison baits in pest control is generally confined to the insect groups Orthoptera (crickets, cockroaches, sometimes locusts), Isoptera (termites) and Hymenoptera (ants), in addition to terrestrial molluscs and vertebrates (especially birds and rodents). These pests are typically gregarious with underground nests. Social insects in nests are particularly difficult to kill by spraying insecticides, and sometimes the nests are hard to locate. Most social insects take food back to the nest, where it is shared by trophallaxis amongst other adults and the larvae. In the case of fungus-growing termites (family Termitidae) and some ants, the collected food material is incorporated into fungus gardens on which fungus grows, and poisons introduced in this manner eventually get distributed throughout the colony.

The earliest poison baits relied upon inorganic stomach poisons such as salts of arsenic, lead, mercury and sodium fluoride, but these were largely replaced by certain organochlorine compounds, and others.

Poison baits generally comprise four separate components:

- *Carrier* or *matrix*: inert material which provides the structure of the bait. this may be an edible or attractive material such as meal (e.g. soyabean, groundnut) or citrus pulp, or a more inert or less attractive substance such as ground rice husks, corn cob grits, clay, vermiculite and gelatine microcapsules.
- *Attractants*: this may be an integral part of the carrier itself (e.g. soyabean meal, citrus pulp), or may be substances such as sugar, molasses or vegetable oils added to the inert carrier base. A recent development in leaf-cutting ant baits is the incorporation of a trail pheromone to attract the worker ants to the baits. Obviously, any chemical that will induce the insects to pick up or eat the bait will make this method of control more successful.
- *Toxicant (poison)*: previously a stomach poison usually, but more recently contact insecticides such as aldrin, dieldrin, and heptachlor were used.
- *Additives*: for specific formulation purposes, such as preservatives, materials to bind the bait together in pellets or blocks, and waterproofing agents. The physical properties of baits can be important, especially for outside use in the tropics with high temperatures, high humidity and torrential monsoon rains. For a broadcast bait to remain effective in the field for several weeks under wet tropical conditions, it must not disintegrate in the rain. A standard ant bait is 'Mirex 450', but the compressed pellets apparently break down in heavy rain and become ineffective. Experiments by Cherrett and others have shown that leaf-cutting ant baits composed of citrus pulp with soyabean oil and aldrin can be rendered fairly waterproof by the addition of a hydrophobic surface deposit of siloxane, which prolongs the effective life of the bait under wet conditions.

Baits are often placed by hand, but clearly this is time-consuming and costly at the present time. Baits for arboreal rats are usually in the form of blocks that can be nailed to the trunk of a palm or the side of a building.

21.8.9 Stickers

Stickers such as methyl cellulose, gelatine, various oils and gums, are used to improve the tenacity of a spray residue on the leaves of the crop, or on to the walls of a grain store. Maximum spray retention is particularly important in the tropics, where rainfall is often monsoonal and torrential. Generally, a fine particle deposit is more tenacious than one of coarse particles. Spreaders usually enhance the tenacity of a deposit and its retention on the plant, although they may retain their wetting properties and thus cause the deposit to be washed off by rain or dew. Some spreaders break down on drying and form insoluble derivatives, and these can greatly enhance retention of the deposit.

21.8.10 Lacquers

In order to achieve a slow release of a pesticide in certain locations, it is possible to formulate some insecticides into lacquer, varnish or paint. The painted area then releases the insecticide slowly over a lengthy period of time. The insecticides used in this manner were mainly organochlorines, and particularly DDT and dieldrin. Incorporation of insecticides into paint is of some value, but the lacquers are of limited use owing to the problem of the lacquer remaining after the pesticide has been dispersed. This practice is of more use against household and stored products pests than crop pests.

21.8.11 Synergists

These are substances which cause a particular pesticide to have an enhanced killing power. They are sometimes called *activators*. The way in which synergists act is not always fully understood, but some operate on a biochemical level inhibiting enzyme systems which would otherwise destroy the toxicant. Usually the synergist itself is not insecticidal. Piperonyl butoxide is a synergist for the pyrethrins and certain carbamates. Some pairs of organophosphorous insecticides have a mutually synergistic action (sometimes called '*potentiation*', especially in American literature). Other synergists include piprotal, propyl isome, sesamin and sesamex. Many of these are particularly effective on the pyrethrins. Some of the IGRs have a synergistic effect with some organophosphorous insecticides.

21.8.12 Insect Growth Regulators

Concern about the impact of pesticides on both health and environment has resulted in the search for alternative control measures for stored-product insect pests. Among such alternatives are insect growth regulators (IGRs), a class of bio-rational compounds that disrupt the normal development of insects. Because of their selectivity of action, these compounds appear to fit the requirements for Third Generation Pesticides, environmentally benign and safer grain protectants. IGRs have been developed commercially, and are being used to control insect pests in agriculture, forestry, public health and stored products. IGRs affect the biology of treated insects; for example, both embryonic and post-embryonic development, reproduction, behaviour and mortality. Abnormal morphogenesis is the observed effect of IGR action on the insects. Many of them are more potent than current insecticides, even against the eggs of many pests.

Compared with the conventional insecticides, IGRs do not exhibit quick knock-down in insects, or cause direct mortality, but the long-term exposure to these compounds largely stops the population growth as a result of the effects mentioned in both the parents and progeny (see Mondal & Parween, 2,000).

On the basis of mode of action, IGRs have been grouped into three categories:

- juvenile hormones (JHs) and their analogues (JHAs), also called juvenoids – they maintain the larval state, and the pupae usually die.
- ecdysone agonists – can be ingested or absorbed through the cuticle; they upset cuticle synthesis in different ways, and the insects die; they also cause female sterility.
- chitin synthesis inhibitors (CSIs) – these upset chitin biosynthesis, and prevent moulting; they act at all stages in the life cycle. Most are ingested as stomach poisons; they are particularly effective against early instars, and can prevent eggs from hatching.

It has long been known that some of the JH chemicals have synergistic activity when used with certain organo-phosphorous insecticides.

ADVANTAGES OF IGRS

The potential of IGRs is of outstanding importance in the field of pest control. They are superior in several aspects to the usual insecticides. The following advantages of IGRs make them efficient broad-spectrum insecticides, and especially safe for use in food stores (Mondal & Parween, 2000):

- IGRs degrade rapidly in the environment; it is reported that degradation half-lives of diflubenzuron and triflumuron in soil were 30 days, and residues were practically undetectable 90 days after application.
- IGRs do not affect beneficial parasites and predators.
- IGRs act only during a few short but sensitive phases of the target insect's life cycle, hence they do not affect non-target organisms.
- IGRs have a wide spectrum of activity on economically important pest species.
- Generally, the insects do not develop resistance to IGRs, and IGRs are effective against insect species which are resistant to conventional insecticides because of their different mode of action.
- In stored product situations where there is a closed environment not subject to continuous reinfestations, juvenoids appear to offer considerable promise for slowing down pest population build-up.
- These compounds can easily be formulated.
- Redistribution of JHA compounds through the treated grains in storage occurs over time, and ultimately even concentration in the grain is achieved.
- The apparent insensitivity of the IGRs to grain moisture may be advantageous in circumstances where grain cannot be dried before treatment, whereas grain moisture has a deleterious effect on the activity of most conventional insecticides.
- The potentiality of IGRs is not seriously affected by temperature.
- IGRs do not alter the physical properties and taste of flour, gluten or bread.
- IGRs do not affect the germination of the treated grains.
- IGRs have no mutagenic and teratogenic effect on warm-blooded animals.
- These compounds show synergism with many conventional insecticides.
- Very small amounts are needed.

DISADVANTAGES OF IGRS

In spite of their many advantages, IGRs have been reported to possess the following disadvantages:

- Prolonging larval life. In some stored-product pests, the larval feeding period was significantly prolonged because of JHA treatment, in comparison with that of the control larvae.
- Timing of application. IGRs are only effective against certain developmental stages of insect pests. Like other (conventional) pesticides, they cannot be used against all life stages. The narrow window of sensitivity, instability in sunlight, and high volatility are the main disadvantages of JHAs.
- CSIs are active against the young larvae only. These compounds should be applied to 1st or 2nd larval instars of the stored-product insects. Moreover, owing to the absence of immediate effects of IGR treatment, stored products in warehouses need long exposure to IGR treatment so that all members of the population are exposed to these compounds during their sensitive periods.
- Cross-resistance. Insecticide-resistant strains of insects may show some level of cross-resistance to the IGRs. However, in many other instances no cross-resistance to IGRs have been reported in stored-product insects.

21.8.13 Pesticide Development

For the last half-century, the main weapon in the human arsenal against insect pests has been chemical insecticides, for they have been effective (high kill, rapid acting, predictable in action) and usually not too expensive (sometimes even quite cheap) to buy. However, we have become overly dependent on their use, and now that their overall usefulness is declining, we are having to rethink our pest control strategies. Part of the problem is that very few new compounds are coming into use; this is partly because we are running our of candidate chemicals, and also because of the costs involved with the new development procedures. Over the last 20-30 years, there have been great changes in the procedure required for the commercial development and official registration of a new pesticide. This has largely been because of the 1986 ‘Control of Pesticides Regulations’ in the UK, and similar legislation in the USA. The official concerns are basically environmental risk (safety) balanced against effectiveness, and the commercial concerns are production problems and cost. The new official requirements are very stringent, and all UK and USA agrochemical companies are having to re-test and re-register all their existing products within a certain time period.

Between 1960 and around 1980, the time required for development of a new pesticide increased from three to five years to ten years or more; the cost from US$ 1 million to 10-20 million; and the number of chemicals tested, from 2,000 to 10,000. The end result is that by 1980 only one or two new insecticides were being produced each year.

21.8.14 Pesticides in Current Use

Pesticides are viewed basically according to their chemical composition, but their mode of action is equally important. As far as food materials in storage are concerned, the main types of pesticides are as follows:

Repellants

These work by keeping the pests (usually insects) away by the use of repellant odours. On a large (commercial) scale, repellants are not used, except for pests such as mosquitoes ('Deet' etc.), biting midges and other medical vectors. But on a domestic scale, they are very popular in some regions. In India, for example, most housewives use bird chillies or neem leaves on top of their rice jars, and these do seem to be effective in keeping insect pests at bay.

Some of the inert organic compounds (see Table 21.1) used for stored grain protection (sand, boric acid, ash etc.) may work partly as physical repellants.

Antifeedants

Neem foliage is recorded to act in this manner.

Smokes

These are finely divided powders mixed with a combustible material, and are effective in an enclosed space, but have no penetrant action. Only a few chemicals are available in this formulation, but they can be useful.

Fumigants

They are volatile substances (liquids or powders) which vaporise under ambient conditions, and the gases fill the sealed stores or silos. They kill the pests via the respiratory system – lungs or tracheoles.

Stomach poisons

These are chemicals which have to be ingested and absorbed through the intestinal tract. The poison is usually mixed with an edible bait to encourage ingestion; used mainly for rodenticides, but also against ants, termites, cockroaches, slugs and snails; on foodstuffs, no bait is needed.

Contact poisons

These chemicals are absorbed directly through the insect cuticle. *Ephemeral poisons* are short-lived (days, usually), and have to be applied directly on to the insect body. The *residual* ones remain active for a long while (weeks or months), and are usually mixed into the surface layers of grain etc. in storage, in order to give it maximum protection.

Residual pesticides

These chemicals are used in two ways – firstly to disinfest the storage premises when they are empty, and secondly as grain (produce) protectants. There is a large number of insecticides available, possessing different properties and showing different levels of effectiveness and persistence, so a wide choice is, in theory, available; but in some tropical countries, there are importation limitations, and so some products may not be available. Contact pesticides are the ones particularly used in this situation.

21.8.15 Effectiveness of Pesticides

The *efficiency* of a pesticide is expressed as the more or less constant proportion of the population killed by the poison, regardless of the actual number of pests involved. The *effectiveness* is more concerned with the number of pests surviving after the treatment, and the damage they do.

Some chemicals achieve a kill rate of 80% of a test pest population, and this is generally regarded as being 'effective' by the pesticide manufacturers. However, other pesticides on sale may only achieve a kill of 70% or 60%. In the literature, it will often be stated that that chemical is 'effective against stored products insects', but often it will not state just how effective. If a kill rate of 80% qualifies a chemical as 'effective', it should be noted that this is under optimum conditions in the laboratory (using a Potter tower, for instance); in the field or produce store, there will be a range of environmental conditions, variation in dose rates and application methods, and the end result can be tremendous variation (usually reduction) in relative effectiveness.

When dealing with the more important pests, a kill rate of 95% or more is required in order to delay the population resurgence, and since the organochlorine compounds (DDT, dieldrin, HCH etc.) are no longer available, this target is not easy to achieve. The situation is exacerbated by the official withdrawal (of approval) in the USA, UK and EU of the important fumigants methyl bromide, ethylene dibromide and ethylene dichloride, on the grounds that they caused cancers in laboratory rats.

Chemicals whose use is banned in the USA, UK and Europe are still being manufactured in some cases, and the existing stock-piles are being sold in the Third World. DDT, HCH, dieldrin etc. are often still very effective in many Third World countries where resistance has not developed to any great extent, and environmental damage is minimized by the high ambient temperatures and high rainfall which successfully degrade the chemicals after use.

The chemicals used in stored products pest control are generally less effective in warm moist tropical climates where it is difficult to keep stored grains dry enough. Many insecticides that work well in Europe and the UK became biologically inactive a few weeks after application in grain store trials in West Africa. Studies have shown that one good insecticide treatment will often keep grain protected (insect free) for 6-12 months in the USA and Europe, and 10-12 months is often achieved. But in the tropics, the time expected is generally 3-6 months. Experiments in East Africa recorded that wheat at a moisture content of

10% required a certain insecticide dosage to kill the most common storage pests, but at 14% a double dose was needed, and at 18% a quadruple dosage was required for the same level of kill. The chemicals used were malathion, pirimiphos-methyl and chlorpyrifos-methyl. It has been found that 25°C is generally the optimum temperature for chemical effectiveness in stored produce. Some insecticides are quite ineffective at a temperature of 5-10°C, but pirimiphos-methyl is usually still good at a low temperature.

For further information on pesticide effectiveness, see Hill (1987).

21.8.16 Disinfestation of Premises

An empty building, silo or container should first be cleaned thoroughly and all debris removed and destroyed. Contact insecticides (such as malathion and pirimiphos-methyl) are best applied as a spray, in water, so as to thoroughly wet the walls and floor, but not to the point of run-off. There are some important differences in persistence of the chemical on different wall surfaces, such as concrete, wood, metal etc.; also, painted surfaces differ. Some pesticides are made into smoke generators (HCH, pirimiphos-methyl) which can be useful in treating high inaccessible places. Others are formulated for fogging in empty buildings. Some of the pesticides will be more persistent than others, and so will have a longer residual effect, and, as mentioned, this can be further affected by the nature of the wall surface.

21.8.17 Grain Protectants

These are the insecticides incorporated into the grain mass to protect it against insect and mite attack. The rate of application is usually calculated so that residual protection is provided right up to the time of consumption. There will clearly be some active insecticide left at the time of use, so these chemicals have to be of low mammalian toxicity, hence the range of candidate chemicals is limited.

With bulk grain storage, the insecticide is usually added as a fine spray during the loading process so as to ensure even application; sometimes as an ultra low volume spray, or using a concentrated solution (malathion). The use of water can be a problem by adding moisture to the grain. Powders and dusts avoid any moisture problems, but they may not be easy to apply uniformly to bulk grain. In small quantities, dusts and powders are easily applied. In small underground pit stores (as used in Ethiopia) there is typically a concentration of *Sitophilus* weevils in the surface 10 cm, so surface treatment may be of particular importance, especially for subsequent applications. Contact insecticides with slight fumigation action are particularly useful in grain, for the vapour stage will penetrate air spaces between the grains – included here is lindane (gamma-HCH), dichlorvos, and pirimiphos-methyl.

Bagged grain in sacks is still the main method of storage in many parts of Asia; if access between the stacked bags is sufficient, the outside of the bags can be sprayed with a concentrated solution of insecticide, and this is generally quite effective with malathion, pirimiphos-methyl and the pyrethroids. Aerosols can also be used effectively between the bags. There is usually a concentration of insects just under the surface of the sack. Often the sacks are too close together for easy access, and then fumigation will be needed. It always helps if the sacks are treated with a persistent insecticide before being filled.

The chemicals generally being used as contact and residual insecticides include gamma-HCH (lindane – now seldom officially recommended), malathion, fenitrothion, chlorpyriphos-methyl, pirimiphos-methyl, and the pyrethroids. The major publication on this topic is by Snelson (1987).

21.8.18 Fumigation

The *Manual of fumigation for insect control* (Monro, 1980) covers this subject in great detail up to the publication date. Because of space restrictions in the present text, fumigation is only reviewed very briefly, with reference to basic features.

Fumigation remains the single most useful method of disinfestations of stored products, and it does not have to be a complicated procedure. Time of exposure is an important factor, for gas penetration of bagged grain, or bulk grain, and the like is remarkably slow. It is feasible to use quite low concentrations of fumigant gas if it is possible to retain the fumigant for a lengthy time. Eggs and pupae are notoriously difficult to kill, even with fumigants, and present day recommendations stipulate longer periods of fumigation than formerly.

Fumigation in a leaky system is a waste of time, so care must be taken to ensure airtight conditions. Whole store fumigation is very effective if the premises can be completely sealed. Silo fumigation is generally effective as gas taps are incorporated into the structure. Stacks of bags can be placed on an impervious sheet on the ground and covered with another, the edges sealed and the interior flooded with methyl bromide. Previous systems just used weights to press the two sheet edges together, and this seldom made an air-tight seal, but new systems have a proper sealing mechanism. On a smaller scale, a recent commercial development includes a reusable fumigation tent closed by plastic zippers that will accommodate a small stack of bags for easy fumigation.

A number of fumigants have been in use worldwide, but two have been far more widely used that any others – these are methyl bromide (CH_3Br) and phosphine (PH_3). Methyl bromide generally does not penetrate produce well, but its circulation can be improved by the addition of CO_2 (either as a solid or a gas). There will be some physical sorption of the fumigant by the materials in the store. Because of the toxicity of methyl bromide, only government approved operators were permitted to use it for fumigation. As a less dangerous and more easily dispensed alternative, the phosphine-generating tablets produced by Detia Co. (Detia tablets, bags etc.) and Degesch Co. (Phostoxin tablets) are very effective, and their use in bulk grain is satisfactory, for the tablets can be distributed throughout the grain using the special applicator tube.

In addition to the structure being air-tight, it is important that the air volume is not too great, otherwise the fumigant gas will be too diluted. It does appear from the literature that only too often in the past, attempts at fumigation have been frustrated by carelessness; the main problems being a structure that was not air-tight, fumigant concentration too low (air volume often too large), and duration of fumigation too short.

Most fumigants are respiratory poisons taken into the insect body via the spiracles and tracheal system, so temperature is a critical factor as it controls the rate of metabolism which in turn controls the rate of respiration (oxygen uptake). Eggs and pupae are difficult to kill (especially at low temperatures) because their metabolic rate is basically low. Temperature is important in that the optimum for pesticide efficiency is 20-25°C; under 5°C, fumigation should not be attempted as the insect respiration rate is just too low, and over the range 6-15°C, insect eggs and pupae may not be killed.

Penetration of bulk grain and bags requires a considerable time; penetration rate is controlled by a combination of fumigant concentration, exposure time, and ambient temperature. As with temperature treatment of bags and bulk grain, the time factor is often grossly underestimated. Detia Co. recommend that with their standard phosphine treatment for bulk grain, the times of exposure should be as follows:

At	5-10°C	time required	10 days
	11-15	time required	5 days
	16-20	time required	4 days
	>20°C	time required	3 days

It is preferable that these times be increased – they should never be decreased.

After the fumigation, airing of buildings and containers is of prime importance to dispel the toxic gases.

In the broadest sense, the term 'fumigation' can be applied to several different techniques, namely:

Vapourising strip – dichlorvos strip hanging from beams etc.; the chemical vapourises from the strip.

Ultra low volume (ULV) fogging – this is a mist effect with tiny droplets in the air, using an ULV preparation (pirimiphos-methyl, most pyrethroids, formaldehyde).

Smoke generator – the pesticide is mixed with a combustible powder; when burning, the chemical is carried up in a smoke cloud (gamma-HCH, pirimiphos-methyl).

Volatile liquids becoming gaseous – the liquid is stored under pressure in a steel cylinder; when the pressure is released, the liquid vapourises (methyl bromide, carbon tetrachloride, carbon disulphide, hydrogen cyanide, ethylene dibromide).

Phosphine sources – the two main producers are Detia Co. with Tablets, Pellets and Bags, and Degesch Co. with Phostoxin Tablets; these are basically aluminium phosphide, which reacts with atmospheric moisture to produce phosphine gas (PH_3); tubular dispensers are available for use in bulk grain.

Reactive crystals – crystals of sodium cyanide react with an acid to generate HCN in gaseous form; calcium cyanide as a powder reacts with atmospheric moisture to release HCN.

Pesticides used at the present time, and others that were particularly successful in the past, for protection of stored grain and other stored products are shown in

Table 21.1, and Appendix 1 gives brief details of their chemical and biological properties. Rodenticides have been mentioned separately in Chapter 10. Various plant products (fresh, dried or as ashes, oils or extracts) are also used for protecting stored products (Chapter 21.9.).

Table 21.1 Pesticides widely used for stored products protection

Pesticide	*Formulation*	*Mode of action*
Inorganic (Inert) compounds		
Ash (Usually wood ash)		
Boric Acid		
Diatomaceous earth		
Sand		
Silica gel (Silica acid)		
Organochlorine compounds		
Gamma-HCH[+] (lindane) (now restricted in use, or banned, in some countries)	e.c., w.p., smoke	contact & localised fumigant action
Organophosphorus compounds		
Azamethiphos	w.p., aerosol	contact & stomach action
Chlorpyrifos-methyl	e.c., ULV	contact action
Dichlorvos	e.c., aerosol, strip	fumigant
Etrimfos	e.c., w.p	contact action
Fenitrothion	e.c., w.p., dusts	contact action
Iodofenphos	e.c., w.p., dusts	contact action
Malathion	e.c., s.c., dusts	contact (fogging)
Methacrifos	e.c., ULV, dusts	contact & vapour action
Phoxim	e.c., w.p., ULV	contact & vapour action
Pirimphos-methyl	e.c., ULV, dusts, smoke	contact (fogging)
Tetrachlorvinphos	e.c., w.p., s.c.	contact action
Trichlorfon	w.p., d.p., baits	contact & stomach action
Carbamates		
Bendiocarb	e.c., w.p., dusts	contact action
Carbaryl	e.c., w.p., dusts	contact action
Pyrethroids		
Bioallethrin	dust, aerosols	contact action
Bioresmethrin	dust, aerosols	contact action
Cypermethrin	e.c., ULV	contact (fogging)
Deltamethrin	e.c., ULV	contact (fogging) residual
Fenvalerate	e.c., ULV	contact (fogging)
Permethrin	e.c., w.p., dust, ULV, smoke	contact (fogging)
Resmethrin	e.c., w.p., ULV	contact (fogging)
Tetramethrin	e.c., d.p., aerosols	contact action (mixtures)

Insect growth regulators		
Diflubenzuron	w.p., ULV}	being developed for use against
Methoprene	e.c., s.c.}	some resistant pests
Fumigants		
Carbon dioxide		
Carbon disulphide+		
Carbon tetrachloride+		
Dichlorvos		
Ethylene dibromide+		
Ethylene dichloride+		
Formaldehyde		mostly used for glasshouse fumigation: against disease
Hydrogen cyanide		mostly used against rodents & vertebrate pests
Methyl bromide+		
Phosphine		tablets, pellets, bags of Aluminium phosphide

Indigenous plant materials
Many different plants are being used in different countries; fresh, dried, or as ashes, or as oils or extractions (see Chapter 21.9)
Rodenticides
Mentioned separately in Chapter 10

+ Use now illegal in the UK and withdrawn from use in Europe for several years.
d.p. = dispersible powder; e.c. = emulsifiable concentrate; s.c. = soluble concentrate; ULV = ultra low volume; w.p. = wettable powder

Some of the chemicals now banned in the UK, Europe and the USA (DDT, HCH, dieldrin etc.) are still being used in other countries and are often still very effective.

21.8.19 Pesticide Resistance

One of the earliest synthetic chemicals used for protection of stored products was lindane; some resistence did develop towards it, but mostly its use was curtailed along with the other organochlorines because of residue problems. Malathion was first used in stored products in the early 1960s, and has been used very widely and extensively for a long time, and resistance has developed in most of the major insect pests in most countries. Since the advent of this resistance, other chemicals have been tested and developed as replacements. Now, some of these new chemicals have been used long enough for resistance to have started in some pests. Fenitrothion and most of the pyrethroids are reported to be non-effective at previous dosages with some of the stored products pests in some regions. Resistance has even been recorded to methyl bromide (formerly thought impossible) and to phosphine!

With pests of field crops, and medical and veterinary insects, it has been found that resistance to pesticides can develop in as few as ten generations when circumstances are conducive. The main danger in pest control efforts, as far as inducing resistance is concerned, is exposure of the pest to sublethal doses. Smallholder farmers do not often use insecticides in their grain stores, but when they do, there is an unfortunate tendency sometimes to use less than the recommended dose because of the cost of the treatment, and thus the partial treatment given is both non-effective and at the same time will encourage development of resistance. One regrettable result was that in 1983, in Bangladesh, resistance to phosphine was recorded – it was attributable to the frequent use of phosphine in leaky structures where the gas concentration was insufficient to kill the pests.

Resistance to pesticides, including rodenticides, is now so widespread, involves so many pests and is developing so rapidly, that further generalisation is inadvisable; suffice it to be aware of the situation, and to seek local advice when planning a pest control programme.

For chemical methods to control insect pests of processed foods, see Gorham (1991), pp. 427-440; and an extensive review of chemicals used as grain protectants is provided by Snelson (1987).

21.9 USE OF INDIGENOUS PLANT MATERIAL (Natural Pest Control)

It has long been known by many tropical peasant farmers that certain plants appear to possess properties that drive away insects or deter them from feeding.

That plants possess qualities which enable them to resist predation (grazing) by phytophagous animals is not surprising. Plants and animals have co-evolved for many millennia, and grazing is the basis of most terrestrial food webs, so it is only to be expected that plants would evolve protective mechanisms, and, of course, that the grazers would in some cases evolve counter mechanisms, and so on. This is a topic of considerable interest in agricultural entomology, and in insect and plant ecology, and much work is in progress.

As far as natural pest control of stored products is concerned, it is only the biochemical aspects of evolved plant protective mechanisms that are relevant. The main groups of chemicals are the terpenes, tannins and certain alkaloids. The tannins are found mostly in horsetails, ferns, gymnosperms and some angiosperms, and they possess definite antibiotic properties. It is thought that the tannins were evolved as a deterrent to grazing reptiles. Alkaloids are of more recent origin and mostly found in angiosperms, and their evolution is thought to be a defence reaction to the grazing of mammals.

It is known that some 2,000 species of plants produce chemicals which have pest-repellent or controlling properties. Some of the chemicals are known to have a repellant effect on the insects, some act as anti-feedants, some are ovicidal, some cause either male or female sterility, and others appear to be either juvenile hormone analogues or mimics. In the majority of cases, however, it is apparently clear that some sort of deterrent effect is associated with the plant material, usually expressed as a lowered infestation or damage level, but the precise mode of action is usually not known.

Peasant farmers and housewives in the tropics have often used some plant material to add to stored foodstuffs to reduce insect pest populations; the earliest recorded cases are hundreds of years ago. At the present time in India, many housewives put some bird chillies into their rice storage jars, or sometimes neem leaves, and they swear that *Sitophilus* infestations are reduced.

Inert (sorptive) dusts have long been used in conjunction with local plant material by peasant farmers. Sand or brick dust may just offer some sort of physical resistance to insect movement. Wood ash is widely used, often as an admixture with sand, but ashes from different plant sources are recorded to have differing effects on the insects.

A layer of plant material on top of an underground grain store has often been recorded as being instrumental in lowering pest populations (often of *Sitophilus* weevils) and reducing damage to the grain. Usually, leaves and twigs, sometimes macerated, and usually to a depth of 5-10 cm, are the plant parts most used. Crushed seeds, or oil extracted from seeds, stems or leaves are also used. The more concentrated plant extracts or materials are often markedly more effective.

Potato tubers in store have been successfully protected against tuber moth infestation in South America by a 10 cm layer of crushed lantana foliage, and also by foliage of *Eucalyptus globulus*.

In the UK, smallholder farmers sometimes interplant *Tagetes* in greenhouse crops, and they claim that the odours from these plants keep away a range of aerial insect pests, especially aphids. If *Tagetes* can repel insects in a greenhouse, it could be expected to be equally effective in the confines of a produce store.

For peasant farmers in the tropics in situations where synthetic pesticides are not readily available, the use of natural control methods are well worth serious consideration. In recognition of this fact, there is now a great deal of research effort being expended in several different parts of the world, including the USA, India, Germany and the UK. As part of the general appropriate technology approach to peasant agriculture, the idea of natural control of storage pests is worth very careful study.

One of the first major publications on the neem tree and other plants was by Schmutter and Ascher (1984).

Some of the common plants that are recorded as being successfully used to combat stored products pests in different parts of the world are shown in Table 21.2. More than 2,000 species of plants are thought to have potential for use in natural control.

Table 21.2 Some common plants used in natural control of stored products pests

Family	*Plant species*
Chenopodiaceae	- *Chenopodium ambrosoides*
Compositae	- *Tagetes minuta*
	- *Tagetes* spp. (African marigold, etc.)
Labiatae	- *Ajuga remota*
Lauraceae	- *Cinnamomum camphora* (Camphor)
Leguminosae	- *Derris* spp. (Derris)
Lobeliaceae	- *Lobelia inflata* (Indian tobacco)
Meliaceae	- *Azadirachta indica* (Neem)
	- *Melia azedarach* (Chinaberry, Persian lilac)
Myrtaceae	- *Eucalyptus globulus*
	- *Melaleuca alterni folia* (Tea tree)
Rutaceae	- *Atalantia monophylla*
Solanaceae	- *Capsicum minimum* (Bird chilli)
	- *Datura* spp. (Thorn apples)
	- *Nicotiana tabacum* (Tobacco)
Verbenaceae	- *Latana camara* (Latana)

For more information, see Golob and Webley (1980), who gave an extensive bibliography up to 1980.

CHAPTER 22

PEST SPECTRA FOR STORED FOODSTUFFS

In many produce store surveys, the total list of species found is published without a clear indication as to which insects were on what products, and what their role was. Thus, at the present time, there is a shortage of information as to the precise food preferences shown by some of the common stored produce pests.

Also, many pest species are polyphagous and opportunistic, and will to some extent feed on whatever food is available in their location, so that little food selection is evident. But some species do show a definite selectivity for particular foods or type of food material, and these are listed below. Space limitations do not permit the listing of other than the major commodities here, and the more important pests.

22.1 PLANT MATERIALS

22.1.1 Cereals

Maize

Some species	Termites	Termitidae etc.		Foraging workers remove grains
Trogoderma granarium	Khapra beetle	Dermestidae		Primary pest; larvae
Prostephanus truncatus	Larger grain borer	Bostrychidae	}	Adults and larvae eat grains
Rhizopertha dominica	Lesser grain borer	Bostrychidae	}	
Lasioderma serricorne	Tobacco beetle	Anobiidae		Larvae damage
Carpophilus hemipterus	Dried-fruit beetle	Nitidulidae		On mouldy grain
Brachypeplus spp.	-	Nitidulidae		Africa only
Urophorus humeralis	Pineapple sap beetle	Nitidulidae		On damaged grains
Cryptolestes spp.	Red grain beetles	Cucujidae		Secondary pests
Oryzaephilus surinamensis	Saw-toothed grain beetle	Silvanidae		On damaged grains
Cathartus quadricollis	Square-necked grain beetle	Silvanidae		In USA a field pest

Cryptophagus spp.	Silky grain beetles	Cryptophagidae	On mouldy grains
Tribolium spp.	Flour beetles	Tenebrionidae	Secondary pests
Gnathocerus maxillosus	Narrow-horned flour beetle	Tenebrionidae	In field and stores
Catolethus spp.	-	Curculionidae	Central America only
Caulophilus oryzae	Broad-nosed grain weevil	Curculionidae	North and South America
Sitophilus oryzae	Rice weevil	Curculionidae	Major primary pest
Sitophilus zeamais	Maize weevil	Curculionidae	Major primary pest
Pyroderces rileyi	Pink scavenger caterpillar	Cosmopterygidae	USA only
Ephestia spp.	Warehouse moths	Pyralidae	} Larvae eat broken grains; secondary
Cadra spp.	Warehouse moths	Pyralidae	} pests
Plodia interpunctella	Indian meal moth	Pyralidae	}
Pyralis spp.	Meal moths	Pyralidae	}
Mussidia nigrivenella	-	Pyralidae	West Africa only
Mus spp.	House mice	Muridae	} Primary pests;
Rattus spp.	Rats	Muridae	} often remove grains from stores

Small grain cereals

At the present time, recorded information does not show much difference between the pests of most small grain cereals in the tropics; in temperate regions, the stored produce cereal pests are relatively few in number. The minor secondary pests eating grain fragments, such as crickets, psocids, cockroaches etc. are not included in the table.

Macrotermes and *Odontotermes* spp.	Termites	Termitidae	} Foraging workers remove grains
Coptotermes spp.	Wet-wood termites	Rhinotermitidae	} from stores
Trogoderma granarium	Khapra beetle	Dermestidae	Primary pest; larvae
Lasioderma serricorne	Tobacco beetle	Anobiidae	Larvae damage grain
Rhizopertha dominica	Lesser grain borer	Bostrychidae	Adults and larvae damage grains
Lophocateres pusillus	Siamese grain beetle	Lophocateridae	Tropical minor pest

Tenebriodes mauritanicus	Cadelle	Trogossitidae	Larvae minor pests
Carpophilus spp.	Dried-fruit beetles	Nitidulidae	On mouldy grains
Cryptolestes spp.	Red grain beetles	Cucujidae	Secondary pests
Cryptolestes ferrugineus	Rust-red grain beetle	Cucujidae	Pest in the UK
Oryzaephilus surinamensis	Saw-toothed grain beetle	Silvanidae	On damaged grain; occurs in UK
Cathartus quadricollis	Square-necked grain beetle	Silvanidae	Minor pest in stores
Ahasverus advena	Foreign grain beetle	Silvanidae	Scavenger mostly
Cryptophagus spp.	Fungus beetles	Cryptophagidae	On mouldy grains
Typhaea stercorea	Hairy fungus beetle	Mycetophagidae	Tropical mostly; on mouldy grain
Latheticus oryzae	Long-headed flour beetle	Tenebrionidae	} Secondary pests; attack damaged
Tribolium spp.	Flour beetles	Tenebrionidae	} grains mostly
Gnathocerus cornutus	Broad-horned flour beetle	Tenebrionidae	}
Sitophilus oryzae	Rice weevil	Curculionidae	Tropical; primary
Sitophilus granaries	Grain weevil	Curculionidae	Temperate; primary
Contarinia sorghicola	Sorghum midge	Cecidomyiidae	Sorghum grains only
Nemapogon granella	Corn moth	Tineidae	Temperate; primary
Endrosis sarcitrella	White-shouldered house moth	Oecophoridae	} On damaged grain mostly; secondary
Hofmannophila pseudospretella	Brown house moth	Oecophoridae	} pests
Anchonoma xerula	Grain moth	Oecophoridae	Japan
Sitotroga cerealella	Angoumois grain moth	Gelechiidae	In field then store
Cadra cautella	Tropical warehouse moth	Pyralidae	Polyphagous; more tropical
Ephestia elutella	Warehouse moth	Pyralidae	Polyphagous; more temperate
Pyralis farinalis	Meal moth	Pyralidae	Cosmopolitan pest
Plodia interpunctella	Indian meal moth	Pyralidae	Subtropics mostly
Corcyra cephalonica	Rice moth	Pyralidae	Humid tropics
Acarus siro	Flour mite	Acaridae	} Abundant in grain
Glycyphagus destructor	Grain mite	Acaridae	} stores (in UK)

Cernula virgata etc.	Small banded	Helicidae	}	Contaminants in harvested grain
Theba spp.	snails	Helicidae	}	
Passer spp.	Sparrows	Ploceidae	}	Granivorous birds
Ploceus spp.	Weavers	Ploceidae	}	invade stores
Columba spp.	Pigeons	Columbidae		Invade grain stores
Streptopelia spp.	Doves	Columbidae		Eat spilled grain mostly
Mus spp.	House mice	Muridae	}	Invade stores to
Rattus spp.	Rats	Muridae	}	steal grain

22.1.2 Cereal Products (flours, meals, bread, biscuits, pasta etc.)

Lepisma sacharina	Silverfish	Lepismatidae		Scavenger
Several species	Cockroaches	Blattidae etc.		General scavengers
Liposcelsis etc.	Booklice	Psocidae		Mouldy produce
Attagenus spp.	Carpet beetles	Dermestidae		Animal food mostly
Lasioderma serricorne	Tobacco beetle	Anobiidae	}	Polyphagous and cosmopolitan
Stegobium paniceum	Biscuit beetle	Anobiidae	}	
Ptinus tectus	Australian spider beetle	Ptinidae	}	More temperate; polyphagous
Other species	Spider beetles	Ptinidae	}	
Rhizopertha dominica	Lesser grain borer	Bostrychidae		Warmer countries
Tenebriodes mauritanicus	Cadelle	Trogossitidae		Widespread
Cryptolestes ferrugineus	Rust-red grain beetle	Cucujidae		Common in the UK
Cryptolestes spp.	Red grain beetles	Cucujidae		Tiny; polyphagous
Oryzaephilus spp.	Grain beetles	Silvanidae		Polyphagous
Ahasverus advena	Foreign grain beetle	Silvanidae		Scavenger mostly
Typhaea stercorea	Hairy fungus beetle	Mycetophagidae		Mouldy produce
Latheticus oryzae	Long-headed flour beetle	Tenebrionidae	}	Feed on a wide range of
Gnathocerus cornutus	Broad-horned flour beetle	Tenebrionidae	}	farinaceous materials
Tenebrio spp.	Mealworm beetles	Tenebrionidae		Temperate; large
Tribolium spp.	Flour beetles	Tenebrionidae		Serious pests

Palorus spp.	Flour beetles	Tenebrionidae	Common minor pests
Sitophilus granarius	Grain weevil	Curculionidae	Temperate
Sitophilus spp.	Maize and rice weevils	Curculionidae	More tropical
Endrosis sarcitrella	White-shouldered house moth	Oecophoridae	} Damage a wide range of produce
Hofmannophila pseudospretella	Brown house moth	Oecophoridae	} by larval feeding
Pyralis farinalis	Meal moth	Pyralidae	} Larvae on wide
Pyralis pictalis	Painted meal moth	Pyralidae	} range of produce
Cadra cautella	Tropical warehouse moth	Pyralidae	Larvae polyphagous
Ephestia elutella	Warehouse moth	Pyralidae	More temperate
Ephestia kuehniella	Mediterranean flour moth	Pyralidae	Subtropical
Plodia interpunctella	Indian meal moth	Pyralidae	Warmer countries
Monomorium pharaonis	Pharaoh ant	Formicidae	Polyphagous
Acarus siro	Flour mite	Acaridae	} Cosmopolitan;
Glycyphagus spp.	House mites etc.	Acaridae	} polyphagous and
Suidasia spp.	-	Acaridae	} important pests
Tyrophagus spp.	Cheese and mould mites	Acaridae	Protein foods
Passer spp.	Sparrows	Ploceidae	Eat flour and bread
Mus spp.	House mice	Muridae	} Ubiquitous and
Rattus spp.	Rats	Muridae	} polyphagous

Farinaceous materials are usually attacked by a wider range of stored product pests than any other commodity stored by man; in point of fact, almost all of the animal species mentioned in this text will feed on these materials to some extent, especially meals, so the total list of pest species is very long.

22.1.3 Pulses

It is evident that there are appreciable differences in feeding preference as far as the insect pests of pulses are concerned, but to date there are insufficient records to warrant a separate approach.

Acanthoscelides obtectus	Bean bruchid	Bruchidae	*Phaseolus* mostly
Bruchus spp.	Pea and bean bruchids	Bruchidae	Different legumes

Callosobruchus spp.	Cowpea bruchids	Bruchidae	Several hosts
Specularius spp.	-	Bruchidae	Mostly field pests
Caryedon serratus	Groundnut beetle	Bruchidae	Groundnut mostly
Zabrotes subfasciatus	Mexican bean beetle	Bruchidae	*Phaseolus* etc.
Lophocateres pusillus	Siamese grain beetle	Lophocateridae	Tropical; uncommon
Apion spp.	Seed weevils	Apionidae	In legume seeds
Melanagromyza spp.	Pulse pod flies	Agromyzidae	Inside pods
Hofmannophila pseudospretella	Brown house moth	Oecophoridae	Larvae eat seeds (groundnuts)
Cydia nigricana	Pea moth	Tortricidae	Larvae eat peas
Laspeyresia glycinivorella	Soya bean pod borer	Tortricidae	Larvae in pods of soya bean
Etiella zinckenella	Pea pod borer	Pyralidae	} Larvae in pods of
Maruca testulalis	Mung moth	Pyralidae	} different legumes
Cernuella virgata	Small banded snail	Helicidae	Contaminate peas for freezing (UK)

22.1.4 Roots and Tubers

Potato (*Solanum tuberosum*)

Aulacorthum solani	Potato aphid	Aphididae	} Infest eyes of potato tubers in
Myzus persicae	Peach potato aphid	Aphididae	} storage, later on sprouting shoots
Rhopalosiphoninus latysiphon	Bulb and potato aphid	Aphididae	}
Melolontha spp.	Chafer grubs	Scarabaeidae	Eat holes in tubers
Agriotes spp.	Wireworms	Elateridae	Small holes eaten
Pnyxia scabiei	Potato scab gnat	Sciaridae	Larvae on tubers
Eumerus spp.	Small bulb flies	Syrphidae	Maggots in tubers
Phthorimaea operculella	Potato tuber moth	Gelechiidae	Larvae bore tubers; major pest
Pyralis spp.	-	Pyralidae	} Larvae found
Euzophera osseatella	Eggplant stem borer	Pyralidae	} occasionally boring in tubers
Agrotis segetum etc.	Common cutworms	Noctuidae	Larvae bite shallow hole in tubers
Deroceras reticulatum and other species	Field slugs	Limacidae	Deep tunnel in tubers

Meloidogne spp.	Root knot nematodes	Heterodidae	Swollen galls on tubers
Cassava			
Prostephanus truncatus	Larger grain borer	Bostrychidae	} Adults bore dried tubers, and larvae
Rhizopertha dominica	Lesser grain borer	Bostrychidae	} develop inside breeding galleries
Dinoderus spp.	Small bamboo borers	Bostrychidae	}
Heterobostrychus spp.	Black borers	Bostrychidae	}
Araeceus fasciculatus	Coffee bean weevil	Anthribidae	Larvae in tubers
Lyctus brunneus	Powder-post beetle	Lyctidae	Larvae bore tubers
Lophocateres pusillus	Siamese grain beetle	Lophocateridae	S.E. Asia mostly
Cryptolestes spp.	Red grain beetles	Cucujidae	Cassava chips mostly
Pyralis manihotalis	Grey pyralid	Pyralidae	Larvae eat tubers
Sweet potato			
Cylas spp.	Sweet potato weevils	Apionidae	Larvae and adults in tunnels in tubers
Yams			
Prionoryctes spp. etc.	Yam beetles	Scarabaeidae	Adults and larvae tunnel tubers
Meloidogyne spp.	Root knot nematodes	Heterodidae	Swollen galls on tubers
Ginger			
Caulophilus oryzae	Broad-nosed grain weevil	Curculionidae	Larvae in rhizome
Eumerus spp.	Small bulb flies	Syrphidae	Larvae bore in rhizomes
Several species	Rhizome flies	Several families	Maggots bore in rhizomes

22.1.5 Vegetables

Brassicas

Brevicoryne brassicae	Cabbage aphid	Aphididae	Mealy aphids in small colonies
Ceutorhynchus pleurostigma	Turnip gall weevil	Curculionidae	Larva in stem gall
Lixus spp.	Cabbage stem weevils	Curculionidae	Larvae gall stem/root
Delia radicum	Cabbage root fly	Anthomyiidae	Larvae in stem and Sprout buttons
Phytomyza spp. etc.	Cabbage leaf miners	Agromyzidae	Larvae mine leaves
Plutella xylostella	Diamond-back moth	Yponomeutidae	Larvae eat leaves and pupate on plant
Hellula spp.	Cabbage webworms	Pyralidae	} Smallish caterpillars eat
Evergestis spp.	Cabbageworms	Pyralidae	} foliage; may be
Crocidolomia binotalis	Cabbage cluster caterpillar	Pyralidae	} found in heart
Pieris spp.	Cabbage butterflies	Pieridae	} Larger caterpillars eat
Mamestra spp.	Cabbage moth etc.	Noctuidae	} leaves and contaminate
Trichoplusia ni	Cabbage semi-looper	Noctuidae	} foliage
Deroceras reticulatum	Field slug	Limacidae	Slugs infest foliage

Onions

Eumerus spp.	Small bulb flies	Syrphidae	Maggots inside bulb
Merodon spp.	Large bulb flies	Syrphidae	Usually single larva in bulb
Delia antiqua	Onion fly	Anthomyiidae	Maggots inside bulb
Rhizoglyphus echinopus	Bulb mite	Acaridae	Mites between scales

Beetroot

Meloidogyne spp.	Root-knot nematodes	Heterodidae	Gall roots

Turnips and Mangels

Ceutorhynchus pleurostigma	Turnip gall weevil	Curculionidae	Larva in root gall

Carrot

Eumerus spp.	Small bulb flies	Syrphidae	Maggots in root
Psila rosae	Carrot fly	Psilidae	Maggots bore root

22.1.6 Fruits

Citrus

Planococcus spp. etc.	Citrus mealybugs	Pseudococcidae	Infest eye and stalk base mostly
Aonidiella spp. etc.	Armoured scales	Diaspididae	On fruit surface; very small
Drosophila spp.	Small fruit flies	Drosophilidae	Infest ripe/over-ripe fruits
Ceratitis spp.	Fruit flies	Tephritidae	} Maggots inside
Dacus spp.	Fruit flies	Tephritidae	} fruit
Prays endocarpa	Citrus rind borer	Yponomeutidae	Larvae mine rind (S.E. Asia)
Cryptophlebia leucotreta	False codling moth	Tortricidae	Caterpillar bores in fruit
Citripestis sagittiferella	Citrus fruit borer	Pyralidae	Caterpillars in fruits (S.E. Asia)
Cryptoblabes gnidiella	-	Pyralidae	In fruits from Mediterranean
Ectomyelois ceratoniae	Locust bean moth	Pyralidae	Larvae in fruits
Paramyelois transitella	Navel orangeworm	Pyralidae	Larvae in fruits in USA
Phyllocoptruta oleivora	Citrus rust mite	Eriophyidae	Mites infest skin surface

Apple

Rhagoletis spp. etc.	Fruit flies	Tephritidae	Maggots in fruits; temperate
Argyresthia spp.	Fruit tortrix moths	Tortricidae	Larvae in fruits
Cydia pomonella	Codling moth	Tortricidae	} Caterpillars in
Cydia prunivora	Lesser appleworm	Tortricidae	} fruits; may pupate in stores

Peach

Drosophila spp.	Small fruit flies	Drosophilidae		Maggots in ripe fruits
Ceratitis spp.	Fruit flies	Tephritidae	}	Maggots inside
Dacus spp.	Fruit flies	Tephritidae	}	developing fruits
Anarsia lineatella	Peach twig borer	Gelechiidae		Larvae inside fruits
Cydia molesta	Oriental fruit moth	Tortricidae		Larvae in fruits
Euzophera bigella	Quince moth	Pyralidae		Larvae in fruit flesh or stone.

Pineapple

Dysmicoccus brevipes	Pineapple mealybug	Pseudococcidae		Bugs infest fruit
Urophorus humeralis	Pineapple sap beetle	Nitidulidae	}	Attracted by sweet sap and found in
Drosophila spp.	Small fruit flies	Drosophilidae	}	tinned fruits
Tapinoma spp. etc.	Sugar ants	Formicidae	}	

Dried fruits *(raisins, currants, dates, figs, apricots etc.)*

Lasioderma serricorne	Tobacco beetle	Anobiidae		Larvae eat fruits
Carpophilus hemipterus	Dried fruit beetle	Nitidulidae	}	Larvae and adults feed on fruits:
Carpophilus spp.	Sap beetles	Nitidulidae	}	major pests
Oryzaephilus surinamensis	Saw-toothed grain beetle	Silvanidae		Adults and larvae eat fruits
Cathartus quadricollis	Square-necked grain beetle	Silvanidae		Minor pest usually
Tribolium castaneum	Red flour beetle	Tenebrionidae	}	Larvae and adults eat dried fruits
Blapstinus spp.	-	Tenebrionidae	}	
Coccotrypes dactyliperda	Date stone borer	Scolytidae		Larvae in unripe fruits, adults emerge in stores
Lonchaea aristella	Fig fly	Lonchaeidae		In dried figs
Nemapogon granella	Corn moth	Tineidae	}	Small caterpillars eat a wide range
Tinea spp.	Clothes moths	Tineidae	}	of dried fruits in
Hofmannophila pseudospretella	Brown house moth	Oecophoridae	}	storage; usually pupate in the
Corcyra cephalonica	Rice moth	Pyralidae	}	stored produce

Arenipses sabella	Greater date moth	Pyralidae	}
Cadra calidella	Dried fruit moth	Pyralidae	}
Cadra cautella	Dried currant moth	Pyralidae	}
Cadra figulilella	Raisin moth	Pyralidae	}
Ephestia elutella	Cocoa moth	Pyralidae	}
Paralipsa gularis	Stored nut moth	Pyralidae	}
Glycyphagus spp.	House mites	Acaridae	Infest dried fruits

Nuts

At present, many records are from unspecified nuts, but it is clear that food preferences exist and may be pronounced. Groundnuts are listed under pulses.

Araecerus fasciculatus	Coffee bean weevil	Anthribidae	Polyphagous
Curculio nucum etc.	Hazelnut weevils	Curculionidae	Larval emergence holes in shell
Archips argyrospilus	Fruit tree leaf roller	Tortricidae	}
Laspeyresia caryana	Hickory shuckworm	Tortricidae	} Larvae in kernels
Cryptophlebia ombrodelta	Macadamia nut borer	Tortricidae	}
Melissopus latiferreanus	Filbertworm	Tortricidae	Walnuts, hazelnuts
Ectomyelois ceratoniae	Locust bean moth	Pyralidae	}
Paralipsa gularis	Stored nut moth	Pyralidae	} Most larvae need
Paramyelois transitella	Navel orangeworm	Pyralidae	} nut shell to be damaged for
Corcyra cephalonica	Rice moth	Pyralidae	} access to kernel
Cadra cautella	Tropical warehouse moth	Pyralidae	}
Plodia interpunctella	Indian meal moth	Pyralidae	On Brazil nuts
Nemapogon granella	Corn moth	Tineidae	Larvae polyphagous
Solenopsis xyloni	Southern fire ant	Formicidae	Almonds in the USA
Acarus siro	Flour mite	Acaridae	On nut kernels
Several species	Rodents	Muridae etc.	Invade stores

22.1.7 *Oil Seeds and Copra*

Oil seeds

Many records refer to oilseeds in general, although some refer specifically to cotton seed, groundnuts, and oil palm kernels.

Trogoderma granarium	Khapra beetle	Dermestidae	}	
Lasioderma serricorne	Tobacco beetle	Anobiidae	}	
Tenebriodes mauritanicus	Cadelle	Trogossitidae	}	Most damage is done by feeding
Phradonoma spp.	-	Dermestidae	}	larvae, but some
Cryptolestes ferrugineus	Rust-red grain beetle	Cucujidae	}	adults also feed on the oilseeds
Oryzaephilus mercator	Merchant grain beetle	Silvanidae	}	
Necrobia rufipes	Copra beetle	Cleridae	}	
Latheticus oryzae	Long-headed flour beetle	Tenebrionidae	}	
Pectinophora gossypiella	Pink bollworm	Glelchiidae		In cotton seed
Promalactis inonisema	Cotton seedworm	Oecophoridae		Cotton seed (Japan)
Aglossa ocellalis	-	Pyralidae		Oil palm (West Africa)

Copra

Dermestes ater	-	Dermestidae	Larval damage mostly
Necrobia rufipes	Copra beetle	Cleridae	Often very abundant
Oryzaephilus surinamensis	Saw-toothed grain beetle	Silvanidae	Abundant, and widespread
Doloessa viridis	-	Pyralidae	Larvae are minor pests

Rape

Recent surveys in the UK revealed almost no insects but a total of 21 species of mite; only two species were particularly abundant.

Acarus siro	Flour mite	Acaridae	}	Pest status and
Glycyphagus destructor	-	Acaridae	}	damage uncertain

22.1.8 Beverages and Spices

Tea is not included here as a separate commodity because of insufficient data at the present time.

Coffee beans

Phradonoma spp.	-	Dermestidae	} Larvae damage
Tribolium castaneum	Red flour beetle	Tenebrionidae	} beans
Araecerus fasciculatus	Coffee bean weevil	Anthribidae	Common pest; serious
Hypothenemus hampei	Coffee berry borer	Scolytidae	Adults bore cherries
Heterobostrychus spp.	Black borers	Bostrychidae	Uncommon; adults bore beans

Cocoa beans

Trogoderma granarium	Khapra beetle	Dermestidae	} Larvae do damage; adults
Lasioderma serricorne	Tobacco beetle	Anobiidae	} seldom
Ptinus tectus	Australian spider beetle	Ptinidae	Adults and larvae feed
Necrobia rufipes	Copra beetle	Cleridae	Widespread pest
Brachypeplus spp.	-	Nitidulidae	Africa only
Carpophilus spp.	Dried fruit beetles	Nitidulidae	Adults and larvae pests
Cryptolestes spp.	Red grain beetles	Cucujidae	Several species
Oryzaephilus mercator	Merchant grain beetle	Silvanidae	Adults and larvae feed
Ahasverus advena	Foreign grain beetle	Silvanidae	Scavenger mainly
Cathartus quadricollis	Square-necked grain beetle	Silvanidae	Minor pest usually
Tribolium spp.	Flour beetles	Tenebrionidae	Adults and larvae pests
Araecerus fasciculatus	Coffee bean weevil	Anthribidae	Adults and larvae pests
Setomorpha rutella	-	Tineidae	West Africa only
Corcyra cephalonica	Rice moth	Pyralidae	Humid tropics mostly
Doloessa viridis	-	Pyralidae	Not common
Cadra cautella	Cocoa moth	Pyralidae	Serious pest
Ephestia elutella	Warehouse moth	Pyralidae	Temperate regions

Chocolate

Stegobium paniceum	Biscuit beetle	Anobiidae	Larval feeding
Corcyra cephalonica	Rice moth	Pyralidae	Humid tropics mostly
Plodia interpunctella	Indian meal moth	Pyralidae	Larval feeding only

Tobacco

Lasioderma serricorne	Tobacco beetle	Anobiidae	Larvae eat leaves; attack cigarettes
Cathartus quadricollis	Square-necked grain beetle	Silvanidae	Minor pest usually
Phthorimaea operculella	Potato tuber moth	Gelechiidae	Larvae mine leaves
Setomorpha rutella	-	Tineidae	West Africa only
Ephestia elutella	Warehouse moth	Pyralidae	More temperate
Glycyphagus domesticus	House mite	Acaridae	All stages infest

Spices

At present, few records actually specify which spices were damaged; there will be major differences because some are dried leaves, others are seeds, and many different plant families are involved.

Coccus hesperidum	Soft brown scale	Coccidae	} Scales found on bay leaves in
Ceroplastes sinensis	Pink waxy scale	Coccidae	} Europe
Lasioderma serricorne	Tobacco beetle	Anobiidae	Sometimes found
Stegobium paniceum	Drugstore beetle	Anobiidae	Coriander seeds etc.
Necrobia rufipes	Copra beetle	Cleridae	Some spices attacked
Oryzaephilus surinamensis	Saw-toothed grain beetle	Silvanidae	Some beans eaten
Palorus subdepressus	Depressed flour beetle	Tenebrionidae	Pepper and ginger attacked
Araecerus fasciculatus	Nutmeg beetle	Anthribidae	Larvae and adults feed on seeds
Corcyra cephalonica	Rice moth	Pyralidae	Humid tropics; on seeds mostly
Plodia interpunctella	Indian meal moth	Pyralidae	Larval feeding

22.1.9 Miscellaneous Plant Materials

Mushrooms *(dried fungi)*

Mycetophila spp.	Mushroom midges	Mycetophilidae	} Part of the
Sciara spp. etc.	Mushroom flies	Sciaridae etc.	} mushroom maggot complex
Nemapogon granella	Corn moth	Tineidae	} Larvae feed on fungal fruiting
Nemapogon spp.	Fungus moths	Tineidae	} bodies
Rhizoglyphus echinopus	Bulb mite	Acaridae	All stages feed on the fungus

Cork *(including corks in wine bottles)*

Hofmannophila pseudospretella	Brown house moth	Oecophoridae	Serious damage recorded
Cadra calidella	Raisin moth	Pyralidae	Larvae recorded eating corks

Book bindings

Lepisma saccharina	Silverfish	Lepismatidae	Slight damage
Periplaneta americana	American cockroach	Blattidae	Severe damage; gnaw bindings
Several species	Booklice	Psocoptera	Slight damage
Hofmannophila pseudospretella	Brown house moth	Oecophoridae	Caterpillars can bore bindings

Dried seaweeds

Hofmannophila pseudospretella	Brown house moth	Oecophoridae	Caterpillars eat materials

Flower bulbs and corms

Eumerus spp.	Lesser bulb flies	Syrphidae	} Maggots in bulbs
Merodon spp.	Large bulb flies	Syrphidae	} of Liliaceae mostly
Stenotarsonemus laticeps	Bulb scale mite	Tarsonemidae	Between bulb scales
Rhizoplyphus echinopus	Bulb mite	Acaridae	Mites between bulb scales

Bean curd

Fannia canicularis	Lesser house fly	Fanniidae	Maggots in the curd

Soya sauce

Drosophila spp.	Vinegar flies	Drosophilidae	Adults attracted by fermentation
Sarcophaga spp.	Flesh flies	Sarcophagidae	Larvae deposited in fluid
Fannia canicularis	Lesser house fly	Fanniidae	Larvae can be expected to infest soya sauce.

22.2 ANIMAL MATERIALS

22.2.1 Fresh Meat

Musca domestica	House fly	Muscidae	}	Eggs laid on
Calliphora spp. etc.	Blowflies	Calliphoridae	}	exposed flesh - termed 'fly
Lucilia spp.	Greenbottles etc.	Calliphoridae	}	blown'; larvae eat flesh; adults feed on surface.
Sarcophaga spp.	Flesh flies	Sarcophagidae		Larviparous
Iridomyrmex spp.	House and meat ants	Formicidae		Often industrial premises infested
Mus musculus	House mice	Muridae	}	Recorded feeding
Rattus spp.	Rats	Muridae	}	on frozen carcasses in stores

Game animals *(birds and mammals)*

Calliphora spp. etc.	Blowflies	Calliphoridae	}	Hanging carcasses 'fly blown'
Lucilia spp.	Greenbottles etc.	Calliphoridae	}	

22.2.2 Dried Meats (biltong, sausages, salami, pressed beef etc.)

Dermestes spp.	Hide beetles	Dermestidae	}	Adults and larvae
Attagenus spp.	Carpet beetles	Dermestidae	}	feed on the meats
Necrobia spp.	Bacon and copra beetles	Cleridae		Somewhat omnivorous
Musca domestica	House fly	Muscidae	}	Eggs laid on
Lucilia spp.	Greenbottles etc.	Calliphoridae	}	exposed
Piophila casei	Cheese skipper	Piophilidae	}	meatstuffs and maggots develop
Pyralis manihotalis	Grey pyralid	Pyralidae		Larvae omnivorous

Iridomyrmex spp.	Meat ants	Formicidae		Infest processing plants
Mus musculus	House mice	Muridae	}	Feed on a wide
Rattus spp.	Rats	Muridae	}	range of meat products

22.2.3 Skins and Hides (basically unprocessed)

Dermestes spp.	Hide beetles	Dermestidae	}	Adults and larvae
Attagenus spp.	Fur and carpet beetles	Dermestidae	}	feed and make holes in hides
Tribolium castaneum	Rust-red flour beetle	Tenebrionidae	}	
Tinea pellionella	Case-bearing clothes moth	Tineidae	}	Caterpillars eat keratin, and hole
Tinea spp.	Clothes moths	Tineidae	}	skins
Tineola bisselliella	Common clothes moth	Tineidae	}	
Monopsis spp.	-	Tineidae	}	
Lardoglyphus konoi	-	Acaridae		Protein feeders

22.2.4 Dried Fish (including fishmeal)

Nauphoeta cinerea	Lobster cockroach	Blaberidae		Found in fishmeals
Dermestes spp.	Hide beetles	Dermestidae		Adults and larvae feed on dried fish/meal
Dermestes frischii	-	Dermestidae		Infest salted fish
Necrobia spp.	Bacon and copra beetles	Cleridae		Larvae eat dried fish
Musca domestica	House fly	Muscidae	}	More abundant in
Calliphora spp.	Blowflies	Calliphoridae	}	fishmeal than on
Lucilia spp.	Greenbottles etc.	Calliphoridae	}	dried fish; salt-sensitive
Sarcophaga spp.	Flesh flies	Sarcophagidae		Larviparous
Piophila casei	Cheese skipper	Piophilidae		Larvae in meal etc.
Tineola bisselliella	Common clothes moth	Tineidae		Larvae prefer meal
Acarus siro	Flour mite	Acaridae	}	Adults and
Lardoglyphus konoi	-	Acaridae	}	nymphs feed; more abundant on
Tyrophagus spp.	-	Acaridae (s.l.)	}	meal

Mus spp.	House mice	Muridae	} Will feed on both
Rattus spp.	Rats	Muridae	} dried fish and fishmeal

(No pests of dried Mollusca yet recorded)

22.2.5 Wool and Furs (including feathers)

Blattella germanica	German cockroach	Epilampridae	Adults and nymphs feed
Dermestes spp.	Carpet beetles	Dermestidae	} Adults and
Attagenus spp.	Fur and carpet beetles	Dermestidae	} larvae feed on keratin in most
Anthrenus spp.	Carpet beetles	Dermestidae	} of its forms
Tinea pellionella	Case-bearing clothes moth	Tineidae	} Caterpillars feed on keratin;
Tinea spp.	Clothes moths	Tineidae	} produce silk and
Tineola spp.	Clothes moths	Tineidae	} may make silken
Monopsis spp.	-	Tineidae	} galleries
Trichophaga spp.	Tapestry moths	Tineidae	Prefer coarser fibres and hairs
Endrosis sarcitrella	White-shouldered house moth	Oecophoridae	} Larvae polyphagous on many different
Hofmannophila pseudospretella	Brown house moth	Oecophoridae	} materials; very abundant pests

(Several species of Pyralinae live in animal and bird nests, and may eat some keratin)

22.2.6 Bones and Bonemeal

Macrotermes spp. (?)	Termites	Termitidae	Africa; eat animal skulls
Phradonoma spp.	-	Dermestidae	Omnivorous
Necrobia spp.	Copra/ bacon beetles	Cleridae	Protein foods
Pyralis manihotalis	Grey pyralid	Pyralidae	Larvae omnivorous
Several species (see Fresh Meat)	Domestic flies	Muscidae and Calliphoridae etc.	Maggots common on fresh bones in stores; sometimes in bonemeal
Lardoglyphus konoi	-	Acaridae	More abundant in bonemeal

22.2.7 Bacon and Hams

Dermestes spp.	Hide beetles	Dermestidae		Adults and larvae feed together
Necrobia ruficollis	Red-necked bacon beetle	Cleridae	}	Larvae burrow in fatty parts mostly;
Necrobia rufipes	Copra beetle	Cleridae	}	adults feed on surface
Lucilia spp.	Greenbottles etc.	Calliphoridae	}	Maggots feed in
Piophila casei	Cheese skipper	Piophilidae	}	the meat and fat
Tyrophagus spp.	Cheese and bacon mites	Acaridae (s.l.)		Several species feed on bacon

22.2.8 Cheeses

Dermestes spp.	Hide beetles	Dermestidae		Several species recorded
Necrobia spp.	Copra and bacon beetles	Cleridae		Larvae tunnel; adults on surface
Piophila casei	Cheese skipper	Piophilidae		Maggots bore in cheeses
Acarus siro	Flour mite	Acaridae	}	Adults and
Glycyphagus spp.	House mites	Acaridae	}	nymphs infest
Tyrophagus spp.	Cheese mites	Acaridae	}	surface and feed there
Mus musculus	House mouse	Muridae	}	Sometimes show
Rattus spp.	Rats	Muridae	}	feeding preferences for cheeses

22.2.9 Oyster Sauce

Musca domestica	Housefly	Muscidae		Adult and larval contamination
Fannia canicularis	Lesser housefly	Fanniidae		Larvae in sauce

22.2.10 Dried Milk

Tribolium castaneum	Rust-red flour beetle	Tenebrionidae		Adults and larvae
Trogoderma glabrum	-	Dermestidae	}	Larvae only feed
Trogoderma variabile	Warehouse beetle	Dermestidae	}	

Acarus farris	-	Acaridae	} All stages found
Tyrophagus putrescentiae	Mould mite	Acaridae	}

22.3. BUILDING AND PACKAGING MATERIALS

22.3.1 Wood (structural timbers, packing cases)

Macrotermes spp.	Termites	Termitidae	Not commonly found
Coptotermes spp.	Wet-wood termites	Rhinotermitidae	Attack damp wood only
Cryptotermes brevis	Dry-wood termites	Kalotermitidae	Uncommon species, in dry wood
Anobium punctatum	Furniture beetle	Anobiidae	Larvae tunnel wood
Xestobium rufovillosum	Death-watch beetle	Anobiidae	Larvae in hard-wood (oak etc.)
Lyctus brunneus	Powder-post beetle	Lyctidae	Often in plywood
Chalcophora spp.	Pine jewel beetles	Buprestidae	Larvae in *Pinus* timber
Hylotrupes bajulus	House longhorn	Cerambycidae	Domestic species in pine timbers
Monochamus spp.	Pine sawyers	Cerambycidae	Larvae in pine timbers
Sipalinus spp.	Pine weevils	Curculionidae	Larvae in *Pinus* timber
Xylocopa spp.	Carpenter bees	Xylocopidae	Breeding tunnels in beams and rafters in tropics

22.3.2 Bamboo

Dinoderus spp.	Small bamboo borers	Bostrychidae	Adults bore stems
Chlorophorus annularis	Bamboo longhorn	Cerambycidae	Larvae bore stem internodes
Xylocopa irridipennis	Bamboo carpenter bee	Xylocopidae	Adults bite holes in internodes

22.3.3 Sacking (Hessian etc.)

Coptotermes spp.	Wet-wood termites	Rhinotermitidae	Damp sacking eaten
Tinea spp.	Clothes moths	Tineidae	} Caterpillars
Tineola spp.	Clothes moths	Tineidae	} polyphagous on
Endrosis sarcitrella	White-shouldered house moth	Oecophoridae	} plant materials mostly; will
Hofmannophila pseudospretella	Brown house moth	Oecophoridae	} spread to produce packing

22.3.4 Straw Dunnage

Pyralis farinalis	Meal moth	Pyralidae	} Larvae of several
Several species	-	(Pyralinae)	} species feed on dried grasses (thatch) in the wild
Glycyphagus destructor	-	Acaridae	Adults and nymphs infest dry grasses

Many records of pests (rats, beetles etc.) being carried to another continent or oceanic islands in straw dunnage in the hold of a ship.

APPENDIX 1

PESTICIDES USED FOR STORED PRODUCTS PROTECTION

The chemicals currently (and recently) used widely are quite few in number, for a basic requirement is for a broad-spectrum insecticide and acaricide which will effectively kill all the insects and mites commonly found in stored grain and other products. For produce stores and warehouses, it is clearly advantageous to be able to have one basic treatment for the whole place, which is why fumigation with methyl bromide or phosphine was so widely practiced.

But there will be occasions when commodities such as dried fish are infested with maggots, or cockroaches or rats are a nuisance, or a tobacco store is infested, and then more specific treatments are needed, and different pesticides may be recommended.

The chemicals listed here are the general ones that are currently being used, or more recently widely used, in stored produce protection against insect and mite pests. Several important pesticides have recently been banned, both in the USA and the UK following reports of carcinogenic properties (methyl bromide, ethylene dibromide etc.), and the last of the organo-chlorine compounds (HCH – Lindane) were withdrawn from use because of environmental risks. But the stockpiles of these chemicals are still being sold abroad, and in some tropical countries are still very effective. In the hot tropics, Lindane does not pose the same environmental risks that it did in the UK and North America because of its more rapid degradation. It must be remembered that in some regions there is widespread resistance shown by many common pests to the major insecticides in use, so local advice should always be sought when contemplating pesticide control measures.

A warning is given if any pesticide is particularly toxic to fish, for many tropical countries either have a freshwater fishing industry, or else farmers rear fish in ponds or paddy fields, and there have been many unfortunate instances in the past of farmers dumping used pesticide containers into the local freshwater bodies. Many on-farm stores are situated in the vicinity of ponds and water courses.

A useful source of information on pesticides as grain protectants is the book by Snelson (1987).

At present, this section does not include substances that are solely fungicidal or bactericidal. *The pesticide manual* (Tomlin, 1994) has to be consulted for these compounds.

INORGANIC (INERT) COMPOUNDS/MINERALS

Peasant farmers in many parts of the world have ancient traditions of adding inert materials to stored grain and seeds to reduce pest infestations. Quite how this operates is not always clear, but certain points seem obvious. Any fine material will have a filter effect, and will fill the interstices between the grains and hamper insect movement. And an abrasive material will scratch the wax from the insect epicuticle, and so increase water loss from the insect body. And some materials clearly contain chemicals that have an effect on the insect pests.

Seeds and seed grain (for next season's planting) are especially valuable to the farmer, and being small in bulk, protection methods can be used with them which would not be feasible or economic for bulk-stored grain to be used as food. For example, the production of special wood ash from selected botanical materials can often only be done on a small scale.

Below are listed some of the usual additives that are being employed by farmers to protect seed and grain in storage, and also a couple of inorganic compounds which are similarly used in recent years.

Ash

This is usually wood ash, and is useful as a physical barrier in the grain, but it can also possess various chemical properties according to its botanical source. Foliage of Neem (and sometimes Lantana) is often burned so that the ash can be added to seeds (and grains) in storage. But rice husk ash is used in Japan and India for rice storage, and cow dung ash in India and Africa with good success rate. Coal ash is also successful.

Boric acid ('Grovex', 'Boric Acid Powder', 'Cockroach Control Agent')

This chemical probably does not strictly belong to this category, but it is an inorganic compound, relatively inert, which has been used successfully to control cockroaches and crickets. Boric acid and Borax were both cleared for use on food premises in the USA (1991).

Diatomaceous earths (Diatomite)

These naturally occurring powders contain tiny silica crystals from the dead diatom bodies, and the abrasive powder so formed can have a wearing effect on the insect cuticle, scratching away the waxy epicuticle so that the insects suffer dehydration and can die of desiccation. But natural deposits of diatomaceous earths are limited in distribution, and only found in a few areas. Generally effective with rice, beans and wheat, and bulk maize, against the common pests.

Sand

Most sands are formed largely of silica crystals of very hard consistency, and fine sands are often added to stored seeds and grains. One problem is that the weight of the grains of sand will tend to result in all the sand falling to the bottom of the container. This method can be successful for stored seeds in small containers, but for bulk grain is less feasible. But sands are widely distributed, and can be found in most areas, and can be reused repeatedly. Cowpeas, groundnuts, beans, peppers etc. mixed with dry sand and sealed in metal drums have been well protected in Africa, India and Indonesia.

Silica gel (Silicic acid) ('Dri-Die', 'Silikil', 'Sprotive Dust')

Often called silica aerogel – the very fine crystals are formulated as a desiccant dust for admixture purposes. The crystals have a hygroscopic effect and absorb atmospheric moisture, and if coating the insect body, will absorb moisture from the insect, leading to death by desiccation. Most silica gels are reusable after heating to drive off the absorbed water.

Approved for use on food premises in the USA (1991).

Kaolin clay

A very fine naturally occurring clay found in large deposits in most countries; used as a dust, sometimes with lime (calcium oxide), when it was either insecticidal or acted as an insect deterrent.

The use of some of these materials, especially the naturally occurring ones, should be investigated more fully, for it would seem that they could be of use in many rural communities in the tropics, where there is of necessity emphasis on low external input sustainable agriculture. Most of these substances can be removed after storage by simply sieving, and then can be reused repeatedly.

ORGANOCHLORINE COMPOUNDS

These are mostly broad-spectrum and persistent poisons, and so in the past several (DDT, dieldrin etc.) were successfully used for the protection of stored products, but in more recent years, concern about environmental contamination and long-term toxicity led to their withdrawal for this purpose, and at the same time many insect pests had developed resistance to them. But in many developing countries, they are still being used, especially lindane.

Gamma-HCH (= Lindane) (= Benzene hexachloride) ('Lindane', 'HCH', 'BHC', 'Gammalin' etc.)

Properties. HCH occurs as five isomers in the technical form, but the active ingredient is the gamma isomer. Lindane is required to contain a minimum of 99% of the gamma isomer. It exhibits a strong stomach poison action (by ingestion), persistent contact toxicity, and some fumigant action, against a wide range of insects. It is non-phytotoxic at insecticidal concentrations, but the technical product causes tainting of many crops, although there is less risk of tainting with lindane.

It occurs as colourless crystals, and is practically insoluble in water; slightly soluble in petroleum oils; soluble in acetone, aromatic and chlorinated solvents. Lindane is stable to air, light, heat and carbon dioxide; unattacked by strong acids, but can be dechlorinated by alkalis.

The acute oral LD_{50} varies considerably according to the conditions of the test – especially the carrier; for rats, it is 88-270 mg/kg; ADI for man 0.01 mg lindane/kg.

Uses. Effective against a wide range of insect pests, including beetle adults and larvae, ants, fly larvae, bugs, caterpillars – in fact, most stored products pests, including mites. It was regarded as being the most potent of all grain protectants in its time.

Caution.

- Harmful to livestock and fish.
- Irritating to eyes and respiratory tract.
- In many countries, it is no longer recommended for use on stored grain or foodstuffs, because of its toxicity and persistence, and insect resistance became widespread; in some countries, its use is banned. But it is still widely used in some tropical countries for grain protection.

Formulations. E.c., w.p., dust, smoke, and many different mixtures with other insecticides and with fungicides.

ORGANOPHOSPHORUS COMPOUNDS

These compounds were developed in World War II as nerve gases, and are among the most poisonous compounds known to man. They are very effective insecticides, but their high level of mammalian toxicity is a serious hazard, and many of the more dangerous chemicals are now withdrawn from general use in most countries. The mode of action in both insects and mammals seems to be inhibition of acetylcholinesterase. Most of the earliest developed compounds are also quite short-lived chemicals, and so have little residual effect. But some of the more recently developed members of this group have both lower mammalian toxicity and greater persistence, and so are of great use in protection of stored products.

Azamethiphos ('Alfacron', 'Snip')

Properties. A broad-spectrum contact and stomach acting insecticide and acaricide, having both a quick knock-down effect and good residual activity.

It forms colourless crystals, insoluble in water, but slightly soluble in some organic solvents; suffers hydrolysis under alkaline conditions.

Acute oral LD_{50} for rats is 1,180 mg/kg.

Uses. Only used at present against household pests, and pests of livestock and hygiene, but it is an effective pesticide.

Caution.

- Toxic to fish.
- Slightly irritating to eyes.

Formulations. W.p. of 10, 100 or 500 g/kg, and an aerosol.

Bromophos ('Brofene', 'Bromovur', 'Nexion', 'Pluridox')

Properties. A broad-spectrum, contact and stomach insecticide; persists on sprayed foliage for 7-10 days. Non-phytotoxic at insecticidal concentration, but suspected of damage to plants under glass. Also used to protect stored products. No systemic action.

It occurs as yellow crystals, relatively insoluble in water, but soluble in most organic solvents, particularly in tetrachloromethane, diethyl ether, and methyl benzene. It is stable in media up to pH9, non-corrosive, and compatible with all pesticides except sulphur and the organometal fungicides.

The acute oral LD_{50} for rats is 3,750-7,700 mg/kg; ADI for man 0.04 mg/kg body weight.

Uses. Effective against a wide range of insects on crops, at concentrations of 25-75 mg/100 1. At rates of 10-20 mg/kg effective on stored grains against most pests – it penetrates grain rapidly. Superior to malathion in many respects. First used in 1968.

Caution.

- Harmful to fish and bees.
- Possibly phytotoxic under glass.

Formulations. E.c. 250, 400 g a.i./l; w.p. 250 g a.i./kg; dusts 20-50 g a.i./kg; atomising concentrate 400 g a.i./l; coarse powder 30 g a.i./kg; granules 50-100 g a.i./kg.

Chlorpyrifos-methyl ('Reldan', 'Dowco-214', 'Cooper Graincote')

Properties. A broad-spectrum insecticide and acaricide; with contact, stomach and fumigant action (vapour).

It forms colourless crystals, insoluble in water but soluble in most organic solvents. Basically stable, but can be hydrolysed under both acid and alkaline conditions.

The acute oral LD_{50} for rats is 1,630-2,140 mg/kg; ADI for man 0.01 mg/kg.

Uses. Mostly used for protection of stored grain and oilseed rape, and against household pests; generally effective against the spectrum of stored products pests.

Apply pre-harvest to empty store and machinery, and admixture with grain after cleaning, drying (to below 14% m.c.) and cooling, for greatest efficiency. First used in 1978.

Caution.

- Dangerous to fish.
- Irritating to eyes and skin.
- Flammable.
- Harvest interval for malting barley – 8 weeks.

Formulations. As an e.c. (240 or 500 g a.i./l or ULV (25 g a.i./l) for the treatment of stored grain and oilseed rape.

Chlorpyrifos ('Dursban') is also used to some extent worldwide, and was cleared for use in food areas in the USA (1991).

Diazinon ('Basudin', 'Diazitol', 'DBD', 'Neocid' etc.)

Properties. A non-systemic broad-spectrum insecticide with some acaricidal action, used against a wide range of insects both in agriculture and veterinary practice. At higher dosages, it may be phytotoxic.

It is a colourless oil, almost insoluble in water, but is miscible with ethanol, acetone, xylene, and is soluble in petroleum oils. It decomposes above 120°C, and is susceptible to oxidation; stable in alkaline media, but is slowly hydrolised by water and dilute acids. It is compatible with most pesticides, but should not be compounded with copper fungicides.

The acute oral LD_{50} for male rats is 108 mg/kg.

Uses. Especially effective against flies and their larvae; also used against mites, thrips, springtails, glasshouse pests, and some bugs (aphids, capsids etc), and stored products pests of all types.

Cleared for use on food premises in the USA (1991).

Caution.

- Dangerous to bees.
- Harmful to fish, livestock, game, wild birds and animals.
- Overdosage may lead to phytotoxicity on some crops.
- Pre-harvest interval for edible crops – usually 2 weeks.
- Pre-access interval for livestock to treated areas – 2 weeks.
- Should no longer be used on rice in S.E. Asia because of its resistance problems.

Formulations. Aerosol solutions; 25% e.c.; 40 and 25% w.p.; 4% dust; 5% granules.

Dichlorvos (= DDVP) ('Nogos', 'Nuvan', 'Dedevap' 'Vapona' etc.)

Properties. A short-lived, broad-spectrum, contact and stomach insecticide with fumigant action. Used on some field crops, and in greenhouses, but mostly in households and for public health. Generally not phytotoxic.

A colourless to amber liquid with b.p. 120°C, slightly soluble in water (1%), but miscible with most organic solvents and aerosol propellants; stable to heat, but hydrolysed in the presence of water; corrosive to iron and mild steel, but non-corrosive to aluminium and stainless steel.

The acute oral LD_{50} for rats is 56-108 mg/kg; ADI for man 0.004 mg/kg.

Uses. Often used for glasshouse fumigation (kills most glasshouse pests), and for space fumigation in produce stores. Because of its low vapour pressure, it is unable to penetrate bulk grain etc., and is no use as a commodity fumigant. As a fumigant, it is effective against flies at low concentrations, and at higher concentrations (0.5-1.0 g a.i./100 m^3), it kills cockroaches and most stored products insects; it has been used successfully in tobacco stores against moths and tobacco beetle. The impregnated resin strips are only used to kill flies and mosquitoes.

It is useful to apply to the empty store as a wet spray or an aerosol. In use since 1966.

Approved as a space treatment on food premises in the USA (1991).

Caution.

- A poisonous substance (classed in the UK as a Part III chemical under the Poisonous Substances in Agriculture Regulations, 1984), and protective clothing should be worn.
- Dangerous to fish and birds.
- Pre-harvest interval – 1 day.
- Pre-access interval to treated area – 12 hours.

Formulations. E.c. of 50 and 100%; aerosols 0.4-1.0%; 0.5% granules; Vapona resin strip; many mixtures as liquids or for fogging.

Etrimfos ('Ekamet', 'Satisfar')

Properties. A broad-spectrum insecticide with contact action, which will also kill some mites; of moderate persistence (1-2 weeks).

It is a colourless oil, insoluble in water, but soluble in most organic solvents; in the pure form unstable, but formulations are stable with a shelf life of about 2 years.

Acute oral LD^{50} for rats is 1,800 mg/kg; ADI for man 0.003 mg/kg.

Uses. Effective against Coleoptera, Lepidoptera, Psocoptera and mites; used to protect grain, rape seed etc. at 3-10 mg/kg for periods up to 1 year.

Recommended as spray for empty stores, but allow sufficient time for pests to emerge from hiding places before grain is introduced. Should be applied as admixture to grain using a suitable sprayer or dust applicator. Grain moisture level should not exceed 16%, and grain should be cooled below 15°C before treatment. Controls malathion-resistant beetles and HCH-resistant mites in the UK.

Caution.

- Dangerous to fish.
- Harmful to livestock, game and wild birds.

Formulations. Dusts of 1 or 2% w/w; e.c. 525 g a.i./l (= 500 g/kg), and oil spray 50 g/kg, and ULV at 400 g/l.

Fenitrothion ('Accothion', 'Dicofen', 'Folithion', 'Sumithion'etc.; many trade names)

Properties. A broad-spectrum contact and stomach insecticide, and a selective acaricide (but of low ovicidal activity), of moderate persistence.

It is a pale brown liquid of bp 145°C, practically insoluble in water, but soluble in most organic solvents, and hydrolised by alkali.

Acute oral LD_{50} for rats is 800 mg/kg; ADI for man 0.003 mg/kg.

Uses. Effective against most stored products beetles, and generally effective against Lepidoptera and some Acarina. Used against many crop pests. Recommended as an interior spray for empty grain stores, but the stores must be cleaned first. First used in 1977. Resistance problems are generally widespread.

Caution.

- Harmful to fish, livestock, game, wild birds and animals.
- Avoid skin contact or inhaling vapours; irritating to the eyes.
- Flammable.

Formulations. E.c. 50%; w.p. 400 g/kg; dusts of 20, 30 or 50 g/kg; and ULV formulation of 1.25 kg/l. Many mixtures, with malathion, tetramethrin, fenvalerate and others. In the UK, the mixture fenitrothion + permethrin + resmethrin ('Turbair Grain Store Insecticide') is formulated only for ULV application in grain stores; presumably these mixtures are more effective when there is danger of resistance.

Idofenphos ('Elocril', 'Nuvan', 'Nuvanol N' etc.)

Properties. A broad-spectrum contact and stomach insecticide and acaricide of moderate persistence, used mostly against stored products pests.

The pure chemical forms colourless crystals, virtually insoluble in water, but soluble in toluene and dichloromethane. Stable in neutral media, but unstable in strong acids and alkalis.

Acute oral LD_{50} for rats is 2,100 mg/kg.

Uses. It is used in crop protection against a wide range of insect pests; in public hygiene, used for fly control; for domestic pest control, especially effective against cockroaches and ants; in produce stores, usually sprayed in empty grain stores.

Caution.

- Irritating to skin and eyes.
- Dangerous to fish.

Formulations. E.c. of 200 g/l; w.p. 500 g a.i./kg; s.c. 500 g/l; and d.p. of 50 g/kg.

Malathion ('Malathion', 'Malatox', 'Cythion' etc.)

Properties. A broad-spectrum contact insecticide and acaricide, generally non-phytotoxic, of brief to moderate persistence, and low mammalian toxicity. Used very widely against crop pests, livestock ectoparasites, and stored products pests for many years – one of the earliest very widely used insecticides, both on crops and in produce stores.

A pale brown liquid of slight solubility in water, miscible with most organic solvents, but low miscibility in petroleum oils. Hydrolysis is rapid at pH above 7.0 and below 5.0; incompatible with alkaline pesticides; corrosive to iron.

Acute oral LD_{50} for rats is 2,800 mg/kg; ADI for man 0.02 mg/kg.

Uses. Generally effective against a wide range of insects and mites, but it has been used in produce stores for many years (since 1960) and resistance is now widespread – no longer recommended in the UK for use in grain stores (1988); but it may still be effective in some countries.

Approved for use on food premises in the USA (1991).

Caution.

- Temporary taint to edible crops can occur, but will pass in 4-7 days.
- Resistance is so widespread that its usefulness is now limited.

Formulations. E.c. from 25-1,000 g a.i./l; dust of 40 g/kg; w.p. of 250 or 500 g/kg; and ULV of 920 g/l.

Methacriphos ('Damfin')

Properties. An insecticide and acaricide with vapour, contact and stomach action, used only for stored products protection by both surface treatment and admixture with grain.

A colourless liquid, slightly soluble in water, but very soluble in most organic solvents.

Acute oral LD_{50} for rats 678 mg/kg; ADI for man 0.0003 mg/kg.

Uses. Only used in grain stores in the UK for protection of wheat, oats and barley. Experiments on stored cereals in Egypt and Indonesia were successful, and malathion-resistant pests were killed.

Recommended to spray empty stores, taking special care to treat cracks and crevices, and also as an admixture treatment.

Caution.

- Irritating to eyes and skin.
- Harmful to animals and wildlife.

Formulations. E.c. of 950 g a.i./l; dust of 20 g/kg; and ULV spray 50 g/l.

Phoxim ('Baythion', 'Volaton' etc.)

Properties. A broad-spectrum contact insecticide, of low mammalian toxicity and brief persistence, also stomach-acting.

A yellow liquid, virtually insoluble in water, but soluble in most organic solvents. Stable in water and to acids, but unstable in alkaline media.

Acute oral LD_{50} for rats is 2,500-5,000 mg/kg; ADI for man 0.001 mg/kg.

Uses. Mostly employed against soil insects, medical insects and stored products pests, but has been successful in foliar application. Effective against a wide range of insects and other arthropods.

Caution.

- Harmful to fish.

Formulations. E.c. of 100, 200, 500 g/l; w.p. 500 g/kg; ULV concentrate of 200 g/kg.

Pirimiphos-methyl ('Actellic', 'Blex', 'Fumite', 'Actellifog')

Properties. A fast-acting, broad-spectrum insecticide and acaricide of limited persistence, with both contact and fumigant action; low mammalian toxicity and non-phytotoxic.

It is a straw-coloured liquid, insoluble in water, but soluble in most organic solvents; decomposed by strong acids and alkalis; does not corrode brass, stainless steel, aluminium or nylon.

Acute oral LD_{50} for rats is 2,050 mg/kg; ADI for man 0.01 mg/kg.

Uses. Recorded as effective aginst the whole spectrum of stored products insect and mite pests. In many countries, now the most widely used stored products pesticide that is not a fumigant. The only insecticide recommended for use on drying fish.

Recommended application to empty buildings by spraying, or fogging, or with smoke generators, and also admixture to grain. Grain should be at less than 18% moisture content; admixture can be of dust or liquid. First used in 1969.

Prostephanus and resistant *Rhizopertha* can be controlled using a mixture with permethrin ('Actellic Super').

Caution.

- Harmful to fish

Formulations. E.c. of 80, 250 or 500 g a.i./l; w.p. of 40% w/w; s.c. 200 g/l; dust of 20 g/kg; and ULV and hot-fogging concentrates. 'Actellic Super' is available as e.c. and dust.

Tetrachlorvinphos ('Gardona', 'Rabond')

Properties. A selective insecticide, effective against many insect pests of major crops, but not Hemiptera and sucking pests; very low mammalian toxicity.

A white crystalline solid, scarcely soluble in water, but soluble in some organic solvents. It is temperature stable, but slowly hydrolyses in water, especially under alkaline conditions.

Acute oral LD_{50} for rats is 4-5,000 mg/kg.

Uses. Widely used on field crops of all types, as well as forestry and pastures, and also for public health, against livestock pests and stored products pests. In Africa, used as a dip for drying fish (against *Dermestes* and *Necrobia* beetles).

Caution.

- Dangerous to fish.

*Formulations.*E.c. of 240 g a.i./l w.p. of 500 and 700 g/kg; s.c. 700 g/l.

Trichlorfon ('Dipterex', 'Tugon', 'Dylox' etc.)

Properties. A contact and stomach insecticide, with a broad range of effectiveness; its activity is related to its conversion metabolically to dichlorvos; brief persistence.

A colourless crystalline powder, soluble in water and some organic solvents; stable at room temperature.

Acute oral LD_{50} for rats is 560-630 mg/kg; ADI for man 0.01 mg/kg

Uses. Effective against many crop pests, and in households; most effective against Diptera and Lepidoptera. In produce stores, used mainly as a space spray.

Caution.

- Harmful to fish

Formulations. W.p. of 500 g a.i./kg; s.p. of 500, 800, 950 g/kg; s.c. 500 g/l; d.p. of 50 g/kg; and ULV formulations of 250, 500 and 750 g/l. A mixture with fenitrothion as a w.p. ('Tugon Fly Bait') contains 10 g/kg.

CARBAMATES

These chemicals were developed as alternatives to the very toxic organophosphorus compounds which were a threat to both human operators and all forms of livestock and wild animals. The mode of action is similar in being inhibition of cholinesterase, in both Mammalia and Insecta, but toxicity levels for the mammals are lower.

Bendiocarb ('Ficam', 'Turcam', 'Garvox' etc.)

Properties. An insecticide with contact and stomach action, useful in stored products and households because of its low odour, and lack of staining and corrosive properties; it has a good residual activity.

It is a colourless solid, slightly soluble in water, and stable to hydrolysis at pH 5, but quite rapidly hydrolysed at pH 7; readily soluble in most organic solvents.

Acute oral LD_{50} for rats is in the range 40-156 mg/kg; ADI for man 0.004 mg/kg.

Uses. Widely used in plant protection; it is effective against cockroaches, ants, flies and fleas, and is used in public health and against storage insects. As mentioned, it is odourless and non-staining, and so is useful in domestic situations.

Approved for use on food premises in the USA (1991).

Caution.

- A toxic chemical, so protective clothing must be worn.
- Dangerous to livestock and wildlife.

Formulations. D.p. of 10 g a.i./kg; w.p. of 200, 500, 760 or 800 g/kg; s.c. 500 g/l; ULV concentrate of 250 g/l; and many others. Several mixtures are also available, some are with synergized pyrethrins.

Carbaryl ('Sevin', 'Carbaryl' etc.)

Properties. A broad-spectrum insecticide with contact and stomach action, and good persistence.

A white crystalline solid, barely soluble in water, but soluble in many organic solvents. Stable to light, heat, and does not readily hydrolyse, and non-corrosive.

Acute oral LD_{50} for rats is 850 mg/kg; ADI for man is 0.01 mg/kg.

Uses. Generally effective against a wide range of insects, especially the biting and chewing types, and can be used in stored grain as well as in empty buildings.

Caution.

- Harmful to fish.

Formulations. W.p. of 500, 800 or 850 g a.i./kg; dusts of 50 or 100 g/kg; s.c.of 220, 300, 400, 440 or 480 g/l, and as true solutions; mixtures with acaricides are available.

PYRETHROIDS

The pyrethrins (as extracted from the flowers of Pyrethrum) have a dramatic knock-down effect on a wide range of insects, but these chemicals are unstable and are rapidly decomposed by sunlight and exposure to air. Their stability and persistence can be enhanced by synergism with chemicals such as piperonyl butoxide; they have very low mammalian toxicity. Because of these qualities, they are still in use, but mostly in public health, especially for mosquito control. Their place in agriculture has generally been taken by the very successful synthetic pyrethroids developed at Rothamsted in the early 1970's. These chemicals have kept the basic insecticidal properties, and in some cases they are greatly enhanced, and they are resistant to photochemical degradation. The value of these third-generation insecticides to world agriculture is inestimable. The only drawback is that they have a higher basic toxicity to mammals, bees, fish etc., and so have to be used very carefully. Most of these chemicals can be successfully synergized.

Bioallethrin ('Biothrin', 'Detmol', 'Esbiol' etc.)

Properties. A potent contact insecticide with a rapid knock-down, effective against a broad range of insects; synergists can be used to delay detoxication.

A viscous amber liquid, almost insoluble in water, but readily soluble in many organic solvents; it occurs as several isomers with slightly different properties.

Acute oral LD_{50} for rats 425-575 mg/kg.

Uses. Used mostly against household insect pests, including mosquitoes; not widely used in produce stores, except for space treatment.

Formulations. Aerosols; powders for dusting; m.l; commonly as a mixture with permethrin and piperonyl butoxide as an aerosol for space treatment.

Bioresmethrin ('Resbuthin', 'Biorex', 'Detmol')

Properties. A potent broad-spectrum insecticide effective against cockroaches, flies, and other stored products pests; with low mammalian toxicity; low toxicity to plants; and short persistence.

The technical product is a viscous brownish liquid which partially solidifies on standing; sunlight accelerates decomposition, and it is hydrolysed by alkali; almost insoluble in water, but soluble in most organic solvents.

Acute oral LD_{50} for rats is 7-8,000 mg/kg.

Uses. Mostly used as a stored grain protectant, but also for cockroach and synanthropic fly control; synergistic factors tend to be low. First used in 1975.

Caution.

- Very toxic to bees; toxic to fish.

Formulations. S.c.and m.l. of 2.5 g/l; aerosols for domestic use at 1 g/l; available in various mixtures.

Cypermethrin ('Cymbush', 'Ripcord', 'Folcord' etc.; many trade names)

Properties. A broad-spectrum synthetic pyrethroid insecticide, with rapid contact and stomach action, equally used for crop protection and against animal ectoparasites (ticks, lice etc.), and in public health.

A generally stable compound; quite persistent; but it occurs as several different isomers with somewhat different properties.

Acute oral LD_{50} values are variable, according to many factors; for rats from 250-4,000 mg/kg; ADI for man 0.05 mg/kg.

Uses. Mostly used for crop protection, against livestock pests, and for public health purposes against flies, cockroaches etc.; a very high activity against Lepidoptera; some formulations in mixtures. It is being used against some organophosphorus-resistant stored products pests. Now being developed for use against timber beetles.

Caution.
- Irritating to skin and eyes.
- Flammable.
- Extremely dangerous to fish.

Formulations.

Most commercial formulations contain a *cis/trans* isomer ratio of 45:55. E.c. of 25-400 g a.i./l; ULV 10-50 g/l; and various mixtures, especially of *cis* isomers of e.c., w.p. etc.

Deltamethrin ('Decis;, 'Butox' etc.)

Properties. A very potent broad-spectrum pyrethroid insecticide, with contact and stomach action, and with good residual activity when used both outdoors and indoors.

The technical product is a colourless crystalline powder, insoluble in water, but soluble in most organic solvents; stable on exposure to both air and sunlight.

Acute oral LD_{50} for rats is 135-5,000 mg/kg according to conditions; ADI for man 0.01 mg/kg.

Uses. Effective against a wide range of insects of all types, and used for crop protection, against livestock ectoparasites, for public health, and for stored products protection.

Caution.
- A Part III Poisonous Substance; protective clothing should be worn.
- Irritating to eyes and skin.
- Flammable.
- Extremely dangerous to fish.

Formulations. E.c. of 25 and 50 g a.i./l; w.p. of 25 or 50 g/kg; dusts of 0.5 and 1.0 g/kg; and ULV concentrates of 4, 5 and 10 g/l; also a series of different mixtures.

Fenvalerate ('Belmark', 'Sumicidin', 'Tirade', 'Pydrin')

Properties. A synthetic pyrethroid insecticide with highly active contact and stomach action; broad-spectrum, and persistent. The technical product is a viscous brown liquid, largely insoluble in water, but soluble in most organic solvents; stable to heat and sunlight; more stable in acid than alkaline media (optimum at pH 4).

Acute oral LD_{50} for rats is 450 mg/kg; ADI for man is 0.02 mg/kg.

Uses. Mostly used in crop protection, but some use in stored products against resistant pests. Also used in public health and animal husbandry against flies and ectoparasites.

Cleared for use in food areas in the USA (1991).

Caution.

- Irritating to eyes and skin.
- Flammable.
- Extremely dangerous to fish.

Formulations. E.c. of 25-300 g a.i./l; ULV of 25-75 g/l; s.c. 100 g/l; and various mixtures.

Permethrin ('Kayo', 'Kalfil', 'Talcord' etc.; many trade names)

Properties. A broad-spectrum contact and stomach action synthetic pyrethroid insecticide; quite persistent. As with some of the other pyrethroids, it gives a good knock-down combined with a persistent protection. The technical product is a brown liquid, insoluble in water, but soluble in most organic solvents; stable to heat, but some photochemical degradation; more stable in acid than alkaline media.

Acute oral LD_{50} for rats varies with the *cis/trans* isomers ratio, carrier, and conditions of use – recorded from 430 to 4,000 mg/kg; ADI for man 0.05 mg/kg.

Uses

Effective against a wide range of insect pests of plants, livestock and stored products; also used to protect wool at 200 mg/kg wool, and to kill cockroaches. In a mixture with pirimiphos-methyl ('Actellic Super') for control of resistant strains of *Rhizopertha* and *Prostephanus.*

Caution.

- Extremely dangerous to fish.
- Flammable.
- Irritating to skin, eyes and respiratory tract.

Formulations. E.c. of 100-500 g a.i./l; w.p. 100-500 g/kg; ULV formulations; dusts and smokes, and aerosols. Not many formulations recommended for stored products use.

Pyrethrum ('Pyrethrum')

Properties. A very broad-spectrum contact insecticide of fairly short persistence (unsynergized), easily degraded on exposure to sunlight. Effective against stored grain pests, especially when synergized with piperonyl butoxide (1:10 ratio); pronounced 'knock-down' effect on flying insects.

Pyrethrum is the oily extract from *Chrysanthemum cinerariaefolium*, and the technical extract contains pyrethrin I and II (very insecticidal), and cinerin I and II (less insecticidal); first recorded use in 1935.

Acute oral LD_{50} for rats is 200-1500 mg/kg; ADI for man 0.4 mg/kg body weight.

Uses. Effective against all types of stored product insects; synergists do not increase its effectiveness against moths. Often used particularly for mosquito control.

Cleared for use in food areas in the USA (1991).

Caution.

- Pyrethrum has a very low acute oral toxicity, and is generally regarded as completely safe to use.

Formulations. Most formulations are in oil as an e.c. or oil solution with piperonyl butoxide in a 1:10 ratio. They can be made into dusts and w.p., but these are not often used.

Resmethrin ('Kilsect', 'Chryson', 'Synthrin' etc.)

Properties. A synthetic pyrethroid insecticide with powerful contact action against a wide range of insects; low phytotoxicity. But it is not synergized by the pyrethrin synergists; often used in mixtures with more persistent insecticides.

A colourless waxy solid; it occurs as a mixture of isomers; insoluble in water, but soluble in most organic solvents. It is decomposed quite rapidly on exposure to light and air, and is unstable in alkaline media.

Acute oral LD_{50} for rats is more than 2,500 mg/kg.

Uses. Widely used against crop pest insects, as well as household and public health insects; often used in mixtures with more persistent insecticides.

Cleared for use in food areas in the USA (1991).

Caution.

- Irritating to eyes, skin and respiratory tract.
- Flammable.

Formulations. E.c., w.p., ULV concentrate; mixtures with fenitrothion, malathion, and tetramethrin are used.

Tetramethrin ('Butamin', 'Duracide', 'Ecothrin', 'Pesguard' etc.)

Properties. A contact insecticide with strong knock-down effect; enhanced kills are obtained using mixtures with other insecticides and synergists.

The technical product is a mixture of isomers, as a pale brown solid or viscous liquid; stable under normal storage conditions; almost insoluble in water, but very soluble in xylene, methanol, and hexane.

Acute oral LD_{50} for rats is more than 5,000 mg/kg.

Uses. At present, only used for public health and against household pests; very effective against cockroaches, flies and mosquitoes.

Formulations. Mostly formulated as mixtures for public health purposes, and many mixtures are available worldwide; also as e.c., d.p., and aerosols.

INSECT GROWTH REGULATORS

Because of the general toxicity of pesticides, and the disruptive effects they can have environmentally, there has been a search for biological compounds that would kill the insect pest without displaying any of the disadvantageous side effects. Chemicals similar to natural insect hormones have shown great promise over recent years; the three groups concerned are classed as:

- juvenile hormones (JHs) involved in metamorphosis and moulting
- ecdysone agonists, which upset cuticle synthesis in several different ways causing the insects to die
- chitin synthesis inhibitors (CSIs), which kill insects at all stages in the life cycle, particularly the early stages.

Collectively, they are known as Insect Growth Regulators (IGRs). The CSIs are not as specific as the juvenile hormones, and thus they are of potentially wider application, and often seem to work as pesticides by interfering with cuticle formation at the time of ecdysis. Several chemical compounds have been developed for this purpose, and have been used in crop protection very successfully. With the spread of insecticide resistance in stored products insects becoming so serious, there is interest in the application of insect growth regulators as alternatives to the more usual insecticides. (See Chapter 21 for more details.)

Diflubenzuron ('Dimilin', 'Empire', 'Larvakil')

Properties. An insect growth regulator active by inhibition of chitin synthesis, so that insect larvae die at ecdysis, and eggs can be prevented from hatching; a CSI chemical.

It is a yellow crystalline solid, of limited solubility.

Acute oral LD_{50} for rats 4,640 mg/kg; ADI for man 0.02 mg/kg.

Uses. Mostly used in agriculture against leaf eating insects (especially Lepidoptera) and some phytophagous mites, but also against fly and mosquito larvae. Research is being done to see if this chemical can be used against some stored products pests that are difficult to control by other means, with considerable success.

Formulations. W.p. of 250 g a.i./kg; concentrates for hot fogging and ULV application.

Methoprene ('Altosid', 'Kabat', 'Diacon', 'Precor' etc.)

Properties. An insect growth regulator of the juvenile hormone type, of short persistence; treated larvae develop apparently normally into pupae, but the pupae die without giving rise to adults.

An amber liquid, insoluble in water but miscible with all common solvents; stable if stored in the dark.

Acute oral LD_{50} for rats is 34,600 mg/kg; ADI for man 0.06 mg/kg.

Uses. Effective against many different insect pests of plants and farm animals, and used in public health and with stored products. In public health, used mostly against Diptera. Successfully used to control *Lasioderma* and *Ephestia elutella* in tobacco warehouses. Its use has been extended against other stored products pests that are difficult to kill with other chemicals.

Formulations. S.c. of 41 and 103 g/l; e.c. of 600 g/l; aerosols for domestic use; and mixtures with insecticides to kill adult insects as well as larvae.

FUMIGANTS

The fumigants by definition act as gases or in some form of vapour phase, and enter the insect body via the spiracles into the tracheal system. The most effective gaseous poisons are very toxic to all forms of life, and their use is often restricted to trained operators. But, of course, the advantage of fumigation is that ideally it kills all the pests present, although eggs and insect pupae, mite hypopi and fungal spores are more difficult to kill, and may not succumb to the vapour.

Some of the most toxic (and effective) chemicals have been shown to possess carcinogenic properties (using laboratory rats), and their use has been either severely curtailed or banned in some countries (USA, UK etc.). And suitable replacements are proving hard to find.

Carbon dioxide (CO_2)

This is clearly not a fumigant in the usual sense, but it is a gas used to control stored products pests on the basis of controlled atmosphere. It is a normal constituent of air in small quantities (usually 0.03%), and is the main by-product of biological aerobic respiration. It has a suffocating effect on living animals, and it interferes with the oxygenation process of haemoglobin in red blood cells. The gas is heavier than air, and so can actually displace oxygen from a suitably designed grain store when introduced from the top. For this purpose, it may be used as a mixture with nitrogen (N_2), which is commercially available as a liquid under pressure. Nitrogen may be used on its own as an inert gas for Modified Atmosphere Packaging (see Chapter 5.9.). Carbon dioxide can be obtained as a liquid under pressure, and also frozen as a solid called 'dry-ice' – at a low temperature, it freezes to a white solid (resembling ice), and vapourises at room temperature, giving off white clouds of cold gas. In many countries it is commercially available as dry-ice.

With controlled atmosphere storage and pest control, the aim is to achieve an atmosphere containing no more than 1% oxygen, at which level all animal life will be killed.

With sealed storage techniques, the normal 15% oxygen content can be used up by the organisms infesting the grain, and eventually the oxygen level will fall to a level inimical to life.

Carbon disulphide (CS_2) (= Carbon bisulphide) ('Weevil-Tox' etc.)

Properties. A liquid with b.p. of 46°C, with flashpoint above 20°C, and it ignites spontaneously at about 100°C; very toxic to man, and can be absorbed through the skin. The vapour is 2.6 times as heavy as air; in tropical countries (at higher ambient temperatures), it vapourises well; it generally penetrates bulk grain well; but in temperate countries, it requires heating to assist vapourisation.

Uses. Often used as a soil sterilant against insects and nematodes, and used in fumigating nursery stock. Previously used in fumigation chambers to treat plant products such as dried peas and beans. Used in some tropical countries (mixed with carbon tetrachloride to reduce fire hazard) to fumigate stored grain; but its toxicity to insects is not high. In most countries, its use has been superseded by methyl bromide, but now may be used more extensively.

Caution.

- Very toxic to man, and can be absorbed through the skin.
- Highly flammable – b.p. of 46°C.
- In the USA in 1986, its use was banned because of the toxicity hazards.

Formulations. For stored grain protection, it can be used alone as the pure chemical, but is more often used as a mixture with carbon tetrachloride to reduce flammability.

Carbon tetrachloride (CCl_4) (= tetrachloromethane)

Properties. A clear liquid, non-flammable, and non-explosive – relatively inert with b.p. of 76°C, miscible with most organic solvents, it readily vapourises by evaporation, but it has a low toxicity to insects. It is a general anaesthetic, and repeated exposure can be dangerous. The vapour is 5 times as dense as air. Usually employed as a mixture to lower flammability of other chemicals, or to aid penetration in bulk grain.

Uses. Its low insecticidal properties limit its usefulness, but it is sometimes used when high concentration or long exposure is possible; one advantage is its low adsorption by treated grain.

Caution.

- Repeated exposure to the vapour can be dangerous.
- In the USA in 1986, it was banned because of toxicity hazards.

Formulations. Used as a pure liquid, by evaporation; or used as mixtures with other fumigants such as carbon disulphide and ethylene dichloride etc.

Dichlorvos

Not truly classified as a fumigant, and already included under the heading of organophosphorus compounds.

Ethylene dibromide ($CH_2Br.CH_2Br$) (= dibromoethane) ('EDB', 'Bromofume', 'Dowfume-W')

Properties. A colourless liquid of b.p. 131°C; highly insecticidal, but dangerous to man. Insoluble in water, but soluble in most organic solvents; stable and not flammable.

Acute oral LD_{50} for rats 146 mg/kg; dermal application will cause severe skin burning.

Uses. Mostly used for protection of stored produce. Formerly, used extensively in the USA and tropics for control of fruit flies (Tephritidae) in fruits and vegetables, and for grain fumigation. It is an important soil fumigant, but it suffers from surface adsorption by many materials, and it does not penetrate well. It is often used for spot treatment in flour mills. It was also used in weak solution as a dip to control fruit fly larvae in fruits. But in 1988 its use in the USA was banned after demonstration of its carcinogenic properties.

Caution.

- A dangerous poisonous substance; carcinogenic; and should only be used with great care and wearing full protective clothing; now banned in the USA.
- In liquid or vapour form, it attacks aluminium and some paints.

Formulations. Available as the pure liquid; for mill use, a mixture with carbon tetrachloride is usual.

Ethylene dichloride ($CH_2Cl.CH_2Cl$) (=dichloroethane) ('EDC')

Properties. A colourless liquid, b.p. 83°C, soluble in most solvents; with moderate insecticidal properties, but dangerous to man; quite flammable; adsorption by grain can be a problem.

Acute oral LD_{50} for rats 670-890 mg/kg.

Uses. An insecticidal fumigant for stored products protection, but not widely used; formulated as mixtures to reduce fire hazard.

Caution.

- Dangerous to man.
- Flammable.

Formulations. Generally only available as a mixture with carbon tetrachloride to reduce the flammability.

Formaldehyde (HCHO) ('Dynoform', 'Formalin')

Properties. A colourless flammable gas (formalin is a solution in water), with a very pungent irritating odour, and very strong fungicidal and bacterial action.

Acute oral LD_{50} for rats 550-800 mg/kg, but acute inhalation LD_{50} 0.82 mg/kg or lower.

Uses. Mostly used for sterilisation of greenhouses and empty buildings, particularly when fungal pathogens or bacteria are a problem; spraying formalin was the former method of application, but, recently, fogging methods have been used very successfully. It is also a useful soil sterilant.

Caution.

- Vapour is very dangerous to man and animals; and 3 days should elapse before entry to treated buildings is permitted.

Formulations. Formalin is a 40% solution in water; methanol may be added to delay polymerization; fogging solutions.

Hydrogen cyanide (HCN) (= hydrocyanic acid) ('HCN', 'Cymag', 'Cyangas')

Properties. A colourless liquid, smelling of almonds; b.p. of 26°C; soluble in water and most organic solvents. A weak acid forming salts which are very soluble in water, and are readily hydrolysed, liberating hydrogen cyanide. Very toxic to insects, but also high mammalian toxicity, either as a vapour or ingested as salts. It does not penetrate grain as rapidly as methyl bromide.

Acute oral LD_{50} for rats is 6.4 mg sodium cyanide/kg.

Uses. Effective against insects and rodents in enclosed spaces, and is used mostly for fumigation of grain and buildings, and for killing rats and rabbits in their burrows. Atmospheric moisture on the salts is usually sufficient, but for rapid vapourisation, an acid is applied. Sometimes used for fumigation of dormant nursery stock against scale insects, less often used for fruit with scale insects; dried plant materials are more successfully treated.

Caution.

- Very dangerous to man and mammals; its use is normally restricted to trained personnel with full protective clothing.
- Danger of fire when used as a liquid.

Formulations. The usual salts are sodium or calcium cyanide; but the liquid is available packed in metal containers with compressed air dispersal.

Methyl bromide (CH_3Br) (= bromomethane) ('Bromogas', 'Dowfume')

Properties. A general poison with moderately high insecticidal and some acaricidal properties, used for space and produce fumigation, and as a soil fumigant against nematodes and fungi; the penetration action is good and rapid.

A colourless liquid with b.p. at 4°C, forming a colourless odourless gas; stable; non-flammable; soluble in most organic solvents; but corrosive to aluminium and magnesium, and a solvent to natural rubber.

Uses. Effective against stored products insects (but less so than HCH or PH_3), and at insecticidal concentrations non-toxic to living plants, fruits etc., so widely used for plant quarantine. With stored products, it has largely replaced HCH in recent years. It is successful as a seed fumigant in bags, and with bulbs against bulb flies, mites and nematodes. Generally used for protection of grain, flours, meals, cereal products, and most stored products. Recently banned in the USA and UK because of alleged carcinogenic properties.

Caution.

- A very poisonous gas, and its use should be restricted to trained personnel.
- Resistance is established in a number of insects in various widespread locations, so local resistance has to be expected.
- Now banned in the USA and UK.

Formulations. Packed as a liquid in glass ampoules (up to 50 ml), or in metal cans and small cylinders for direct use; chloropicrin is sometimes added (up to 2%) as a warning gas.

Phosphine (PH_3) (= hydrogen phosphide) ('Detia Gas', 'Phostoxin', 'Gastoxin') (Aluminium phosphide)

Properties. Phosphine is a colourless odourless gas, highly insecticidal, and a potent mammalian poison, very effective as a fumigant.

It is produced usually from aluminium phosphide, which is in the form of yellowish crystals; stable when dry, but reacting with water or atmospheric moisture to release phosphine, leaving a small harmless residue of aluminium oxide. The gas is spontaneously flammable, so the usual formulation also releases carbon dioxide and ammonia to reduce the fire hazard; the gas reacts with all metals, especially with copper.

Uses. One of the most toxic fumigants to stored produce insects; normally only used for stored products protection, but sometimes used to kill rats in their burrows. Resistant stages of mites (egg, hypopus) may not be killed, so mite populations may recur some time after treatment (predatory mites are invariably killed). Seed germination is not affected by phosphine. Gas penetration is quite good, and tablet dispensers are available for placement of tablets in bulk grain.

Caution.

A very potent poison, so care should be taken – ideally only to be used by trained personnel.

- Exposure time is generally 3-10 days – adequate exposure is vital, followed by adequate airing.
- Silo workers in the USA have suffered chromosomal damage, probably from gas release before tablet placement was completed.
- Resistance is now established in a few locations in some insects, probably following inefficient fumigation and carelessness.

Formulations. 'Phostoxin' tablets (3g) or pellets (0.6 g); 'Detia' pellets and tablets, and bags (34 g powder). The bags are for space fumigation, and may be joined into a long strip. Long tubular dispensers are available for use in bulk grain, and both pellets and tablets can be added directly to a grain stream in the store. Degesch now also make a plate and a strip formulation for space fumigation – the strip being a series of joined plates. 'Detiaphos' is a magnesium phosphide formulation made for the more rapid release of the phosphine, and for a smaller final residue.

Fumigant mixtures

Many fumigants are marketed in mixtures with other compounds; the main reasons for this being:

- To reduce flammability risk.
- Some liquid-type fumigants on their own do not penetrate bulk grain well, whereas a mixture may be far more effective at penetration.
- Highly volatile fumigants (such as methyl bromide) may diffuse downwards too rapidly, so that addition of a less volatile chemical (ethylene dibromide) will help to achieve a more uniform treatment, especially under warm tropical conditions.
- The toxic ingredient, after dilution, may spread more uniformly. Carbon tetrachloride is only moderately insecticidal, but it aids the dispersal of other chemicals such as ethylene dibromide in bulk grain.

To find an ideal fumigant for the treatment of bulk grain is not easy, for it needs to be a broad-spectrum insecticide and acaricide that will spread quickly and evenly throughout the bulk grain, and will not suffer too much surface adsorption; it has to be highly lethal to insects; but inevitably resistance will develop, even though ideally it should be a very slow process with fumigation, as higher levels of insect kill are usual if the procedure has been followed carefully.

The three most extensively used fumigants worldwide have been ethylene dibromide, methyl bromide, and phosphine; but now the first two are banned in the USA as carcinogens, and resistance is developing rapidly to the third, so the search for suitable control programmes continues.

Pesticide mixtures

There is now an increasing tendency to formulate pesticide mixtures for stored products protection. This has originated partly following the development of resistance to the main pesticides being used (malathion etc.) and partly because of the gradual withdrawal of the organochlorine compounds with their useful persistence and broad-spectrum activity.

Chemical mixing is generally an attempt to achieve an adequate combination of appropriate toxicity (usually broad-spectrum), persistence, hazard reduction, and reduced likelihood of resistance development. Results to date have been mixed. One very successful case was the use of 'Drione Dust' to control *Blattus orientalis* in England – it was a mixture of 40% silica-gel, with pyrethrins (1%), and synergized by piperonyl butoxide (10%); the treatment was effective for six months.

Rodenticides

Rats and mice will be killed when buildings and produce stores are fumigated with methyl bromide or phosphine, and so rodenticides are not included here in detail; but the chemicals used specifically for rodent control are mentioned in Chapter 10.

Fungicides and Bactericides

As already mentioned, it is not feasible to include details of the chemicals that are specifically used against fungi and bacteria.

Recommended sources include:

- *BCPC pest and disease control handbook* (for the UK). London: BCPC. (latest ed.).
- *The pesticide manual: a world compendium.* London: BCPC & RSC. (Tomlin, 1994). (10th ed., or latest ed.)
- Snowdon, A.L. (1990). *A colour atlas of post-harvest diseases and disorders of fruits and vegetables.* Vol. 1 *General introduction and fruits.* Vol. 2 *Vegetables.* London: Wolfe Scientific.
- Other titles are listed in the References.

APPENDIX 2

Group for Assistance on Systems Relating to Grain After-harvest (GASGA)

In 1971, a seminar was held at Ibadan, Nigeria, on *The Storage of Grains particularly in the Humid Tropics*, organised jointly by the Ford Foundation, International Institute of Tropical Agriculture (IITA) and IRAT. At the meeting, a large group of international representatives decided to organise GASGA to disseminate information and expertise in relation to third world postharvest grain losses; the group of eight organisations is primarily linked with aid donor operations.

A newsletter and various other publications are produced, some being printed by GTZ; at the present time, the Secretariat for the group is provided jointly by IRAT and ODNRI.

The eight member organisations of GASGA were as follows:

The Commonwealth Scientific and Industrial Research Organisation, Canberra, Australia (CSIRO).
Food and Agriculture Organisation of the United Nations, Rome, Italy (FAO).
Deutsche Gesellschaft für Technische Zusammenarbeit (GTZ) GmbH, Eschborn, Federal Republic of Germany (GTZ).
The International Development Centre, Ottawa, Canada (IDRC).
L'Institut de Recherches Agronomiques Tropicales et des Cultures Vivrières, Paris, France (IRAT).
Koninklijk Instituut voor de Tropen, Amsterdam, The Netherlands (KIT).
The Food and Feed Grain Institute of Kansas State University, Manhattan, Kansas, USA (KSU).
Overseas Development Natural Resources Institute, Chatham Maritime, Chatham, Kent, England (ODNRI).

APPENDIX 3

SOME STANDARD ABBREVIATIONS

Units and general abbreviations

a.e.	acid equivalent	m.c.	moisture content
a.i.	active ingredient	mg	milligram
b.p.	boiling point	min	minute
°C	degrees Celsius	ml	millilitre
cm	centimetre	mm	millimetre
c.s.	concentrated solution	pH	hydrogen ion concentration
cu.m	cubic metre	post-em	post-emergence
cv.	cultivar	ppm	parts per million
cwt	hundredweight (112 lb)	pre-em	pre-emergence
d.p.	dispersible powder	RH	relative humidity
e.c.	emulsifiable concentrate	s	second
g	gram	s.c.	soluble concentrate
h	hour	sp.	species
ha	hectare	spp.	species (plural)
h.v.	high volume	ssp.	subspecies
i.r.	infra-red	$	dollar
kg	kilogram	s.u.l.v.	special ultra-low volume
km	kilometre	u.l.v. (ULV)	ultra-low volume spray
£	pound sterling	u.v. (UV)	ultra-violet
l	litre	var.	variety
lb	pound (weight) (16 oz)	vol.	volume
LC_{50}	median lethal concentration	w.p.	wettable powder
LD_{50}	median lethal dose	w.s.p.	water-soluble powder
l.v.	low volume	w/v	weight for volume
m	metre	w/w	weight for weight

Miscellaneous abbreviations

ADI	Acceptable daily intake	(v)CJD	(variant)Creutzfeld-Jacob Disease
Bth	*Bacillus thuringiensis*		
BC	Biological control	COP	Code of Practice
BSE	Bovine spongiform encephalopathy	CSI	Chitin synthesis inhibitor
		EAG	Electro-antennagram
Cda	Controlled droplet application	FIFO	First-in first-out principle

GM	Genetically modified (food)
GV	Granulosis virus
HACCP	Hazard analysis critical control points
HMSO	Her Majesty's Stationery Office, UK
ICM	Integrated crop management
IGR	Insect growth regulators
IPM	Integrated pest management
JH	Juvenile hormone
JHA	Juvenile hormone analogue
MAP	Modified atmosphere packaging
MSG	Monosodium glutamate
OC	Organochlorine compounds
OP	Organophosphorous compounds
PHV	Polyhedrosis virus
PM	Pest management
SIRM	Sterile insect release method (= SMRT)
SMRT	Sterile male release treatment

APPENDIX 4

ACRONYMS FOR SOME ORGANISATIONS

AACC	American Association of Cereal Chemists
ACIAR	Australian Centre for International Agricultural Research
ADAS	Agricultural Development and Advisory Service (formerly NAAS), MAFF, UK
ARC	Agricultural Research Council, UK
BCPC	British Crop Protection Council
BM (NH)	British Museum (Natural History), London, UK (now NHM)
BPCA	British Pest Control Association
BSI	British Standards Institute
BSRS	Biological Sites Recording Scheme, UK
CABI	Commonwealth Agricultural Bureaux International, UK
CDC	Centres for Disease Control, USA
CIAT	Centre for International Tropical Agriculture, Cali, Colombia
CIBC	Commonwealth Institute of Biological Control (Headquarters), London, UK
CIE	Commonwealth Institute of Entomology (now IIE), London, UK
CIHI	Commonwealth Institute of Helminthology International, St Albans, UK
CMII	Commonwealth Mycological Institute International, Kew, UK
COPR	Centre for Overseas Pest Research (later TDRI, now TRI)
CSIRO	Commonwealth Scientific and Industrial Research Organisation, Canberra, Australia
CSL	Central Science Laboratory, Slough, UK
DEFRA	Department for the Environment, Food and Rural Affairs, UK (formerly MAFF)
EEC	European Economic Community
EFFoST	European Federation of Food Science and Technology
EU	European Union
EPA	Environmental Protection Agency, Washington, USA
EPPO	European Plant Protection Organisation, Paris, France
FAO	Food and Agricultural Organisation of the United Nations, Rome, Italy
FDA	Food and Drug Agency, USA
FSA	Food Standards Agency, UK
GASGA	Group for Assistance on Systems Relating to Grain After-harvest
GTZ	German Agency for Technical Co-operation
HGCA	Home-Grown Cereals Authority, London, UK
IAC	International Agricultural Centre, Wageningen, Netherlands

IAEA	International Atomic Energy Agency, Vienna
ICI	Imperial Chemical Industries, UK
ICIPE	International Centre for Insect Physiology and Ecology, Nairobi, Kenya
ICRISAT	International Crops Research Institute for the Semi-arid Tropics, Hyderabad, India
IEHO	Institute of Environmental Health Officers
IFIS	International Food Information Service, Reading, UK
IFR	Institute of Food Research, Norwich, UK
IFST	Institute of Food Science and Technology, UK
IIE	International Institute of Entomology, London, UK
IITA	International Institute of Tropical Agriculture, Ibadan, Nigeria
INRA	French National Institute of Agricultural Research
IRRI	International Rice Research Institute, Manila, Philippines
ISO	International Standards Organisation
IUFoST	International Union of Food Science and Technology
MAFF	Ministry of Agriculture, Fisheries and Food (now DEFRA), UK
NAAS	National Agricultural Advisory Service (now ADAS), MAFF, UK
NAPPO	North American Plant Protection Organisation, USA
NASA	National Aeronautic and Space Administration, USA
NHM	Natural History Museum, London, UK (formerly BM (NH))
NRI	Natural Resources Institute, University of Greenwich
NVRS	National Vegetable Research Station, Wellesbourne, UK
ODA	Overseas Development Administration, London
ODM	Ministry of Overseas Development, UK
ODNRI	Overseas Development Natural Resources Institute, London, UK (now NRI)
PHLS	Public Health Laboratory Service, UK
PIL	Pest Infestation Laboratory, MAFF, Slough, UK
RESL	Royal Entomological Society of London, UK
RSH	Royal Society of Health, UK
RTI	Royal Tropical Institute, Amsterdam, Netherlands
TDRI	Tropical Development and Research Institute (now NRI), London, UK
TPI	Tropical Products Institute (now NRI), London, UK
US AID	United States Aid for International Development, USA
USDA	United States Department of Agriculture, USA
WHO	World Health Organisation, Geneva, Switzerland
WTO	World Trade Organisation, Geneva, Switzerland

REFERENCES

Adams, J.M. (1977). *A bibliography on post-harvest losses in cereals and pulses with particular reference to tropical and subtropical countries.* (G 110, Tropical Products Institute). London: Her Majesty's Stationery Office.

Agricultural Development and Advisory Service. (1986). *Insect and mite pests in food stores.* (P. 483, Ministry of Agriculture, Fisheries & Food). Alnwick: Ministry of Agriculture, Fisheries & Food.

Agricultural Development and Advisory Service. (1987). *Insects in farm-stored grain.* (P. 368, Ministry of Agriculture, Fisheries & Food). Alnwick: Ministry of Agriculture, Fisheries & Food.

Agriculture International. (1989). Post harvest – storing, handling, processing the crops when grown. *Agriculture International Supplement, 47,* 236-250.

Aitken, A.D. (1975). *Insect travellers* (Vol. 1 Coleoptera, Vol. 2 Other Orders). (Ministry of Agriculture, Fisheries & Food Technical Bulletin 31). London: Her Majesty's Stationery Office.

Ames, G., Clucas, I. & Paul, S.S. (1991). *Post-harvest losses of fish in the tropics.* London: Natural Resources Institute.

Avidoz, A. & Harpaz, I. (1969). *Plant pests of Israel.* Jerusalem: Israel University Press.

Barbosa-Canovas, G.V. & Vega-Mercado, H. (1995). *Drying of foods.* London: Chapman & Hall.

Baur, F.J. (Ed.). (1984). *Insect management for food storage and processing.* St. Paul, MA: American Association of Cereal Chemists.

Blatchford, S.M. & Wye, A.J. (1971). *Crop storage bibliography (with particular reference to the storage of durable agricultural produce in tropical and sub-tropical countries).* (G. 64, Tropical Products Institute). London: Her Majesty's Stationery Office.

Bouchat, I.R. (Ed.). (1987). *Food and beverage mycology.* (2nd ed.). New York: Van Nostrand Reinhold.

Bousquet, Y. (1990). *Beetles associated with stored products in Canada: an identification guide.* Canadian Government Publication Centre.

Boxall, R.A. (1987). *A bibliography on post-harvest losses in cereals and pulses with particular reference to tropical and sub-tropical countries.* (G. 197, Tropical Development and Research Institute). London: Her Majesty's Stationery Office.

British Museum (Natural History). (1951). *Clothes moths and house moths* (Economic Series No. 14). London: British Museum (Natural History).

Burkholder, W.E. (1985). Pheromones for monitoring and control of stored-product insects. *Annual Review of Entomology, 30,* 257-272.

Busvine, J.R. (1966). *Insects and hygiene.* London: Methuen.

Cadogan, A. (Ed.). (2000). *Biological nomenclature.* (3rd ed.). London: Institute of Biology.

Caresche, L., Cotterell, G.S. , Peachey, J.E., Rayner, R.W. & Jacques-Felix, H. (1969). *Handbook for phytosanitary inspectors in Africa.* Lagos, Nigeria: Organization of African Unity.

Champ, B.R. & Dyte, C.E. (1976). *Report of the Food & Agriculture Organization global survey of pesticide susceptibility of stored grain pests.* (Food & Agriculture Organization Plant Production and Protection Series). Rome: Food & Agriculture Organization.

Champ, B.R. & Highley, E. (Eds.). (1985). *Pesticides and humid tropical grain storage systems.* (Australian Centre for International Agricultural Research Proceedings No. 14). Canberra: Australian Centre for International Agricultural Research.

Chandler, A.C. & Read, C.P. (1961). *Introduction to parasitology* (with specific reference to the parasites of man). (10th ed.). New York: Wiley.

Cheng, T.C. (1986). *General parasitology.* (2nd ed.). Orlando, Fa: Academic Press.

Cliver, D.O. (1983) *Manual of food virology.* (VPH/83.46). Geneva: World Health Organisation.

Clucas, I.J. (1982) *Fish handling, preservation and processing in the tropics: Part* 2. (G. 145). London: Tropical Development & Research Institute.

Corbet, A.S. & Tams, W.H.T. (1943). Keys for the identification of the Lepidoptera infesting stored food products. *Proceedings of the Zoological Society of London,* (*B*), *113*, 55-148.

Cornwell, P.B. (1968). *The cockroach.* Vol. 1. (The Rentokil Library). London: Hutchinson.

Cornwell, P.B. (1976). *The cockroach.* Vol. 2. (The Rentokil Library). New York: St Martins Press.

Dales, M.J. (1994). *Controlling insect pests of stored products using insect growth regulators and insecticides of microbial origin.* (Bulletin 64, Natural Resources Institute). London: Overseas Development Agency.

Dalgleish, J. McN. (1990). *Freeze-drying for the food industries.* London: Elsevier.

Deutsche Gesellschaft für Technische Zusammenarbeit. (1980). *Post harvest problems.* (Documentation of an OAU/GTZ Seminar; Lome, March 1980). Eschborn: Deutsche Gesellschaft für Technische Zusammenarbeit.

Deutsche Gesellschaft für Technische Zusammenarbeit. (1985). *Control of infestation by Trogoderma granarium and Prostephanus truncatus to stored products.* (Proceedings of an International Seminar held in Lome, Togo, December 1984). Eschborn: Deutsche Gesellschaft für Technische Zusammenarbeit.

Deutsche Gesellschaft für Technische Zusammenarbeit. (1986). *Technical cooperation in rural areas plant and post-harvest protection. Facts and figures 1986.* (Schr. Der Deutsche Gesellschaft für Technische Zusammenarbeit, No. 194). Eschborn: Deutsche Gesellschaft für Technische Zusammenarbeit.

Dick, K. (1988). *A review of insect infestation of maize in farm storage in Africa, with special reference to the ecology and control of Prostephanus truncatus.* (Bulletin No. 18). London: Overseas Development & Natural Resources Institute.

Ebeling, W. (1975). *Urban entomology.* University of California, Division of Agricultural Science.

Food & Agriculture Organisation/International Atomic Energy Agency. (1975). *Requirements for the irradiation of food on a commercial scale*. (Proceedings of a panel, Vienna, 18-22 March 1974). Vienna: International Atomic Energy Agency.

Forsyth, J. (1966). *Agricultural insects of Ghana*. Accra, Ghana: Ghana University Press.

Goater, B. (1986). *British pyralid moths*. (A guide to their identification). Colchester, Essex: Harley Books.

Golob, P. & Hodges, R. (1982). *Study of an outbreak of Prostephanus truncatus (Horn) in Tanzania*. (G 164, Tropical Products Institute). London: Her Majesty's Stationery Office.

Golob, P. & Webley, D.J. (1980). *The use of plants and minerals as traditional protectants of stored products*. (Tropical Products Institute, G138). London: Tropical Products Institute.

Gorham, R. (Ed.). (1991). *Ecology and management of food-industry pests*. (Food and Drugs Administration Technical Bulletin No. 4). Arlington Va: Association of Official Analytical Chemists.

Gould, G.W. (1995). *New methods of food preservation*. London: Chapman & Hall.

Hagstrum, D.W., Reed, S., & Kenkel, P. (1999). Management of stored wheat insect pests in the USA. *Integrated Pest Management Reviews, 4*, 127-142.

Haines, C.P. (1977). *The potato tuber moth, Phthorimaea operculella (Zeller): a bibliography of recent literature and a review of its biology and control on potatoes in the field and in store*. (G. 112, Tropical Products Institute). London: Tropical Products Institute.

Harrigan, W.F. (1988). Laboratory methods in food microbiology. (3rd ed.). London: Academic Press.

Harris, K.L. & Lindblad, C.J. (Eds.). (1978). *Postharvest grain loss assessment methods*. St Paul, Ma: American Association of Cereal Chemists.

Harris, W.V. (1971). Termites – their recognition and control. Longman: London.

Hasan, M. & Khan, A.R. (1998). Control of stored-product pests by irradiation. *Integrated Pest Management Reviews, 3*, 15-29.

Hayes, P.R. (1992). *Food microbiology and hygiene*. (2nd ed.). London: Chapman & Hall.

Heath, J. & Maitland Emmet, A. (Eds.). (1985). The moths and butterflies of Great Britain and Ireland, Vol. 2 Cossidae – Heliodinidae. Colchester, Essex: Harley Books.

Hickin, N.E. (1974). *Household insect pests*. (The Rentokil Library). London: Hutchinson.

Hickin, N.E. (1975). *The insect factor in wood decay*. (3rd ed.). (The Rentokil Library). London: Hutchinson.

Hill, D.S. (1987). *Agricultural insect pests of temperate regions and their control*. Cambridge: Cambridge University Press.

Hill, D.S. (1990). *Pests of stored products and their control*. London: Bellhaven Press.

Hill, D.S. (1997). *The economic importance of insects*. London: Chapman & Hall.

Hinton, H.E. (1943). The larvae of the Lepidoptera associated with stored products. *Bulletin of Entomological Research, 34*, 163-212.

Hocking, B. (Ed.). (1965). *Armed forces manual on pest control.* (3rd ed.). (Defence Research Board, Department of National Defence, Canada). Ottawa: Queen's Printer Canada.

Hodges, R.J. (1979). *A review of the biology and control of the Rice Moth Corcyra cephalonica Stainton (Lepidoptera: Galleriinae).* (G. 125, Tropical Products Institute). London: Her Majesty's Stationery Office.

Hodges, R.J. (1986). The biology and control of *Prostephanus truncatus* (Horn) (Coleoptera; Bostrichidae) – a destructive storage pest with an increasing range. *Journal of Stored Product Research, 22,* 1-14.

Home-Grown Cereals Authority. (1992). *Integrated pest control strategy for stored grain; surface pesticide treatments of aerated commercial and farm stores to control insects and mites.* (Home-Grown Cereals Authority Project Report No. 57.). London: Home-Grown Cereals Authority, Hamlyn House.

Hopft, H.S., Morley, G.E.J. & Humphries, J.R.O. (Eds.). (1976). *Rodent damage to growing crops and to farm and village storage in tropical and subtropical regions.* (Results of a postal survey 1972-3). London: Tropical Products Institute.

Hughes, A.M. (1976). *The mites of stored food.* (2nd ed.). (Ministry of Agriculture, Fisheries & Food, Technical Bulletin No. 9). London: Her Majesty's Stationery Office.

Hui, Y. (Ed.). (1994). *Foodborne disease handbook.* (Vols 1-3). New York: Marcel Dekker.

Jay, J.J. (1986). *Modern food microbiology.* (3rd ed.). New York: Van Nostrand Reinhold.

Land, C.M. van't (1991). *Industrial drying equipment.* New York: Marcel Dekker.

Lawson, T.J. (Ed.). (1987). *Stored products pest control.* (British Crop Protection Council Monograph No. 37). Thornton Heath: British Crop Protection Council.

Macrae, R., Robinson, R.K., & Sadler, M.J. (Eds). (1993). *Encyclopedia of food science, food technology and nutrition.* (Vols 1-8). San Diego: Academic Press.

McFarlane, J.A. (1989). *Guidelines for pest management research to reduce stored food losses caused by insects and mites.* (Bulletin No. 22). London: Overseas Development & Natural Resources Institute.

Mathlouthy, M. (1994). *Food packaging and preservation.* London: Chapman & Hall.

Matthews, G.A. (1979). *Pesticide application methods.* London: Longmans.

Meehan, A. (1989). *Rats and mice – their biology and control.* East Grinstead: Rentokil.

Ministry of Agriculture, Fisheries & Food. (1989). *A retailers guide to: European Community quality standards for fresh fruit and vegetables.* (Ministry of Agriculture, Fisheries & Food; Agriculture Department Leaflet 893). London: Ministry of Agriculture, Fisheries & Food.

Ministry of Agriculture, Fisheries & Food. (1989). *Pesticides 1989.* (Pesticides approved under the Control of Pesticides Regulations 1986). London: Her Majesty's Stationery Office.

Mondal, K.A.M.S.H. & Parween, S. (2000). Insect growth regulators and their potential in the management of stored-product insect pests. *Integrated Pest Management Reviews, 5* (4), 255-295.

Monro, H.A.V. (1980). *Manual of fumigation for insect control.* (2nd ed.). Rome: Food & Agriculture Organisation.

Mound, L. (Ed.). (1989). *Common insect pests of stored food products.* (7th ed.). (British Museum (Natural History), Econ. Series No. 15). London: British Museum (Natural History).

Mountney, G.J. & Gould, W.A. (1988). *Practical food microbiology and technology.* (3rd ed.). New York: Van Nostrand Reinhold.

Mourier, H. & Winding, O. (1997). *Collins guide to wild life in house and home.* London: Collins.

Munroe, J.W. (1966). *Pests of stored products.* (The Rentokil Library). London: Hutchinson.

National Academy of Sciences. (1978). *Postharvest food losses in developing countries.* Washington DC: National Academy of Sciences.

Overseas Development & Natural Resources Institute. (1989). *Pesticide index.* (1989 ed.). London: Overseas Development & Natural Resources Institute.

Ordish, G. (1952). *Untaken harvest.* London: Constable.

Parkin, E.A. (1956). Stored product entomology (the assessment and reduction of losses caused by insects to stored foodstuffs). *Annual Review of Entomology, 1*, 223-40.

Parshad, V.R. (1999). Rodent control in India. *Integrated Pest Management Reviews, 4*, 97-126.

Phillips, C.A. (1997). Preserving food: a change in the atmosphere. *Biologist, 44*, 297-300.

Public Health & Industrial Pesticides Council, (1989) *The Phipco manual* (A guide to the requirements on the storage, supply, sale, and use of public health and industrial pesticides by formulators, distributors and commercial users). Sheringham: Public Health & Industrial Pesticides Council.

Pinniger, D. (1989) *Insect pests in museums.* London: Institute of Archaeology Publications.

Pitt, J.L. & Hocking, A.D. (1985). *Fungi and food spoilage.* North Ryde: Academic Press.

Prakash, I. (Ed.). (1989). *Rodent pest management.* London: Wolfe Medical Publications.

Prevett, P.F. (1990). Stored-product entomology – the challenge of the next decade. *Bulletin of Entomological Research, 80,* 3-6.

Rahman, M.A. & Hill, D.S. (*in press*) A comparison of food safety code violations among southeast Asian, Hispanic and Caucasian retail food outlets in central California, USA.

Richards, O.W. & Davies, R.G. (1977). *Imms' general textbook of entomology.* (10th ed.). Vol. 1 *Structure, physiology and development.* Vol. 2 *Classification and biology.* London: Chapman & Hall.

Roberts, T.J. (1981). *Hand book of vertebrate pest control in Pakistan.* Karachi: Pakistan Agricultural Research Council.

Robinson, R.K. (Ed.) (1999). *Encyclopaedia of food microbiology.* (Vols 1-3). London: Academic Press.

Robinson, W.H. (1996). *Urban entomology: Insect and mite pests in the human environment.* London: Chapman & Hall.

Russell, N.J. & Gould, G.W. (Eds.). (1991). *Food preservatives.* Oxford: Blackie.

Sadler, M.J. (Ed.). (1999). *Encyclopaedia of human nutrition.* (Vols 1-3). London: Academic Press.

Scher, B. (Ed.). (1999). *Controlled release delivery systems for pesticides. New York: Marcel Dekker.*

Schmidt, G.D. & Roberts, L.S. (2000). *Foundations of parasitology.* (6th ed.). Boston: McGraw-Hill.

Schmutterer, H. (1990). Properties and potential of natural pesticides from the Neem Tree, *Azadirachta indica. Annual Review of Entomology, 35,* 271-297.

Schmutterer, H. & Ascher, K.R.S. (Eds.). (1984). *Natural pesticides from the Neem Tree and other tropical plants.* Eschborn: Deutsche Gesellschaft für Technische Zusammenarbeit.

Schal, C. & Hamilton, R.L. (1990). Integrated suppression of synanthropic cockroaches. *Annual Review of Entomology, 35,* 521-551.

Schulten, G.G.M. (1981). *Post-harvest improvements for the eighties and nineties.* Proceedings of the 10th Annual Seminar of the Group for Assistance on Systems Relating to Grain After-harvest.

Scopes, N. & Ledieu, M. (1983). *Pest and disease control handbook.* (2nd ed.). London: British Crop Protection Council.

Smith, K.V.G. (Ed.). (1973). *Insects and other arthropods of medical importance.* London: British Museum (Natural History).

Snelson, J.T. (1987). *Grain protectants.* (Australian Centre for International Agricultural Research Monograph No. 3). Canberra: Australian Centre for International Agricultural Research.

Snowdon, A.L. (1990). *A colour atlas of post-harvest diseases and disorders of fruits and vegetables.* Vol. 1 *General introduction and fruits.* Vol. 2 *Vegetables.* London: Wolfe Scientific.

Southgate, B.J. (1979). Biology of the Bruchidae. *Annual review of entomology, 24,* 449-473.

Tomlin, C. (Ed.). (1994). *The pesticide manual* (A world compendium). (10th ed.). London: British Crop Protection Council & Royal Society of Chemistry.

Tropical Development & Research Institute. (1984). *Insects and arachnids of tropical stored products: their biology and identification.* (A training manual). Compiled by Dobie, P., Haines, C.P., Hodges, R.J. & Prevett, P.F. Slough, England: Tropical Development & Research Institute.

Tropical Development & Research Institute. (1986). *Group for assistance on systems relating to grain after-harvest seminar on fumigation technology in developing countries.* London: Overseas Development Agency.

United States Department of Agriculture. (1966). *The yearbook of agriculture, 1966: protecting our food.* Washington DC: United States Government Printing Office.

United States Department of Agriculture. (1986). *Stored grain insects.* (United States Department of Agriculture Handbook No. 500). Washington DC: United States Government Printing Office.

Wohlgemuth, R. Harnisch, R., Thiel, R., Buchholz, H. & Laborius, A. (1987). *Comparing tests on the control and long-term action of insecticides against stored product pests under tropical climate conditions.* Eschborn: Deutsche Gesellschaft für Technische Zusammenarbeit.

Worthing, C.R. & Walker, S.B. (1987). *The pesticide manual.* (A world compendium). (8th ed.). London: British Crop Protection Council.

The major publisher of books on food science in the widest sense was, for many years, Chapman & Hall, of London, but they have now been acquired by Kluwer Academic Publishers, of the Netherlands. The last Chapman & Hall catalogue to be printed was in 1995, and contained 85 pages of titles all relating to "Food Science & Technology".

INTERNATIONAL JOURNALS ON FOOD SCIENCE

Advances in Food and Nutrition Research. Vol. 1. (1956). London: Academic Press.

Biodeterioration Abstracts. Vol. 1. (1987). Wallingford: CABI.

Biotechnology and Food Process Engineering. Vol. 1. (1990). New York: Marcel Dekker.

Food Safety. Vol. 1. (1993). New York: Marcel Dekker.

Food Science & Technology Today. Vol. 1. (1999). London: IFST.

International Journal of Food Science and Technology. Vol. 1. (1987). London: IFST.

Journal of Stored Products Research. Vol. 1. (1964). Oxford: Pergamon Press.

Nutrition and Food Science. Vol. 1. (1993). Bradford: McB University Press.

Postharvest Biology and Technology. Vol. 1. (1991). Amsterdam: Elsevier Science.

Postharvest News and Information. Vol. 1. (1990). Wallingford: CABI.

Trends in Food Science and Technology. Vol. 1. (1990). Oxford: Elsevier Trends.

Most countries, and in the USA most States, have their own journal on Public Health/Hygiene which does include papers on food pests, food contamination, and the like.

INDEX

GPSR Compliance
The European Union's (EU) General Product Safety Regulation (GPSR) is a set of rules that requires consumer products to be safe and our obligations to ensure this.

If you have any concerns about our products, you can contact us on

ProductSafety@springernature.com

In case Publisher is established outside the EU, the EU authorized representative is:

Springer Nature Customer Service Center GmbH
Europaplatz 3
69115 Heidelberg, Germany

www.ingramcontent.com/pod-product-compliance
Ingram Content Group UK Ltd.
Pitfield, Milton Keynes, MK11 3LW, UK
UKHW020134160926
475607UK00003B/29
* 9 7 8 1 4 0 2 0 0 7 3 5 4 *